PHYSICS

TOWER, SMITH, TURTON,
AND
COPE

Y   R   RY   B   YRB

(See p.441)

THREE-COLOR PRINTING

Y. Yellow impression; negative made through a blue-violet filter.   *R.* Crimson impression; negative made through a green filter.   *RY.* Crimson on yellow. *B.* Blue impression; negative made through a red filter.   *YRB.* Yellow, crimson, and blue combined; the final product.   (Courtesy of Phototype Engraving Co., Philadelphia.)

PHYSICS

BY

WILLIS E. TOWER, M. Sci. (Univ. of Illinois)
HEAD OF THE DEPARTMENT OF PHYSICS, ENGLEWOOD
HIGH SCHOOL, CHICAGO

CHARLES H. SMITH, M. E. (Cornell)
HEAD OF THE DEPARTMENT OF PHYSICS AND ASSISTANT
PRINCIPAL, HYDE PARK HIGH SCHOOL, CHICAGO

CHARLES M. TURTON, A. M. (Syracuse)
HEAD OF THE DEPARTMENT OF PHYSICS, BOWEN
HIGH SCHOOL, CHICAGO

IN COLLABORATION WITH

THOMAS D. COPE, Ph.D. (Pennsylvania)
ASSISTANT PROFESSOR OF PHYSICS, UNIVERSITY
OF PENNSYLVANIA

BASED UPON

PRINCIPLES OF PHYSICS

BY

TOWER, SMITH and TURTON

WITH 7 PLATES AND 448 OTHER ILLUSTRATIONS

PHILADELPHIA
P. BLAKISTON'S SON & CO.
1012 WALNUT STREET

PREFACE

In the preparation of this text, the *pupil*, his experience, needs, and interests have been constantly kept in mind. The order of topics, illustrations, and problems have been selected with the purpose of leading the *pupil* into a clear understanding of the physical phenomena continually taking place about him.

The recommendations and conclusions reached by the "New Movement in the Teaching of Physics" have been incorporated into the book as a whole. These conclusions indicate that the most efficient teaching in physics involves a departure from the quantitative, mathematical methods of presentation that were in general use a dozen or more years ago, toward a method better adapted to the capabilities, interests, and requirements of the young people in our physics classes.

The older methods are effective with a portion of the student body which has the greater mathematical ability and training, but they discourage a large majority of the pupils who are not gifted or prepared for severe mathematical analysis. For this reason, many of the more difficult mathematical demonstrations often given in physics texts are omitted. Most of the problems involve only the units employed in practical every-day measurements.

The portions of Mechanics that are ordinarily so difficult for the average pupil are not taken up until he has covered considerable ground with which he is more or less familiar and not until he has become somewhat accustomed to the methods of study and the technical terms of the subject.

The pupil comes to the study of physics with a great number of experiences and impressions of physical phenomena continually occurring about him. In recognition of this fact, it has been thought best to consider first the explanation of common things well known to all pupils, such as the diffusion of gases, evaporation of liquids, expansion of bodies when heated, and capillary action. Since the molecular theory of matter is now supported by so many conclusive evidences, we have not hesitated to make free use of it in the early chapters. The applications of this theory are extremely helpful in explaining every-day phenomena. Our experience shows that beginners in physics understand and apply this theory without difficulty.

The illustrations and drawings have been selected from a pedagogical rather than a spectacular point of view. Practically all of them are new. The problems and exercises have been selected for the distinct purpose of illustrating the principles taught in the text and for their practical applications.

Many direct applications to common every-day experiences are given in order to connect the subject matter with the home environment and daily observation of physical phenomena. Some phenomena are mentioned without detailed explanation as it is felt that the presentation of these subjects in this manner is better for this grade of student than a complete analysis.

Some of the special features of the text may be briefly summarized as follows:

(A) *Simplicity of presentation* is emphasized. The methods of attack, the illustrations and examples employed in developing the subjects are particularly adapted to beginners in physics.

(B) The text is divided into some *seventy-seven sections,* each containing material enough for one recitation.

(C) Each of these sections is summarized by a list of *important topics* which point out to the pupil the principles and subject matter requiring most careful attention. The lists of important topics are also of assistance to the teacher in assigning recitations.

(D) The *problems and practical exercises* emphasize physical principles as distinguished from mathematical training. A list of exercises is placed at the end of the several sections. They are in sufficient number to permit testing at many points and of a choice of problems by teachers.

The authors wish to express their appreciation for suggestions and helpful criticisms to many who have read the text in manuscript or proof. Especially to Professor A. P. Carman of the University of Illinois and his associate, Professor F. R. Watson, who have gone carefully over the whole text; and to Mr. Chas. M. Brunson, Scott High School, Toledo, Ohio, Mr. Frank E. Goodell, North High School, Des Moines, Iowa, and to Mr. Walter R. Ahrens, Englewood High School, Chicago, for assistance in reading the proofs. Also to Mr. W. H. Collins, Jr., Bowen High School, Chicago, who supervised the preparation of drawings for the diagrams and figures; and to many firms and individuals that have courteously furnished material for illustrations.

WILLIS E. TOWER.
CHARLES H. SMITH.
CHARLES M. TURTON.

ON THE STUDY OF PHYSICS

When a pupil begins the study of Physics he has in his possession many bits of knowledge which are fundamental in the science. He has learned to throw a ball and can tell how a thrown ball moves. He has drawn out nails with a claw hammer. He has seen wood float and iron sink. He has sucked liquids up through straws. In his mother's kitchen, he sees water as ice, liquid, and steam. On a wintry day he reads the temperature on a thermometer. He sees sparks fly from car wheels when the brakes are applied. He has played with a horseshoe magnet, and has found the north by means of a compass. The telephone, the electric light and the motor he sees, and perhaps uses, many times a day. He dresses before a mirror, focuses his camera, watches the images at a moving picture show, and admires the colors of the rainbow. He has cast stones into water to watch the ripples spread, has shouted to hear the echo, and perhaps plays some musical instrument. These, and a thousand other things, are known to the intelligent and normal boy or girl who has reached the age at which the study of Physics is properly begun.

To a great extent even the terms used in the science are familiar to the beginner. He speaks of the horsepower of an engine, reads kilowatt-hours from the meter in the cellar, and may know that illuminating gas costs one dollar per thousand "cubic feet." "Ampere" and "volt" are words he frequently hears and sees.

When he takes up the study of Physics, the attitude of

the student toward these familiar things and words must undergo a change. Casual information about them must be changed to sound knowledge, purposely acquired. Hazy notions about the meanings of words must be replaced by exact definitions. Bits of knowledge must be built into a structure in which each fact finds its proper place in relation to the others.

The only agent which can accomplish these changes is the student himself. He must consciously and purposely seek the truth and must reflect upon it until he sees it in its relation to other truth. Upon him, and upon him alone, rests the final responsibility for the success or failure of his study.

But the student is not without assistance. In his teacher he finds a guide to stimulate, to direct, and to aid his efforts, and a critic to point out wherein his efforts have failed and wherein they have succeeded. Weights, measures, and other apparatus are furnished to enable him to answer for himself questions which have arisen in his studies.

In addition to these the student has his text book, his teacher for his hours of private study. A good text book is an inspiring teacher in print. It directs attention to things familiar to the student through long experience, and inspires him to make a closer scrutiny of them. It invites him to observe, to analyze, to compare, to discover likenesses and differences in behavior. It questions him at every turn. Its ever repeated challenge reads, "Weigh and consider." It furnishes him needed information that he cannot otherwise acquire. It satisfies his desire to know, "By whom, where, when, and how was this first discovered?"

The student of Physics must never forget that he is studying not pages of text but the behavior and properties of iron, water, mica, moving balls, pumps, boiling liquids, compressed air, mirrors, steam engines, magnets, dynamos, violins, flutes, and a host of other things. His studies should, whenever possible, be made first hand upon the things themselves. The text is an aid to study, never a substitute for the thing studied.

It is an excellent plan for each student to select some one thing for special study, the telephone for example. By observation, experiment, and reading, he may acquire a large amount of valuable information about such a subject while pursuing his course in Physics. Every part of the science will be found to bear some relation to it.

The student who takes up the study of Physics in the way suggested will find himself at the end of a year of study in possession of much new and valuable knowledge about the physical world in which he lives. By virtue of this knowledge he will be better able to enjoy the world, to control it, and to use it.

THOMAS D. COPE.

PHILADELPHIA.

CONTENTS

xiii

PHYSICS

CHAPTER I

INTRODUCTION AND MEASUREMENT

(1) INTRODUCTION

1. Physics, an Explanation of Common Things.—Many students take up the study of physics expecting to see wonderful experiments with the "X" rays, wireless telegraphy, dynamos, and other interesting devices. Others are dreading to begin a study that to them seems strange and difficult, because they fear it deals with ideas and principles that are beyond their experience and hard to comprehend.

Each of these classes is surprised to learn that *physics is mainly an explanation of common things.* It is a study that systematizes our knowledge of the forces and changes about us; such as the pull of the earth, the formation of dew, rain and frost, water pressure and pumps, echoes and music, thermometers and engines, and many other things about us with which people are more or less familiar. Physics is like other school subjects, such as mathematics and language, in having its own peculiar vocabulary and methods of study; these will be acquired as progress is made in the course.

The most useful habit that the student of physics can

form is that of connecting or relating each *new idea* or *fact* that is presented to him to *some observation* or *experience* that will illustrate the new idea. This relating or connecting of the new ideas to one's own personal experience is not only one of the best known means of cultivating the memory and power of association, but it is of especial help in a subject such as physics, which deals with the systematic study and explanation of the facts of our every-day experience.

2. Knowledge—Common and Scientific.—This leads to the distinction between *common knowledge* and *scientific knowledge*. We all possess common knowledge of the things about us, gained from the impressions received by our senses, from reading, and from the remarks of others. *Scientific* knowledge is attained when the bits of common knowledge are connected and explained by other information gained through study or experience. That is, common knowledge becomes scientific, when it is *organized*. This leads to the definition: *Science is organized knowledge.*

Common knowledge of the forces and objects about us becomes scientific only as we are able to make accurate measurements of these. That is, science is concerned not only in *how* things work, but even more in *how much* is involved or results from a given activity. For example, a scientific farmer must be able to compute his costs and results in order to determine accurately his net profits. The business man who is conducting his business with efficiency knows accurately his costs of production and distribution.

This book is written in the hope that it will make more scientific the student's common knowledge of the forces and changes in the world about him and will give him many ideas and principles that will help him to acquire the habit of looking from effects to their natural causes and thus tend to develop what is called the *scientific habit of thought*.

3. Hypothesis, Theory, and Law.—Three words that are frequently used in science may be mentioned here: *hypothesis*, *theory*, and *law*. An hypothesis is a supposition advanced to explain some effect, change, or condition that has been observed. For example, the Nebular Hypothesis of which many high-school students have heard, is an attempt to explain the origin of the sun, the earth, the planets, and other solar systems.

A theory is an hypothesis which has been tested in a variety of ways and which seems to fit the conditions and results so that it is generally accepted as giving a satisfactory explanation of the matter in question. The Molecular Theory of Matter which states that matter of all kinds is composed of very small particles called molecules (see Art. 6), is a familiar example of a theory.

A theory becomes a law when it may be definitely proved. Many laws are expressed in mathematical language, *e.g.*, the law of gravitation. (See Art. 88.) Many of the laws of physics are illustrated by laboratory experiments, which show in a simple way just what the law means.

Exercises

Explain what is meant by the following terms and expressions:
1. Common knowledge.
2. Scientific knowledge.
3. Science.
4. Topics in physics.
5. Scientific habit of thought.
6. Value of relating new ideas to former experiences.
7. Hypothesis.
8. Theory.
9. Law.

(2) THE STATES OF MATTER

4. Physics Defined.—In the study of any science or field of knowledge, it is helpful to have a basis for grouping or classifying the facts studied. In physics we are to study the objects, forces, and changes about us, to understand them and their relations to one another. Accordingly, physics, dealing with the material world about us, is often defined as *the science of matter and energy, matter* being *anything that occupies space* and *energy* the *capacity for doing work*. This definition of physics while not strictly accurate is sufficiently comprehensive for our present purpose.

5. The Three States of Matter.—Our bodies are *matter* since they occupy space. Further, they possess *energy* since they are able to do work. In beginning the study of physics it will simplify our work if we study one of these topics before the other. We will therefore begin with matter and consider first its three states.

Some bodies are *solid;* as ice, iron, wax. Others are *liquid;* as water, mercury, oil. Still others are in the state of gas; as steam, air, and illuminating gas. Further we notice that the same substance may be found in any one of the three states. For example water may be either ice, water or steam; that is, either a solid, a liquid, or a gas.

Most persons have heard of *liquid air* and possibly some know of *ice air, i.e.,* air cooled until it not only liquefies, but is solidified. On the other hand, iron may be melted and, if heated hot enough, may be turned into iron vapor. In fact most substances by heating or cooling sufficiently may be changed into any one of the three states.

Before defining the three states, let us consider the *structure* of matter. This may help us to answer the question: How is it possible to change a hard solid, such as ice, into a liquid, water, and then into an invisible gas like steam? This is explained by the molecular theory of matter.

6. The Molecular Theory of Matter.—It is believed that all bodies are made up of very small particles called *molecules*, and that these instead of being packed tightly together like square packages in a box, are, strange as it may seem, very loosely packed even in solids and do not *permanently* touch their neighbors. The size of these molecules is so minute that it has been estimated that if a drop of water could be magnified to the size of the earth, the molecules magnified in the same proportion would be in size between a baseball and a football. The air and all other gases are believed to be made up of molecules in *rapid motion*, striking and rebounding continually from one another and from any objects in contact with the gas.

7. States of Matter Defined.—These ideas of the structure of matter assist us in understanding the following definitions: *A solid is that state of matter in which the molecules strongly cling together and tend to keep the same relative positions.* (This of course follows from the tendency of a solid to retain a definite form.) *A liquid is that state of matter in which the molecules tend to cling together yet move about freely.* Hence a liquid takes the form of any vessel in which it is placed. *A gas is that state of matter in which the molecules move about freely and tend to separate indefinitely.* Hence a gas will fill any space in which it is placed.

8. Effect of Heat on Matter.—It is further believed that when a body is heated, that the action really consists in making its molecules move or vibrate faster and faster as the heating progresses. This increase of motion causes the molecules to push apart from one another and this separation of the molecules causes an expansion of the body whether it be solid, liquid, or gas. Fig. 1 shows the expansion of air in an air thermometer. Fig. 2 shows the expansion of a solid on heating.

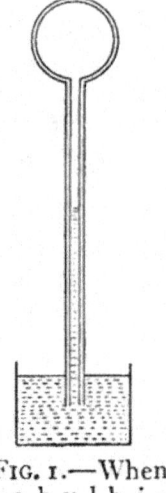

FIG. 1.—When the bulb is heated, the air within expands forcing down the water in the tube.

9. **Physical and Chemical Changes.** A change of state such as the freezing or boiling of water is called a *physical* change, for this change has not affected the identity of the substance. It is water even though it has become solid or gaseous. Heating a platinum wire red hot is also a physical change for the wire on cooling is found to be the same substance as before. Further if salt or sugar be dissolved in water the act of *solution* is also a physical change since the identical substance (salt or sugar) is in the solution and may be obtained by evaporating the water.

If some sugar, however, is heated strongly, say in a test-tube, it is found to blacken, some water is driven off

(a)   (b)

FIG. 2 (a) represents a straight bar made of a strip of brass and a strip of iron riveted together and attached to a handle. Upon heating the compound bar in a gas flame, the brass expands faster than the iron causing the bar to bend toward the latter as in Fig. 2 (b).

and on cooling *some black charcoal is found in the tube instead of the sugar.* This action which has resulted in a *change in the nature of the substance* treated is called a *chemical* change. To illustrate further, if some magnesium wire is heated strongly in a flame, it burns, giving off an intense light and when it cools one finds it changed to a light powdery substance like ashes. Chemical changes, or those that change the nature of the substance affected, are studied in *chemistry*. In *physics* we have to do only with physical changes, that is, with those changes that do not affect the nature of the substance.

Important Topics

1. Physics defined.
2. The three states of matter; solid, liquid, gas.
3. Molecular theory of matter.
4. Physical and chemical changes.

Exercises

Write out in your own words your understanding of:
1. The structure of matter.
2. Some of the differences between solids, liquids, and gases.
3. How to change solids to liquids and gases and *vice versa*.
4. The reason for the changes of size of a body on heating.
5. Why cooling a gas tends to change it to a liquid or a solid.
6. The actual size of molecules.

Which of the following changes are chemical and which physical?

Give reasons.

1. Melting of ice.
2. Burning of a candle.
3. Production of steam.
4. Falling of a weight.
5. Drying of clothes.
6. Making an iron casting
7. Decay of vegetables.
8. Sprouting of seeds.
9. Flying an aeroplane.
10. Growth of a plant.
11. Grinding of grain.
12. Sawing a board.
13. Pulverizing stone.
14. Making toast.
15. Sweetening tea or coffee with sugar.
16. Burning wood or gas.

(3) The Metric System

10. The Metric System.—In order to study the three states of matter with sufficient exactness it is necessary to employ a system of measurement. The system universally employed by scientists is called *The Metric System*. In many respects it is the most convenient for all purposes. Every student should therefore become familiar with it and learn to use it. At the present time, not only do scientists everywhere use it, but many countries have adopted it and use it in common measurements. It was legalized in the United States in 1866. The metric system was originated by the French Academy of Sciences during the latter part of the 18th century. There were so many different systems of weights and measures in use, each country having a system of its own, that commerce was much hindered. It was therefore decided to make a system based upon scientific principles. The length of the earth's quadrant passing from the equator to the pole was determined by surveying and computation. One-ten-millionth of this distance was selected as the unit of length and called a *meter*. Accurate copies of this meter were made and preserved as standards.

Later surveys have shown that the original determination of the earth's quadrant was not strictly accurate; so that after all the meter is not exactly one-ten-millionth of the earth's quadrant.

11. The Standard Meter.—The standard unit of *length* in the metric system is the *meter*. It is the distance, at the temperature of melting ice, between two transverse parallel lines ruled on a bar of platinum (see Fig. 3), which is kept in the Palace of the Archives in Paris. Accurate copies of this and other metric standards are also kept at the Bureau of Standards at Washington, D. C. Fig. 4 shows the relation between the inch and the centimeter (one-hundredth of a meter).

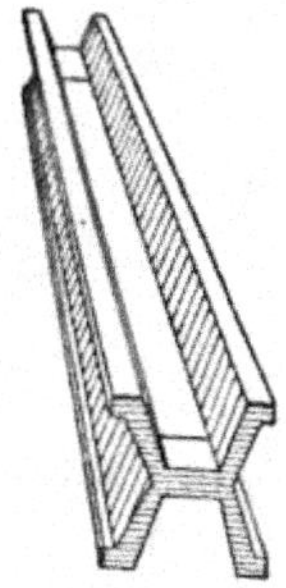

FIG. 3.—The standard meter.

12. Units and Tables in the Metric System.—The metric unit of *area* commonly used in physics is the *square centimeter*.

The standard unit of *volume* or capacity is the *liter*. It is a cube one-tenth of a meter on each edge. It is equal to 1.057 quarts. It corresponds, therefore, to the quart in English measure.

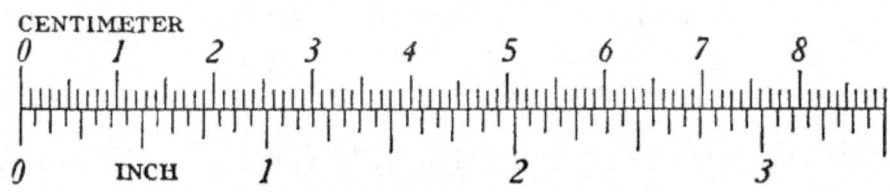

FIG. 4.—Centimeter and inch scales.

The standard unit of *mass* is the *kilogram*. It is the mass of 1 liter of pure water at the temperature of its greatest density, 4°C. or 39.2°F.

The three principal units of the metric system, the *meter*, the *liter*, and the *kilogram*, are related to one another in a simple manner, since the liter is a cube one-tenth of a meter in each dimension and the kilogram is the mass of a liter of water. (See Fig. 5.)

The metric system is a *decimal* system that is, one unit is related to another unit in the ratio of *ten* or of some power of ten. This is indicated by the following tables:

Metric Table of Length

10 millimeters   (mm.) equal 1 centimeter.
10 centimeters   (cm.) equal 1 decimeter.
10 decimeters   (dm.) equal 1 meter.
10 meters   (m.)   equal 1 dekameter.
10 dekameters   (Dm.) equal 1 hectometer.
10 hectometers   (hm.) equal 1 kilometer.
10 kilometers   (km.) equal 1 myriameter.

The measures commonly used are the *centimeter*, *meter* and *kilometer*.

Metric Table of Mass (or Weight)

10 milligrams   (mg.) equal 1 centigram.
10 centigrams   (cg.) equal 1 decigram.
10 decigrams   (dg.) equal 1 gram.
10 grams   (g.)   equal 1 dekagram
10 dekagrams   (Dg.) equal 1 hectogram.
10 hectograms   (hg.) equal 1 kilogram.
10 kilograms   (kg.) equal 1 myriagram.

The masses commonly used are the *milligram*, *gram* and *kilogram*.

Notice in these tables the similarity to 10 mills equal 1 cent, 10 cents equal 1 dime, 10 dimes equal 1 dollar, in the table of United States money.

Other tables in the metric system are built upon the same plan. Learn the prefixes in order thus: milli, centi, deci, deka, hecto, kilo, myria. The first three prefixes are Latin numerals and represent divisions of the unit. The last four are Greek numerals and represent multiples. In these tables, milli means $\frac{1}{1000}$, centi means 1/100, deci means 1/10, deka means 10, hecto, 100, kilo, 1000, myria, 10,000. Two other prefixes are sometimes used, *micro* which means 1/1,000,000; as microfarad or microvolt, and *meg* which means 1,000,000, as megohm meaning 1,000,000 ohms.

13. Advantages of the Metric System.—*First*, it is a decimal system; *second*, the same form and prefixes are used in every table; *third*, the standards of length (meter), volume (liter), and mass (kilogram) bear a simple relation to one another. This simple relation between the three standard units may be given thus: *first*, the liter is a cubic decimeter, and *second*, the kilogram is the mass of a liter of water. (See Fig. 5) Since the liter is a cubic deci-

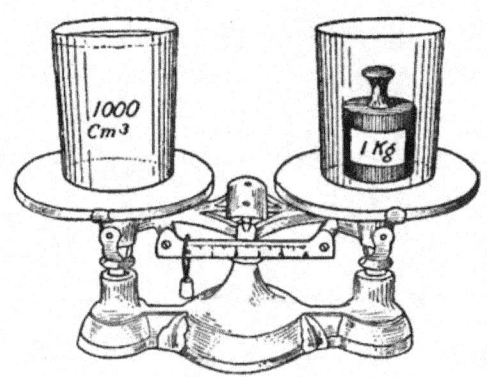

FIG. 5.—One liter of the water has a mass of one kilogram.

meter, the length of one side is 10 cm. The liter therefore holds 1000 ccm. (10 × 10 × 10). Therefore, 1 liter = 1 cu. dm. = 1000 ccm. and since 1 liter of water has a mass of 1 kg. or 1000 g., then 1000 ccm. of water has a mass of 1000 g., or 1 *ccm. of water has a mass of* 1 g.

The following table of equivalents gives the relation between the most common English and metric units. Those marked (*) should be memorized.

(*)	1 meter	= 39.37 inches.	1 cu. in.	= 16.387 ccm.
(*)	1 inch	= 2.54 cm.	1 cu. ft.	= 28315 cm.
	1 foot	= 30.48 cm.	1 cu. m.	= 1.308 cu. yd.
	1 mile	= 1.609 km.	(*) 1 liter	= 1.057 qt.
	1 sq. in.	= 6.45 sq. cm.	(*) 1 kg.	= 2.204 lbs.
	1 sq. cm.	= .155 sq. in.	1 g.	= 15.44 grains.
	1 sq. m.	= 1.196 sq. yd.	1 lb.	= 0.4536 kg.
	1 acre	= 0.405 ha.	1 oz.	= 28.35 g.
	1 hectare	= 2.45 acres.	1 g.	= 0.0353 oz.

THE C. G. S. SYSTEM. Scientists have devised a plan for expressing any measurement in terms of what are called the *three funda-*

mental units of *length, mass,* and *time.* The units used are the *centi-meter,* the *gram* and the *second.* Whenever a measurement has been reduced to its equivalent in terms of these units, it is said to be expressed in *C. G. S.* units.

Important Topics

1. The metric system; how originated.
2. Units; meter, liter, kilogram.
3. Metric tables.
4. Advantages of the metric system.
5. Equivalents.
6. The C. G. S. system.

Exercises

1. Which is cheaper, milk at 8 cents a quart or 8 cents a liter? Why?
2. Which is more expensive, cloth at $1.00 a yard or at $1.00 a meter? Why?
3. Which is a better bargain, sugar at 5 cents a pound or 11 cents a kilogram? Why?
4. Express in centimeters the height of a boy 5 ft. 6 in. tall.
5. What is the length of this page in centimeters? In inches?
6. What is the mass of a liter of water? Of 500 ccm.? Of 1 ccm.?
7. From Chicago to New York is 940 miles. Express in kilometers.
8. A 10-gallon can of milk contains how many liters?
9. What will 100 meters of cloth cost at 10 cents a yard?
10. What will 4 kg. of beef cost at 15 cents a pound?
11. What will 5½ lbs. of mutton cost at 40 cents a kilogram?
12. How can you change the state of a body? Give three methods.
13. Correct the statement 1 ccm. = 1 g.
14. How many liters in 32 quarts?

CHAPTER II

MOLECULAR FORCES AND MOTIONS

(1) Evidences of Molecular Motion in Gases

14. Size of Molecules.—The difference between solids, liquids, and gases has been explained as due to the different behavior of molecules in the three states of matter. That is, in solids they cling together, in liquids they move freely, and in gases they separate. At this time we are to consider the *evidences* of molecular motion in gases. It must be kept in mind that molecules are exceedingly small. It has been said that if a bottle containing about 1 ccm. of ordinary air has pierced in it a minute opening so that 100,000,000 molecules (a number nearly equal to the population of the United States) pass out every second, it would take, not minutes or hours, but nearly 9000 years for all of the molecules to escape. The number of molecules in 1 ccm. of air at 0°C. and 76 cm. pressure has been calculated by Professor Rutherford to be 2.7×10^{19}. It is evident that such minute particles cannot be seen or handled as *individuals*. We must judge of their size and action by the results obtained from experiments.

15. Diffusion of Gases.—One line of evidence which indicates that a gas consists of moving particles is the rapidity with which a gas having a strong odor penetrates to all parts of a room. For example, if illuminating gas is escaping it soon diffuses and is noticed throughout the room. In fact, the common experience of the diffusion of gases having a strong odor is such that we promptly recognize that it is due to motion of some kind.   The gas having the odor consists of little particles that are continually hitting their neighbors and are being struck and buffeted in turn until the individual molecules are widely scattered.   When cabbage is boiled in the kitchen soon all in the house know it.   Other illustrations of the *diffusion* of gases will occur to anyone from personal experience, such for instance as the pleasing odor from a field of clover in bloom.

The following experiment illustrates the rapid diffusion of gases.

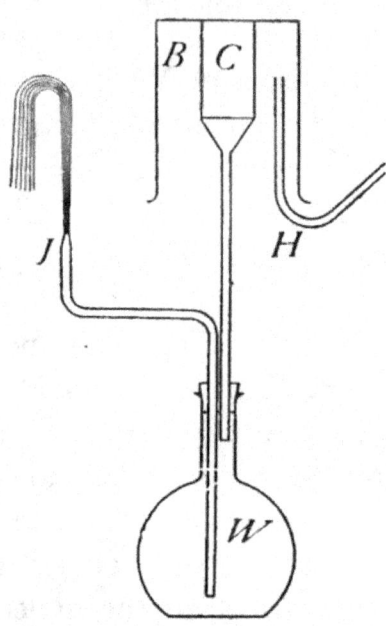

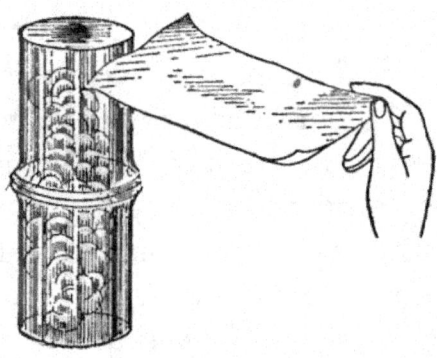
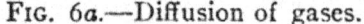

FIG. 6a.—Diffusion of gases.   FIG. 6b.—Effusion of gases.

Take two tumblers (see Fig. 6a), wet the inside of one with a few drops of strong ammonia water and the other with a little hydrochloric acid.   Cover each with a sheet of clean paper.   Nothing can now be seen in either tumbler.   Invert the second one over the first with the paper between, placing them so that the edges will match. On removing the paper it is noticed that both tumblers are quickly filled with a cloud of finely divided particles, the two substances having united chemically to form a new substance, ammonium chloride.

On account of their small size, molecules of air readily pass through porous solids, cloth, unglazed earthenware, etc.   The following experiment shows this fact strikingly. (See Fig. 6b.)

A flask containing water is closed by a rubber stopper through which pass the stem of a glass funnel and a bent glass tube that has been drawn out to a small opening (*J*). The funnel has cemented in its top an inverted porous clay jar (*C*), over the top of the latter is placed a beaker (*B*). A piece of flexible rubber tubing (*H*) leading from a hydrogen generator is brought up to the top of the space between the jar and the beaker. When hydrogen gas is allowed to flow into the space between *C* and *B*, the level of the water in *W* is seen to lower and a stream of water runs out at *J* spurting up into the air.

On stopping the flow of hydrogen and removing *B*, the water falls rapidly in *J* and bubbles of air are seen to enter the water from the tube. (The foregoing steps may be repeated as often as desired).

This experiment illustrates the fact that the molecules of some gases move faster than those of some other gases. Hydrogen molecules are found to move about four times as fast as air molecules. Hence, while both air and hydrogen molecules are at first going in opposite directions through the walls of *C*, the hydrogen goes in much faster than the air comes out. In consequence it accumulates, creates pressure, and drives down the water in *W* and out at *J*. On removing *B*, the hydrogen within the porous cup comes out much faster than the air reënters. This lessens the pressure within, so that air rushes in through *J*. This experiment demonstrates not only the fact of molecular motion in gases but also that molecules of hydrogen move much faster than those of air. (This experiment will work with illuminating gas but not so strikingly.)

Careful experiments have shown that the speed of ordinary air molecules is 445 meters or 1460 ft. per second; while hydrogen molecules move at the rate of 1700 meters or 5575 ft. or more than a mile per second.

16. Expansion of Gases.—Gases also possess the property of indefinite expansion, that is, if a small quantity of gas is placed in a vacuum, the gas will expand immediately to fill the entire space uniformly. This is shown by an experiment with the air pump. On raising the piston the air follows instantly to fill up the space under it. As the air is removed from the receiver of an air pump the air remaining is uniformly distributed within.

17. How Gases Exert Pressure.—It is further found that air under ordinary conditions exerts a pressure of about 15 lbs. to the square inch. In an automobile tire the pressure may be 90 lbs. and in a steam boiler it may be 200 lbs. or more to the square inch.

How is the pressure produced? The molecules are not packed together solidly in a gas, for when steam changes to water it shrinks to about $\frac{1}{1600}$ of its former volume. Air diminishes to about $\frac{1}{800}$ of its volume on changing to liquid air. The pressure of a gas is not due then to the gas filling all of the space in which it acts, but is due rather to the *motion* of the molecules. The blow of a single molecule is imperceptible, but when multitudes of molecules strike against a surface their combined effect is considerable. In fact, this action is known to produce the pressure that a gas exerts against the walls of a containing vessel. Naturally if we compress twice as much gas into a given space there will be twice as many molecules striking in a given time, which will give twice as much pressure.

If gas is heated, it is found that the heat will cause a swifter motion of the molecules. This will also make the molecules strike harder and hence cause the gas to expand or exert more pressure.

17a. Brownian Movements.—Direct photographic evidence of the motion of molecules in gases has been obtained by studying the behavior of minute drops of oil suspended in stagnant air. Such drops instead of being at rest are constantly dancing about as if they were continually receiving blows from many directions. These motions have been called *Brownian Movements* (see Fig. 7).

It has been proved that these movements are due to the blows that these small drops receive from the swiftly moving molecules of the gas about them. If the drops are made smaller or the gas more dense, the movements increase in intensity. These effects are especially marked at a pressure of 0.01 of an atmosphere.

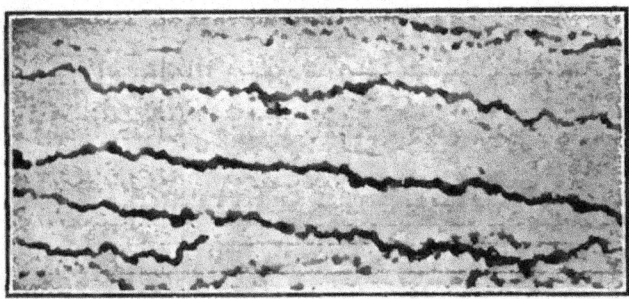

Fig. 7.—Photograph of Brownian movement. This record is prepared by the aid of Siedentopf's ultra-microscope and a plate moving uniformly across the field from left to right.

Important Topics

It is assumed that air and all gases are made up of molecules in rapid motion; that this motion is dependent upon temperature and pressure. Evidence of this is shown by (a) diffusion, (b) expansion, (c) pressure. Brownian Movements.

Questions

1. What is the molecular (kinetic) theory of gases?
2. What three kinds of evidence help to confirm the theory?
3. What have you seen that seems to show that a gas consists of molecules in motion?
4. How many meters long is a 10-ft. pole?
5. A 50-kg. boy weighs how many pounds?
6. What are three advantages of the metric system?
7. What will 12 qts. of milk cost at 8 cents a liter?
8. A cube 1 meter each way will contain how many cubic centimeters? How many liters? What will a cubic meter of water weigh?

2

(2) MOLECULAR MOTION IN LIQUIDS

18. Diffusion of Liquids.—From the evidence given in Arts. 14–17, (a) of diffusion of odors, (b) of the continued *expansion* of air in the air pump, and (c) of the pressure exerted by a gas in all directions, one may realize without difficulty that a *gas consists of small particles in rapid motion*. Let us now consider some of the evidence of molecular motion in liquids. If a little vinegar is placed in a pail of water, all of the water will soon taste sour. A lump of sugar in a cup of tea will sweeten the entire contents. This action is somewhat similar to the diffusion of gases but it takes place much more slowly. It is therefore believed that the motion of liquid molecules is much slower than that of gas molecules.

Again, if a dish of water is left standing in the open air in fine weather, within a few days the dish will become dry though no one has taken anything from it. We say the water has *evaporated*. What was liquid is now *vapor*. If we were to observe carefully any dish of water we would find that it continually loses weight on dry days. That is, there is a constant movement of the molecules of water into the air. This movement of the molecules is explained as follows. There appear to be in the dish of water some molecules that by moving back and forth acquire a greater velocity than their neighbors; when these reach the surface of the liquid, some vibration or movement sends them flying into the air above. They are now vapor or gas molecules, flying, striking, and rebounding like the air molecules. Sometimes on rebounding, the water molecules get back into the water again. This is especially apt to happen when the air is damp, *i.e.*, when it contains many water molecules. Sometimes the air over a dish becomes *saturated*, as in the upper part of a corked bottle containing water. Although molecules are continually leaving the surface of the water they cannot escape from the bottle, so in time as many molecules must return to the water from the space above as leave the water in the same time. When this condition exists, the air above the water is said to be *saturated*. On very damp days the air is often saturated. The explanation above shows why wet clothes dry so slowly on such a day (See Arts. 166–7 on Saturation.)

19. Cooling Effect of Evaporation. We have seen that warming a gas increases its volume. This expansion is due to the increased motion of the warmed molecules. Now the molecules that escape from a liquid when it evaporates are naturally the fastest moving ones, *i.e.*, the hottest ones. The molecules remaining are the slower moving ones or colder molecules. The liquid therefore becomes colder as it evaporates, unless it is heated. This explains why water evaporating on the surface of our bodies cools us. In evaporating, the water is continually losing its warm, fast moving molecules. The *cooling effect* of evaporation is therefore an evidence of molecular motion in liquids.

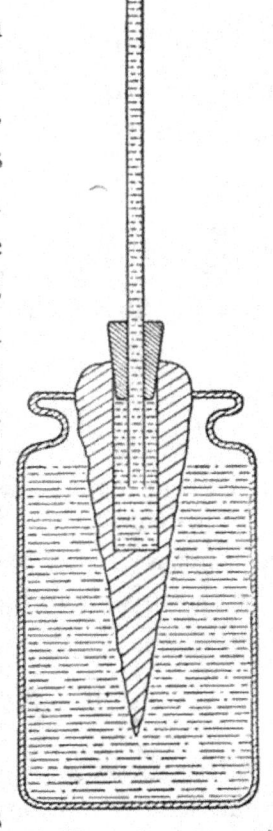

FIG. 8.—Osmosis shown by carrot placed in water.

20. Osmosis.—If two liquids are separated by a membrane or porous partition, they tend to pass through and mix. This action is called osmose, or *osmosis*.

Such a movement of liquid molecules in osmosis may be illustrated by filling a beet or carrot that has had its interior cut out to form a circular opening (see Fig. 8) with a thick syrup. The opening is then closed at the top with a rubber stopper through which passes a long glass tube.

If the carrot is immersed in water, as in Fig. 8, a movement of water through the porous wall to the interior begins at once. Here, as in the experiment of the hydrogen and air passing through the porous cup, the lighter fluid moves faster. The water collecting in the carrot rises in the tube. This action of liquids passing through porous partitions and mingling is called *osmosis*.

Gases and liquids are alike in that each will *flow*. Each is therefore called a *fluid*. Sometimes there is much resistance to the flow of a liquid as in molasses. This resistance is called *viscosity*. Alcohol and gasoline have little viscosity. They are *limpid* or *mobile* Air also has some viscosity. For instance, a stream of air always drags some of the surrounding air along with it.

Important Topics

1. Liquids behave as if they were composed of small particles in motion.
2. This is shown by (1) Diffusion, (2) Solution, (3) Evaporation, (4) Expansion, (5) Osmosis.

Exercises

1. Give an example or illustration of each of the five evidences of molecular motion in liquids.
2. When is air saturated? What is the explanation?
3. Why does warming a liquid increase its rate of evaporation?
4. Air molecules are in rapid motion in all directions. Do they enter a liquid with a surface exposed to the air? Give reason.
5. What are some of the inconveniences of living in a saturated atmosphere?
6. Fish require oxygen. How is it obtained?

(3) Molecular Forces in Liquids

21. Cohesion and Adhesion.—In liquids "the molecules move about freely yet tend to cling together." This tendency of molecules to cling together which is not noticeable in gases is characteristic of *liquids* and especially of *solids*. It is the cause of the viscosity mentioned in the previous section and is readily detected in a variety of ways. For instance, not only do liquid molecules cling together to form drops and streams, but they cling to the molecules of solids as well, as is shown by the wet surface of an object that has been dipped in water. The attraction of like molecules for one another is called *cohesion*, while the attraction of *unlike molecules* is called *adhesion*, although the force is the same whether the molecules are alike or unlike. It is the former that causes drops of water to form and that holds iron, copper, and other solids so rigidly together. The adhesion of glue to other objects is well known. Paint also "sticks" well. Sometimes the "joint" where two boards are glued together is stronger than the board itself. The force of attraction between molecules has been studied carefully. The attraction acts only through very short distances. The attraction even in liquids is considerable and may be measured. The cohesion of water may be shown by an experiment where the force required to pull a glass plate from the surface of water is measured.

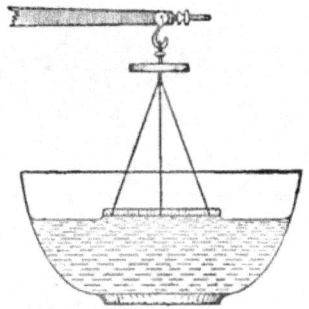

Fig. 9.—The water is pulled apart.

Take a beam balance and suspend from one arm a circular glass plate, Fig. 9. Weigh the plate and its support. Adjust the glass plate so that it hangs horizontally and just touches the surface of clean water, the under side being completely wet. Now find what additional weight is required to raise the glass plate from the water.

Just as the plate comes from the water its under side is found to be wet. That is, *the water was pulled apart,* and the plate was not pulled from the water. The cohesion of the water to itself is not so strong as its adhesion to the glass.

The cohesion of liquids is further shown by the form a drop of liquid tends to take when left to itself. This is readily seen in small drops of liquids. The spherical shape of drops of water or mercury is an example. A mixture of alcohol and water in proper proportions will just support olive oil within it. By carefully dropping olive oil from a pipette into such a mixture, a drop of the oil, an inch or more in diameter suspended in the liquid, may be formed. It is best to use a bottle with plane or flat sides, for if a round bottle is used, the sphere of oil will appear flattened.

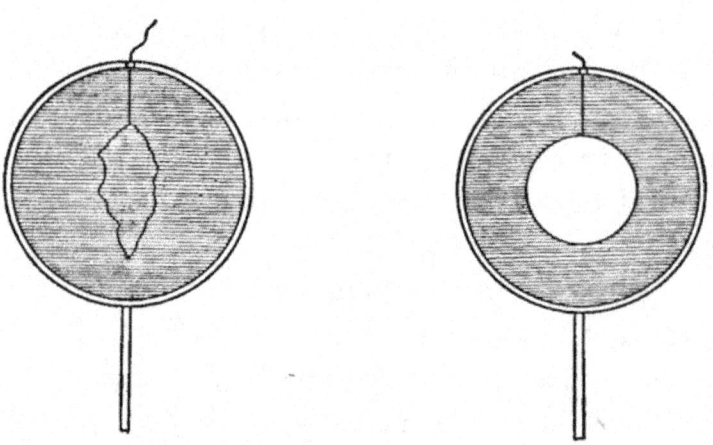

FIG. 10 *a.* FIG. 10 *b.*
FIGS. 10 *a* AND *b.*—Surface tension of a liquid film.

22. Surface Tension.—The cohesion of liquids is also indicated by the tendency of films to assume the smallest possible surface. Soap bubble films show this readily. Fig. 10 *a* represents a circular wire form holding a film in which floats a loop of thread. The tension of the

film is shown in Fig. 10 *b* by the circular form of the loop
after the film within it has been pierecd by a hot wire,
Fig. 11 shows a rectangular wire form with a "rider."
The tension in the film draws the rider forward.

A soap bubble takes its
spherical shape because
this form holds the con-
fined air within the

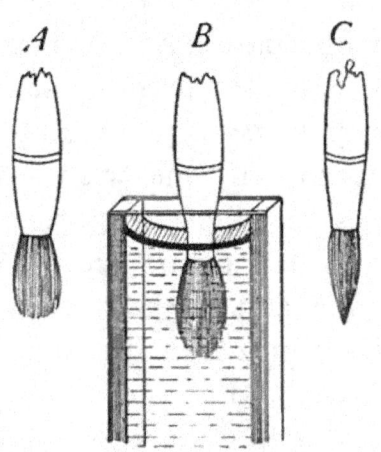

FIG. 12.—Surface tension causes
the pointed shape.

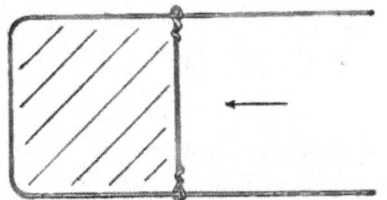

FIG. 11.—The rider is drawn forward.

smallest possible surface. A drop of liquid is spherical
for the same reason. Many illustrations of the tension
in films may be given. Users of water colors notice that
a dry camel's-hair brush is bushy. (Fig. 12 *A*). When
in water it is still bushy. (Fig. 12 *B*.)
But when it is taken from the water
and the excess is shaken from it, it is
pointed as in Fig. 12 *C*. It is held to
the pointed shape by the tension of
the liquid film about the brush.

FIG. 13.—A needle
depresses the surface
when floating.

The surface of water acts as if covered by a film which
coheres more strongly than the water beneath it. This is
shown by the fact that a steel needle or a thin strip of
metal may be floated upon the surface of water. It is supported by the surface film. (See Fig. 13.) If the film
breaks the needle sinks. This film also supports the little
water bugs seen running over the surface of a quiet pond in

summer. The surface film is stronger in some liquids than in others. This may be shown by taking water, colored so that it can be seen, placing a thin layer of it on a white surface and dropping alcohol upon it. Wherever the alcohol drops, the water is seen to pull away from it, leaving a bare space over which the alcohol has been spread. This indicates that the alcohol has the weaker film. The *film of greasy benzine is stronger* than the film of the pure material. If one wishes to remove a grease spot and places pure benzine at the center of the spot, the stronger film of the greasy liquid will pull away from the pure benzine, and spread out, making a larger spot than before, while if pure benzine is placed *around the grease spot,* the greasy liquid at the center pulls away from the pure benzine, drawing more and more to the center, where it may be wiped up and the grease entirely removed.

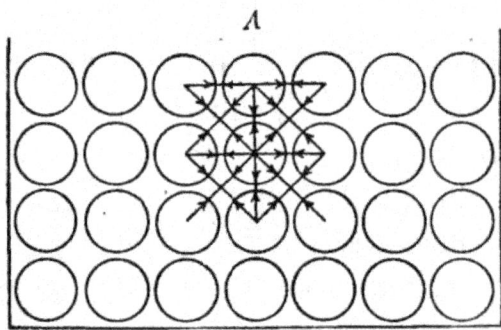

FIG. 14.—The molecule at *A* is held differently from one within the liquid.

23. Explanation of the Surface Film.—Beneath the surface of a liquid each molecule is attracted by all the other molecules around it. It is attracted equally in all directions. Consequently the interior molecules move very easily over each other in any direction. A molecule at the surface, as at *A*, Fig. 14, is not attracted *upward* by other liquid molecules. Its freedom of motion is thereby hindered with the result that a molecule at the surface behaves differently from one beneath the surface. The surface molecules act as if they form an elastic skin or membrane upon the liquid surface.

24. Capillarity.—A striking action of the surface film of a liquid is seen in the rise of liquids in tubes of small bore when the liquid *wets* them. If the liquid *does not wet* the tube, as when mercury is placed in glass, the liquid is depressed. It is found in general that: *Liquids rise in capillary tubes when they wet them and are depressed in tubes which they do not wet; the smaller the diameter of the tube the greater the change of level.* (See Fig. 15.) This action is explained as follows: The molecules of a liquid have an attraction for each other and also

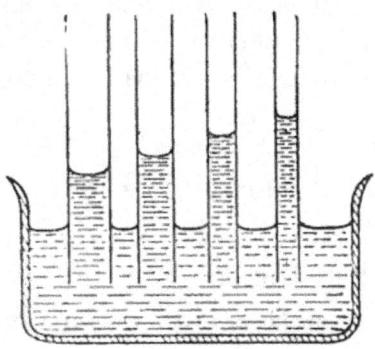

FIG. 15.—Capillary attraction in tubes.

for the sides of a tube. The former is called "cohesion for itself," the latter is called "adhesion for the sides of the containing vessel." If the cohesion for itself is greater than the adhesion for the side of the containing vessel, the liquid is pulled away from the side and is depressed. If the adhesion is greater, the liquid is elevated. This action is called "capillary action" from the Latin word (*capillus*) signifying hair, since it shows best in fine hairlike tubes.

There are many common illustrations of capillary action: oil rising in a wick; water rising in a towel or through clothes; ink in a blotter, etc. The minute spaces between the fibers composing these objects act as fine tubes. If cloth is treated with a preparation which prevents water from adhering to its fibers, the material will not be wet when water is poured upon it, because the water will not run in between the fibers; a surface film spreads over the cloth so that no water enters it. *Cravenette cloth* has been treated in this way and hence is waterproof.

The action of this film may be shown by the following experiment. Dip a sieve of fine copper gauze in melted paraffin, thus coating each wire so that water will not adhere to it. Water may now be poured into the sieve, if a piece of paper is first laid in it to break the force of the water. On carefully removing the paper the surface film of the water will prevent the passage of the water through the sieve.

25. Capillary Action in Soils.—The distribution of moisture in the soil depends largely upon capillary action. When the soil is compact the minute spaces between the soil particles act as capillary tubes, thus aiding the water to rise to the surface. As the water evaporates from the surface more of it rises by capillary action from the damper soil below. Keeping the soil loose by cultivation, makes the spaces between the particles too large for much capillary action, thus the moisture is largely prevented from rising to the surface.

In the semi-arid regions of the West *"dry farming"* is successfully practised. This consists in keeping the surface covered with a "dust mulch" produced by frequent cultivation. In this way the moisture is kept below the surface, where it can be utilized during the hot dry summer by the roots of growing plants.

Important Topics

1. Attractive forces between liquid molecules.
2. Cohesion (like molecules); adhesion (unlike molecules).
3. Special effects of this force are classified as (a) capillary action, and (b) surface tension.

Exercises

1. What evidence of capillary action have you seen outside of the laboratory?
2. What is the explanation for capillary action?
3. Where are surface films found?
4. What are three common effects of surface films?
5. Explain why cravenette cloth sheds water.
6. If a circular glass disc 10 cm. in diameter requires 50 grams of force to draw it from the water, what is the cohesion of water per square centimeter?
7. What is the weight in grams of 1 ccm. of water? of a liter of water?
8. Name five examples of adhesion to be found in your home.
9. Under what conditions will a liquid wet a solid and spread over it.
10. When will it form in drops on the surface?
11. Explain the proper procedure for removing a grease spot with benzine.
12. What difference is there between a liquid and a fluid?
13. Why cannot a "soap bubble" be blown from pure water?
14. Which are larger, the molecules of steam or those of water? Why?
15. Why is the ground likely to be damp under a stone or board when it is dry all around?
16. Why does any liquid in falling through the air assume the globule form?
17. Give three examples of capillary attraction found in the home. Three out of doors.
18. Why does cultivation of the soil prevent rapid evaporation of water from the ground?

(4) EVIDENCES OF MOLECULAR FORCES IN LIQUIDS AND SOLIDS

26. Solutions.—A crystal of potassium permanganate is placed in a liter of water. It soon dissolves and on shaking the flask each portion of the liquid is seen to be colored red. The dissolving of the permanganate is an illustration of the attraction of the molecules of water for the molecules of the permanganate. We are familiar with this action in the seasoning of food with salt and sweetening with sugar.

Water will dissolve many substances, but in varying degrees, *i.e.*, of some it will dissolve much, of others, little, and some not at all. Further, different liquids have different solvent powers. Alcohol will dissolve resin and shellac, but it will not dissolve gum arabic, which is soluble in water. Benzine dissolves grease. Beeswax is not dissolved by water, alcohol or benzine, but is soluble in turpentine.

It is found that the *temperature* of the liquid has a marked effect upon the amount of substance that will dissolve. This is an indication that the *motions* of the molecules are effective in solution. It appears that dissolving a solid is in some respects similar to evaporation, and just as at higher temperatures more of the liquid evaporates, because more of the molecules will escape from the liquid into the air above, so at higher temperatures, more molecules of a solid will detach themselves through greater vibration and will move into the liquid.

Further, just as an evaporating liquid may saturate the space above it so that any escape of molecules is balanced by those returning, so with a dissolving solid, the liquid may become *saturated* so that the solution of more of the solid is balanced by the return of the molecules from the liquid to the solid condition.

27. Crystals and Crystallization.—This return from the liquid to the solid state, of molecules that are in solution, is especially noticeable when the solution is cooling or evaporating and hence is losing its capacity to hold so much of the solid. On returning to the solid, the molecules attach themselves in a definite manner to the solid portion, building up regular solid forms. These regular forms are *crystals.* The action that forms them is called *crystallization.*

Each substance seems to have its own *peculiar form of crystal* due to the manner in which the molecules attach themselves to those previously in place. The largest and most *symmetrical crystals are* those in which the molecules are deposited slowly with no disturbance of the liquid. Beautiful crystals of alum may be obtained by dissolving 25 g. of alum in 50 ccm. of hot water, hanging two or three threads in the solution and letting it stand over night. The thread fibers provide a foundation upon which crystals grow.

When a *solution* of a solid evaporates, the molecules of the *liquid* escape as a gas, the molecules of the *solid* remain accumulating as crystals. This principle has many uses: (a) sea water is purified by *evaporating the water and condensing the vapor,* which of course forms pure water. (b) Water is forced down to *salt* beds where it dissolves the salt. The brine is then raised and evaporated, leaving the salt in the evaporating pans.

28. Absorption of Gases by Solids and Liquids.—If a piece of heated charcoal is placed in a test-tube containing ammonia gas, inverted in mercury, the ammonia is seen to disappear, the mercury rising to take its place. The ammonia has been absorbed by the charcoal, the gas molecules clinging closely to the solid. The charcoal being very porous presents a large surface to the action of the gas.

This experiment indicates that attraction exists beween gas molecules and other molecules. Many porous substances have this power of absorbing gases. We have all noticed that butter has its flavor affected by substances placed near it.

That *liquids absorb gases* is shown by slowly heating

cold water in a beaker. Small bubbles of air form on the sides and rise before the boiling point is reached. Ammonia gas is readily absorbed in water, the bubbles disappearing almost as soon as they escape into the water from the end of the delivery tube. *Household ammonia* is simply a solution of ammonia gas in water. On warming the solution of ammonia the gas begins to pass off; thus, warming a liquid tends to drive off any gas dissolved in it.

Soda water is made by forcing carbon dioxide gas into water under strong pressure. When placed in a vessel open to the air the pressure is lessened and part of the gas escapes. The dissolved gas gives the characteristic taste to the beverage.

Important Topics

1. The solution of solids is increased by heating.
2. The solution of gases is decreased by heating.
3. Pressure increases the quantity of gas that can be dissolved in a liquid.
4. The attraction (cohesion) of molecules of a dissolved solid for each other is shown by crystallization.

Exercises

1. How do fish obtain oxygen for breathing?
2. Why does warming water enable it to dissolve more of a salt?
3. Why does warming water lessen the amount of a gas that will stay in solution?
4. Will water absorb gases of strong odor? How do you know?
5. Name three solvents. Give a use for each.
6. What liquids usually contain gases in solution? Name some uses for these dissolved gases.
7. What is the weight of a cubic meter of water?
8. Name three substances obtained by crystallization.
9. How is maple sugar obtained?
10. Name five crystalline substances.

(5) Evidence of Molecular Forces in Solids

29. Differences between Solids and Gases.—In studying gases, it is seen that they behave as if they were composed of small particles in rapid motion, continually striking and rebounding, and separating to fill any space into which they are released. This action indicates that there is practically no attractive force between such molecules.

Between the molecules of a solid, however, the forces of attraction are strong, as is shown by the fact that a solid often requires a great force to pull it apart; some, as steel and iron, show this property in a superlative degree, a high-grade steel rod 1 cm. in diameter requiring nearly 9 tons to pull it apart. Tests show that the breaking strengths of such rods are directly proportional to their areas of cross-section. That is, twice the area has twice the breaking strength.

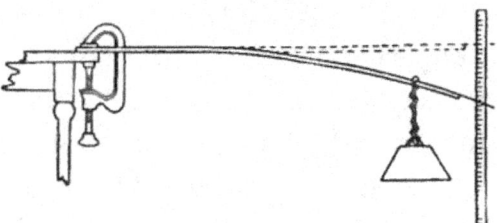

FIG. 16.—Elasticity of bending.

30. Elasticity.— Fully as important as a knowledge of the breaking strengths of solids, is the knowledge of what happens when the forces used are not great enough to break the rods or wires.

Take a wooden rod (as a meter stick) and clamp one end to the table top, as in Fig. 16. At the other end hang a weight. Fasten a wire to this end so that it projects out in front of a scale. Add successively several equal weights and note the position of the wire each time. Remove the weights in order, noting the positions as before. The rod will probably return to the first position.

This simple experiment illustrates a characteristic of solids: that of changing shape when force is applied and of returning to the original shape when the force is removed. This property is called *elasticity*.

Tests of elasticity are made by subjecting wire of different materials but of the same dimensions to the same tension. The one changing least is said to have the greatest *elastic force* or elasticity. If greater forces are applied to the wire and then removed, one will finally be found that will permanently stretch the wire so that it will not return exactly to the former length. The wire has now passed its *elastic limit* and has been permanently stretched.

Just as there are great differences between the *elastic forces* of different substances, so there are great differences in the *limits of elasticity*. In some substances the limit is reached with slight distortion, while others are *perfectly elastic* even when greatly stretched. India rubber is an example of a body having *perfect* elasticity through wide limits. Glass has great *elastic force* but its *limit* of *elasticity* is soon reached. Substances like India rubber may be said to have great "*stretchability*," but little elastic force. In physics, elasticity refers to the elastic force rather than to ability to endure stretching.

31. Kinds of Elasticity.—*Elasticity may be shown in four ways: compression, bending or flexure, extension or stretching, twisting or torsion.* The first is illustrated by squeezing a rubber eraser, the second by an automobile spring, the third by the stretching of a rubber band, the fourth by the twisting and untwisting of a string by which a weight is suspended.

There are two kinds of elasticity: (1) elasticity of form or shape; (2) elasticity of volume. Gases and liquids possess elasticity of volume, but not of shape, while solids may have both kinds. Gases and liquids are perfectly elastic because no matter how great pressure may be applied, as soon as the pressure is removed they regain their former volume. No solid possesses perfect elasticity, because sooner or later the limit of elasticity will be reached.

32. Hooke's Law.*—On examining the successive movements of the end of the rod in Art. 30, we find that they are approximately equal. Carefully conducted experiments upon the elasticity of bodies have shown that the changes in shape are *directly proportional* to the forces applied, provided that the limit of elasticity is not reached. This relation, discovered by Robert Hooke, is sometimes expressed as follows: "*Within the limits of perfect elasticity, all changes of size or shape are directly proportional to the forces producing them.*"

33. Molecular Forces and Molecular Motions.—If a solid is compressed, on releasing the pressure the body regains its former shape if it has not been compressed too far. This indicates that at a given temperature the "molecules of a solid tend to remain at a fixed distance from each other, and resist any attempt to decrease or increase this distance." This raises the question, Why does not the cohesion pull the molecules tightly together so that compression would be impossible? The reason is that heat affects the size of solid bodies. On lowering the temperature, bodies do contract, for as soon as the temperature is lowered the vibration of the molecule is lessened. On raising the temperature the molecules are pushed farther apart.

The size of a body, then, is the result of a balance of opposing forces. The attractive force between the molecules pulling them together is *cohesion*, while the force which pushes them apart is due to the motions of the molecules. Raising the temperature and thus increasing

---

* A law is a statement of a constant mode of behavior. It is often expressed in mathematical language.

the motion causes expansion; lowering the temperature decreases the molecular motion and so causes contraction. If an outside force tries to pull the body apart or to compress it this change of size is resisted by either cohesion or molecular motion.

34. Properties of Matter.—Many differences in the physical properties of solids are due to differences between the cohesive force of different kinds of molecules. In some substances, the attraction is such that they may be rolled out in very thin sheets. Gold is the best example of this, sheets being formed $\frac{1}{300,000}$ of an inch thick. This property is called *malleability*. In other substances the cohesion permits it to be drawn out into fine threads or wire. Glass and quartz are examples of this. This property is called *ductility*. In some, the cohesion make the substance excessively *hard*, so that it is difficult to work or scratch its surface. The diamond is the hardest substance known. Some substances are *tough*, others *brittle*. These are tested by the ability to withstand sudden shocks as the blow of a hammer.

Important Topics

1. Molecular forces in solids; (*a*) adhesion, (*b*) cohesion.
2. Elasticity, Hooke's Law.
3. Contraction on cooling.
4. Malleability, ductility, hardness, brittleness, etc.

Exercises

1. Give an illustration of Hooke's Law from your own experience.
2. What devices make use of it?
3. Do solids evaporate. Give reasons.
4. When iron is welded, is cohesion or adhesion acting?
5. When a tin basin is soldered, is cohesion or adhesion acting?
6. Sometimes a spring is made more elastic by *tempering* and made soft by *annealing*. Look up the two terms. How is each accomplished?

7. Review the definitions: solid, liquid, and gas. Why do these definitions mean more to you now than formerly?

8. If a wire is stretched 0.3 cm. on applying 4 kg. of force, what force will stretch it 0.75 cm? Explain.

9. How long will it take under ordinary conditions for a gas molecule to cross a room? Give reasons for your answer.

10. What is meant by the elastic limit of a body?

11. Without reaching the elastic limit, if a beam is depressed 4 mm. under a load of 60 kg., what will be the depression under a load of 400 kg.? Of 600 kg.?

12. Name three substances that possess elasticity of volume.

13. Give three examples of each; elasticity of (1) compression, (2) stretching, (3) torsion, (4) flexure

Review Outline: Introduction and Molecules

Physics; definition, topics considered, physical and chemical changes.

Science; hypothesis, theory, law. Knowledge; common, scientific.

Matter; three states, molecular theory. Mass, weight, volume.

Metric system; units, tables, equivalents, advantages.

Evidences of molecular motions; gases (3), liquids (5), solids (3).

Evidences of molecular forces; liquids (3), solids (many) special properties such as: elasticity, tenacity, ductility, hardness, etc.

Hooke's law; applications.

CHAPTER III

MECHANICS OF LIQUIDS

(1) The Gravity Pressure of Liquids

35. Pressure of Liquids against Surfaces.—The sight of a great ship, perhaps built of iron and floating on water, causes one to wonder at the force that supports it. This same force is noticed when one pushes a light body, as a cork, under water. It is quite evident in such a case that a force exists sufficient to overcome the weight of the cork so that it tends to rise to the surface. Even the weight of our bodies is so far supported by water that many persons can float.

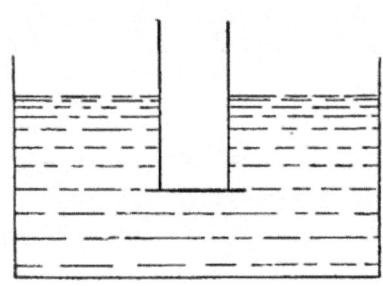

Fig. 17.—Water forces the card against the chimney.

The following experiment provides a means of testing this force:

If an empty can is pushed down into water, we feel at once the force of the liquid acting against the object and tending to push it upward. It may be noticed also that so long as the can is not completely submerged the deeper the can is pushed into the water the greater is the upward force exerted by the liquid.

We may test this action in various ways: a simple way is to take a cylindrical lamp chimney, press a card against its lower end and place it in the water in a vertical position. The force of the water will hold the card firmly against the end of the chimney. (See Fig. 17.) The amount of force may be tested by dropping shot into the tube until the card drops off. At greater depths more shot will be required, showing that the force of the water increases with the depth. Or one may pour water into the chimney. It will then be found that the card does not drop until the level of the water inside the chimney is the same as on the outside. That is, before the card will fall off, the water must stand as high within the chimney as without no matter to what depth the lower end of the chimney is thrust below the surface of the water.

36. Law of Liquid Pressure.—As there is twice as much water or shot in the chimney when it is filled to a depth of 10 cm. as there is when it is filled to a depth of 5 cm. the force of the water upward on the bottom must be twice as great at a depth of 10 cm. as at a depth of 5 cm. Since this reasoning will hold good for a comparison of forces at any two depths, we have the law: "*The pressure exerted by a liquid is directly proportional to the depth.*"

The amount of this force may be computed as follows: First, the card stays on the end of the tube until the *weight* of water from above equals the force of the water from below, and second, the card remains until the water is at the same *height* inside the tube as it is outside. Now if we find the weight of water at a given depth in the tube, we can determine the force of the water from below. If for instance the chimney has an area of cross-section of 12 sq. cm. and is filled with water to a depth of 10 cm., the volume of the water contained will be 120 ccm. This volume of water will weigh 120 g. This represents then, not only the weight of the water in the tube, but also the force of the water against the bottom. In a similar way one may measure the force of water against any horizontal surface.

37. Force and Pressure.—We should now distinguish between *force* and *pressure*. Pressure refers to the force acting against *unit area*, while force refers to the action against the whole surface. Thus for example, the atmospheric *pressure* is often given as 15 pounds to the square inch or as one kilogram to the square centimeter. On the other hand, the air may exert a *force* of more than 300 pounds upon each side of the hand of a man; or a large ship may be supported by the *force* of thousands of tons exerted by water against the bottom of the ship.

In the illustration, given in Art. 36, the upward *force* of the water against the end of the tube at a depth of 10 cm. is computed as 120 grams. The *pressure* at the *same* depth will be 10 grams per sq. cm. What will be the pressure at a depth of 20 cm.? at a depth of 50 cm.? of 100 cm.? Compare these answers with the law of liquid pressure in Art. 36.

38. Density.—If other liquids, as alcohol, mercury, etc., were in the jar, the chimney would need filling to the same level outside, with the *same* liquid, before the card would fall off. This brings in a factor that was not considered before, *that of the mass* of a cubic centimeter of the liquid.* This is called the *density* of the liquid. Alcohol has a density of 0.8 g. per cubic centimeter, mercury of 13.6 g. per cubic centimeter, while water has a density of 1 g. per cubic centimeter.

39. Liquid Force against Any Surface.—To find the force exerted by a liquid against a surface we must take into consideration the area of the surface, and the height and the *density* of the liquid above the surface. The following law, and the formula representing it, which concisely expresses the principle by which the force exerted by a liquid against any surface may be computed, should be memorized:

The force which a liquid exerts against any surface, equals the area of the surface, times its average depth below the surface of the liquid, times the weight of unit volume of the liquid.

Or, expressed by a formula, $F = Ahd$. In this formula, "F" stands for *the force which a liquid exerts against any*

---

*The *mass* of a body is the *amount of matter in it*, the *weight* is the *pull of the earth upon it*.

surface, "A" *the area of the surface,* "H," for *the average depth (or height) of the liquid pressing on the surface,* and "d", for *the weight of unit volume of the liquid.* This is the first illustration in this text, of the use of a formula to represent a law. Observe how accurately and concisely the law is expressed by the formula. When the formula is employed, however, we should keep in mind the law expressed by it.

We must remember that a liquid presses not only downward and upward but sideways as well, as we see when water spurts out of a hole in the side of a vessel. Experiments have shown that at a point the pressure in a fluid is the same in all directions, hence the rule given above may be applied to the pressure of a liquid against the side of a tank, or boat, or other object, provided we are accurate in determining the *average depth of the liquid;* The following example illustrates the use of the law.

For Example: If the English system is used, the area of the surface should be expressed in square feet, the depth in feet and the weight of the liquid in pounds per cubic foot. One cubic foot of water weighs 62.4 lbs.

Suppose that a box 3 ft. square and 4 ft. deep is full of water. What force will be exerted by the water against the bottom and a side?

From the law given above, the force of a liquid against a surface equals the product of the *area* of the surface, the *depth* of the liquid and its weight per unit volume, or using the formula, $F = Ahd$. To compute the downward force against the bottom we have the area, 9, depth, 4, and the weight 62.4 lbs. per cubic foot. $9 \times 4 \times 62.4$ lbs. $= 2246.4$ lbs. To compute the force against a side, the area is 12, the average depth of water on the side is 2, the weight 62.4, $12 \times 2 \times 62.4$ lbs. $= 1497.6$ lbs.

Important Topics

1. Liquids exert pressure; the greater the depth the greater the pressure.

2. Difference between force and pressure.
3. Rules for finding upward and horizontal force exerted by a liquid. $F = Ahd$.
4. Weight, mass, density.

Exercises

1. What is the density of water?
2. What force is pressing upward against the bottom of a flat boat, if it is 60 ft. long, 15 ft. wide and sinks to a depth of 2 ft. in the water? What is the weight of the boat?
3. If a loaded ship sinks in the water to an average depth of 20 ft., the area of the bottom being 6000 sq. ft., what is the upward force of the water? What is the weight of the ship?
4. If this ship sinks only 10 ft. when empty, what is the weight of the ship alone? What was the weight of the cargo in Problem 3?
5. What is the liquid force against one side of an aquarium 10 ft. long, 4 ft. deep and full of water?
6. What is the liquid force on one side of a liter cube full of water? Full of alcohol? Full of mercury? What force is pressing on the bottom in each case?
7. What depth of water will produce a pressure of 1 g. per square centimeter? 10 g. per square centimeter? 1000 g. per square centimeter?
8. What depth of water will produce a pressure of 1 lb. per square inch? 10 lbs. per square inch? 100 lbs. per square inch?
9. What will be the force against a vertical dam-breast 30 meters long, the depth of the water being 10 meters?
10. A trap door with an area of 100 sq. dcm. is set in the bottom of a tank containing water 5 meters deep. What force does the water exert against the trap door?
11. What is the force on the bottom of a conical tank, filled with water, the bottom of which is 3 meters in diameter, the depth 1.5 meters?
12. If alcohol, density 0.8 were used in problem 11, what would be the force? What would be the depth of alcohol to have the same force on the bottom as in problem 11?
13. What is the pressure in pounds per square inch at a depth of 1 mile in sea water, density 1.026 grams per cc.?

14. Find the force on the sides and bottom of a rectangular cistern filled with water, 20 ft. long, 10 ft. wide, and 10 ft. deep?
15. Find the force on the bottom of a water tank 14 ft. in diameter when the water is 15 ft. deep, when full of water.
16. Find the force on one side of a cistern 8 ft. deep and 10 ft. square, when full of water.
17. Find the force on a vertical dam 300 ft. long and 10 ft. high, when full of water.
18. Find the pressure at the bottom of the dam in question 17.
19. Why are dams made thicker at the bottom than at the top?
20. A ship draws 26 ft. of water, i.e., its keel is 26 ft. under water. What is the liquid force against a square foot surface of the keel? Find the pressure on the bottom.

(2) Transmission of Liquid Pressure

40. Pascal's Principle.—Liquids exert pressure not only due to their own weight, but when confined, may be made to transmit pressure to considerable distances. This is a matter of common knowledge wherever a system of waterworks with connections to houses is found, as in cities.

The transmission of liquid pressure has a number of important applications. The principle underlying each of these was first discovered by Pascal, a French scientis of the seventeenth century. Pascal's Principle, as it is called, may be illustrated as follows:

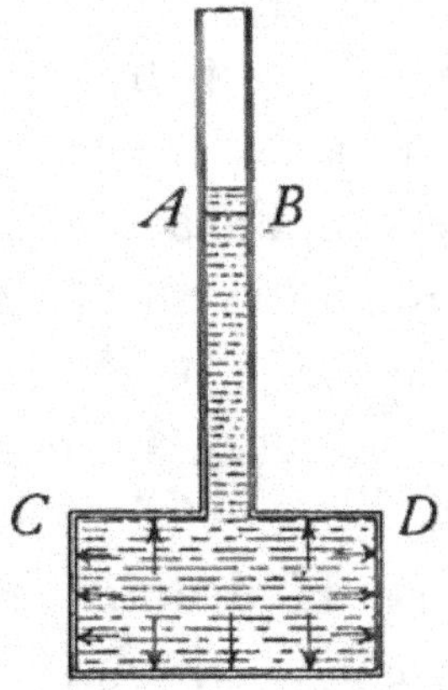

Fig. 18.—The force increases with the depth.

Suppose a vessel of the shape shown in Fig. 18, the upper part of which we may assume has an area of 1 sq. cm., is filled with water up to the level AB. A pressure will be exerted upon each square centimeter of area depending upon the depth. Suppose that the height of AB above CD is 10 cm., then the force upon 1 sq. cm. of CD is 10 g., or if the area of CD is 16 sq. cm., it receives a force of 160 g.

If now a cubic centimeter of water be poured upon *A B* it will raise the level 1 cm., or the head of water exerting pressure upon *CD* becomes 11 cm., or the total force in *CD* is 16×11 g., *ie.*, each square centimeter of *CD* receives an additional force of 1 g. *Hence the force exerted on a unit area at A B is transmitted to every unit area within the vessel.*

The usual form in which this law is expressed is as follows: *Pressure applied to any part of a confined liquid is transmitted unchanged, in all directions, and adds the same force to all equal surfaces in contact with the liquid.*

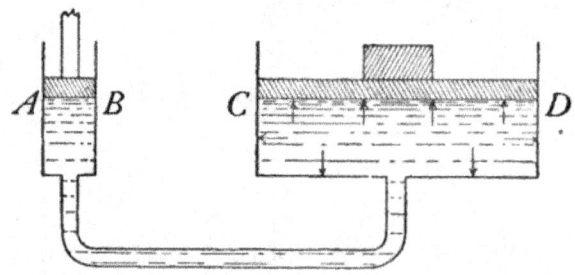

FIG. 19.—The force is proportional to the area.

The importance of this principle, as Pascal himself pointed out, lies in the fact that by its aid we are able to exert a great force upon a large area by applying a small force upon a small area of a confined liquid, both areas being in contact with the same liquid. Thus in Fig. 19 if the area of the surface *CD* is 2000 times the area of the surface *AB*, then 1 lb. applied to the liquid on *AB* will exert or sustain a force of 2000 lbs. on *CD*.

41. Hydraulic Press.—An important application of Pascal's principle is the *hydraulic press*. See Fig. 20. It is used for many purposes where great force is required, as in pressing paper or cloth, extracting oil from seeds, lifting heavy objects, etc. Many high school pupils have been seated in a *hydraulic chair* used by a dentist or barber. This chair is a modified hydraulic press.

The hydraulic press contains two movable pistons, P and p (see Fig. 20). The larger of these, P, has a cross-sectional area that may be 100 or 1000 times that of the smaller. The smaller one is moved up and down by a lever; on each upstroke, liquid is drawn in from a reservoir, while each down-stroke forces some of the liquid into the space about the large piston. Valves at V and V' prevent the return of the liquid. If the area of P is 1,000 times

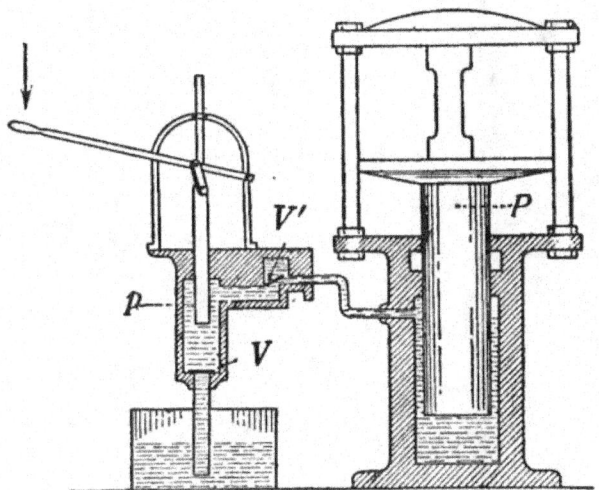

FIG. 20.—Cross-section of a hydraulic press.

that of p, then the force exerted by P is 1000 times the force employed in moving p. On the other hand, since the liquid moved by the small piston is distributed over the area of the large one, the latter will move only $\frac{1}{1000}$ as far as does the small piston. The relation between the motions of the two pistons and the forces exerted by them may be stated concisely as follows: The *motions* of the two *pistons* of the hydraulic press are inversely proportional to the forces exerted by them. The *cross-sectional areas* of the two pistons are, on the other hand, directly proportional to the forces exerted by them.

An application of Pascal's principle often employed in

cities is the hydraulic elevator. In this device a long plunger or piston extends downward from the elevator car into a cylinder sunk into the earth, sometimes to a depth of 300 ft. Water forced into this cylinder pushes the piston upward and when the water is released from the cylinder the piston descends.

Fig. 21 represents another form of hydraulic elevator, where the cylinder and piston are at one side of the elevator shaft. In this type, to raise the elevator, water is admitted to the cylinder pushing the piston downward.

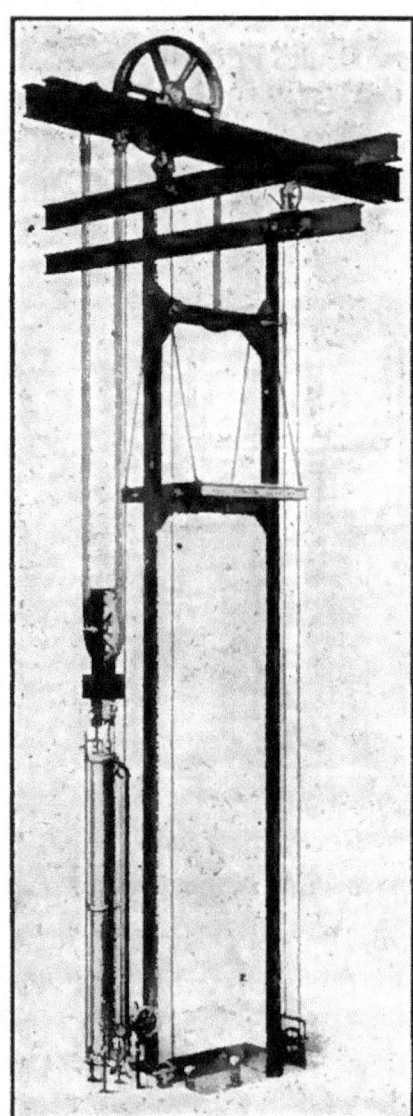

FIG. 21.—A hydraulic freight elevator.

42. Artesian Wells.—

Sometimes a porous stratum containing water in the earth's crust is inclined. Then if there are impervious strata (see Fig. 22), both above and below the water-bearing one, and the latter comes to the surface so that rain may fill it, a well sunk to the water-bearing stratum at a point where it is below the surface will usually give an artesian well, that is, one in which the water rises to or above the surface. Many are found in the United States.

43. Standpipes and Air Cushions.—Many who have

lived in cities where water is pumped into houses under pressure know that the water pressure is changed when

FIG. 22.—Conditions producing an artesian well.

several faucets are opened at the same time. Again, if several persons are using a hose for sprinkling, the pressure

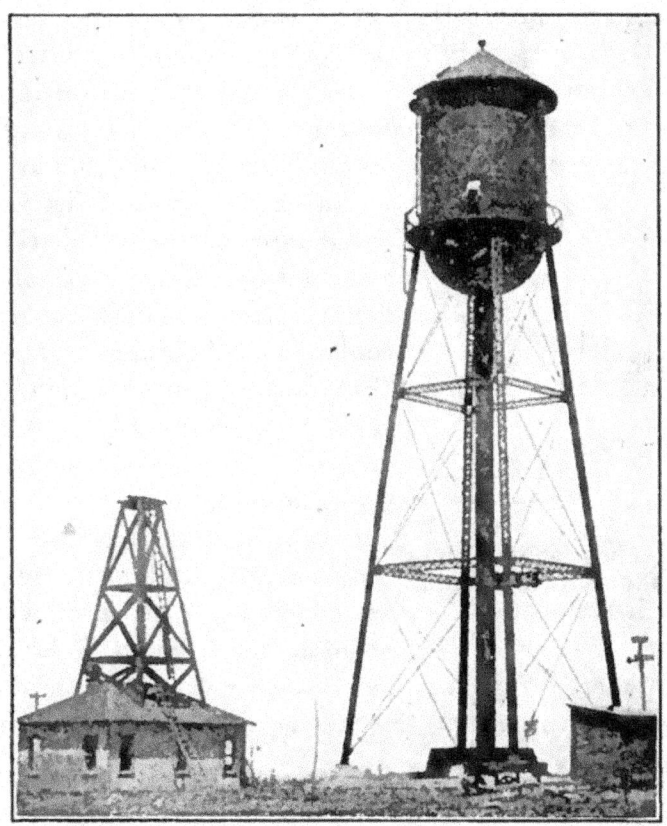

FIG. 23.—A standpipe.

may be lessened so as to be insufficient to force the water above the first floor. In order to allow for these changes some flexibility or spring must be introduced somewhere into the water-pipe system. Water is nearly incompressible and if no means were employed to take care of the pressure changes, the sudden stopping and starting of the flow would cause serious jars and start leaks in the pipes. Two common devices for controlling sudden changes in the water pressure are the *standpipe* and the *air cushion.*

The *standpipe* is simply a large vertical tube connected to the water mains from which and into which water readily flows. When many faucets are opened the water lowers; when most faucets are closed the water rises, giving a simple automatic control of the surplus water and a supply of water for a short time during a shutdown of the pumps. Standpipes are often used in towns and small cities. Fig. 23 represents the standpipe at Jerome, Idaho.

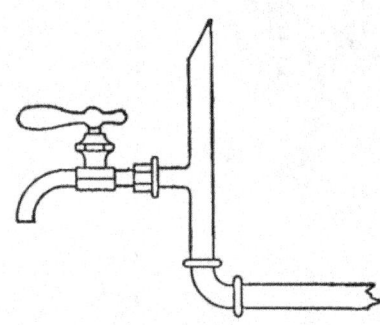

FIG. 24.—The short pipe above the faucet contains air forming an air cushion.

The *air cushion* (Fig. 24) is a metal pipe or dome filled with air attached to a water pipe where sudden changes in pressure are to be controlled. At many faucets in a city water system such an air cushion is employed. It contains air; this, unlike water, is easily compressible and the confined air when the tap is suddenly closed receives and checks gradually the rush of water in the pipe. Even with an air cushion, the "pound" of the water in the pipe when a tap is suddenly closed is often heard. If air cushions were not provided, the "water hammer" would frequently crack or break the pipes.

Important Topics

1. Pascal's law.
2. Hydraulic press.
3. Artesian wells.
4. Standpipes and air cushions.

Exercises

1. Where have you *seen* an air cushion?   Describe it and its use.
2. Where have you *seen* an hydraulic press?   Why and how used?
3. Where have you *seen* hydraulic elevators?   What moves them?
4. Where do you know of liquids under pressure?   Three examples.
5. What is the pressure in water at a depth of 1500 cm.   Express in grams per square centimeter and in kilograms per square centimeter.
6. What head[1] of water is required to give a pressure of 200 g. per square centimeter?   2 kg. per square centimeter?

   [1] "Head" is a term used to express the vertical height of water in pipes.

7. What *pressure* will be produced by a "head" of water of 20 meters?
8. If 1728 cn. in. of water are placed in a vertical tube·1 sq. in. in cross section to what height would the water rise?   It would give how many feet of *head?*
9. What would the water in problem 8 weigh?   What pressure would it produce at the bottom, in pounds per square inch? From this, compute how many feet of "head" of water will produce a pressure of 1 lb. per square inch.
10. Using the result in problem 9, what "head" of water will produce a pressure of 10 lbs. per square inch?   100 lbs. per square inch?
11. From the result in 9, 100 ft. of "head" of water will produce what pressure?   1000 ft. of "head?"
12. If the diameter of the pump piston in a hydraulic press in 2 cm. and that of the press piston 50 cm. what will be the force against the latter if the former is pushed down with a force of 40 kg.?

(3) ARCHIMEDES' PRINCIPLE

44. A Body Supported by a Liquid.—Among the applications of the force exerted by a liquid upon a surface, Archimedes' Principle is one of the most important.

Most persons have noted that a body placed in water is partly or wholly supported by the force of the water upon it.   A stone held by a cord and lowered into water is felt to have a part of its weight supported, while a piece of cork or wood is wholly supported and floats.

The human body is almost entirely supported in water, in fact, many people can easily float in water. It was the consideration of this fact that led the Greek philosopher Archimedes to discover and state the principle that describes the supporting of a body in a liquid.

45. Archimedes' Principle.—"*A body immersed in a liquid is pushed up by a force equal to the weight of the liquid that it displaces.*" The proof for this law is simply demonstrated. Suppose a cube, *abcd*, is immersed in water (Fig. 25). The upward force on *cd* is equal to the weight of a column of water equal to *cdef*. (See Art. 39.) The downward force upon the top of the cube is equal to the weight of the column of water *abef*. Then the net upward force upon the cube, that is, the upward force upon the bottom less the downward force upon the top, or the buoyant force exerted by the liquid is exactly equal to the weight of the displaced water *abcd*.

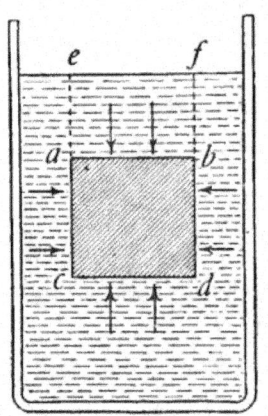

FIG. 25.—Theoretical proof of Archimedes' principle.

46. Law of Floating Bodies.—This same reasoning may be applied to any liquid and to any body immersed to any depth below the surface of the liquid. If the body weighs more than the displaced liquid it will sink. If it weighs less than the displaced liquid it will float or rise in the water. A block of wood rises out of the water in which it floats until its own weight just equals the weight of the water it displaces. From this we have the law of floating bodies.

A floating body displaces its own weight of the liquid in which it floats.

To test the law of floating bodies, take a rod of light wood 1 cm. square and 30 cm. long (Fig. 26). Bore out one end and fill the opening with lead and seal with paraffin so that the rod will float vertically when placed in water. Mark upon one side of the rod a centimeter scale, and dip the rod in hot paraffin to make it waterproof. Now find the weight of the stick in grams and note the depth to which it sinks in water in centimeters. Compute the weight of the displaced water. It will equal the weight of the rod.

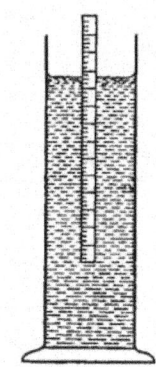

FIG. 26.—A floating body displaces its own weight of water.

47. Applications of Archimedes' Principle. There are numerous applications of Archimedes' Principle and the law of floating bodies.

(a) **To Find the Weight of a Floating Body: Problem.**—A boat 20 ft. long and with an average width of 6 ft. sinks to an average depth of 3 ft. in the water. Find the weight of the boat. What weight of cargo will sink it to an average depth of 5 ft.?

Solution.—The volume of the water displaced is 20 × 6 × 3 cu. ft. = 360 cu. ft. Since 1 cu. ft. of water weighs 62.4 lbs., 360 × 62.4 lbs. = 22,464 lbs., the weight of water displaced. By the law of floating bodies this is equal to the weight of the boat. When loaded the volume of water displaced is 20 ft. × 6 × 5 ft. which equal 600 cu. ft. 600 × 62.4 lbs. = 37,440 lbs. This is the weight of the water displaced when loaded. 37,440 lbs. − 22,464 lbs. = 14,976 lbs., the weight of the cargo.

(b) **To Find the Volume of an Immersed Solid: Problem.**—A stone weighs 187.2 lbs. in air and appears to weigh 124.8 lbs. in water. What is its volume?

Solution.—187.2 lbs. − 124.8 lbs. = 62.4 lbs., the buoyant force of the water. By Archimedes' Principle, this equals the weight of the displaced water which has a volume of 1 cu. ft. which is therefore the volume of the stone.

(c) **To Find the Density of a Body:** The density of a body is defined as the mass of unit volume.

We can easily find the mass of a body by weighing it, but the volume is often impossible to obtain by measurements, especially of irregular solids.

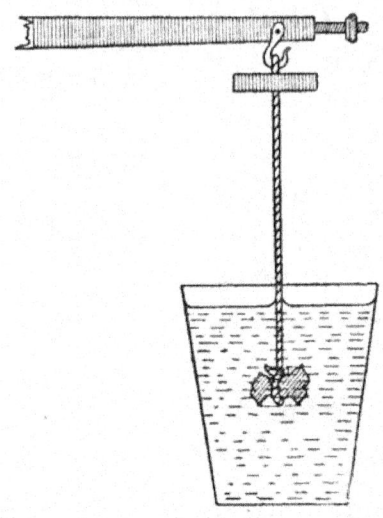

FIG. 27.—A method of weighing a body under water.

Archimedes' Principle, however, provides a method of finding the volume of a body accurately by weighing it first in air and then in water (Fig. 27), the apparent loss in weight being equal to the weight of the displaced water. One needs only to find the volume of water having the same weight as the loss of weight to find the volume of the body.

If the metric system is used, 1 ccm. of water weighs 1 g., and the volume is numerically the same as the loss of weight.

Important Topics

1. Archimedes' Principle.
2. Law of floating bodies.
3. The applications of Archimedes' Principle are to determine (a) the weight of a floating body; (b) the volume of an immersed solid, and (c) the density of a body.

Exercises

1. Look up the story of Archimedes and the crown. Write a brief account of it.
2. Why is it easier for a fat man to float in water than for a lean one?

3. A fish weighing 1 lb. is placed in a pail full of water. Will the pail and contents weigh more than before adding the fish? Why?

4. Why can a large stone be lifted more easily while under water than when on the land?

5. Why does the air bubble in a spirit level move as one end of the instrument is raised or lowered?

6. Why does a dead fish always float?

7. A ship is built for use in fresh water. What will be the effect on its water line when passing into the ocean?

8. Why can small bugs walk on water while large animals cannot?

9. If an object weighing 62.4 lbs. just floats in water, what weight of water does it displace? What volume of water is displaced? What is the volume of the body?

10. What is the volume of a man who just floats in water if he weighs 124.8 lbs.? If he weighs 187.2 lbs.?

11. An object weighing 500 g. just floats in water. What is its volume? How much water does a floating block of wood displace if it weighs 125 lbs.? 125 g.? 2 kg.? 2000 kg.?

12. A flat boat 10 × 40 ft. in size will sink how much in the water when 10 horses each weighing 1250 lbs. are placed on board?

13. A ship 900 ft. long and 80 ft. average width sinks to an average depth of 25 ft. when empty and 40 ft. when loaded. What is the weight of the ship and of its load?

14. Will a 1000 cc. block sink or float in water if it weighs 800 g.? If it weighs 1200 g.? Explain.

15. If a 1000 c.c. block of metal weighing 1200 g. is placed in the water in mid ocean what will become of it?

16. Prove Archimedes' Principle by use of the principles of liquid pressure.

17. An irregular stone, density 2.5, g. per ccm. displaces 2 cu. ft. of water What is its weight? Its apparent weight in water?

18. Will the depth to which a vessel sinks in water change as she sails from Lake Ontario into the Atlantic Ocean? Why?

19. If the density of sea water is 1.0269 g. per cubic centimeter and that of ice 0.918 g. per ccm., what portion of an iceberg is above water?

20. In drawing water from a well by means of a bucket, why is less force used when it is under water than when entirely above?

21. A stone which weighs 300 lbs. can be lifted under water with a force of 150 lbs. What is the volume of the stone?

22. The average density of the human body is 1.07 grams per c.c. How much water will a man who weighs 150 lbs. displace when diving? How much when floating?

(4) Density and Specific Gravity

48. Density.—The density of a substance is often used as a test of its purity. Archimedes in testing King Hiero's crown to find out if it were made of pure gold, determined first its density. It is by such tests that the purity of milk, of alcohol, of gold, and a great variety of substances is often determined.

Knowledge of methods of finding density is of value to everyone and should be included in the education of every student. *The density of a substance is the mass of unit volume of the substance.* In the metric system, for example, the density of a substance is the mass in grams per 1 ccm. Taking water, 1 ccm. weighs 1 gr. or its density is therefore 1 g. to the cubic centimeter. A cubic centimeter of aluminium weighs 2.7 g. Its density therefore is 2.7 g. per ccm.

49. Specific Gravity.—*Specific gravity is the ratio of the weight of any volume of a substance to the weight of an equal volume of water.* Its meaning is not quite the same as that of density, since specific gravity is always a *ratio, i.e.,* an *abstract* number, as 2.7. Density of a substance is a *concrete* number, as 2.7 grams per ccm. In the metric system the density of water is one gram per cubic centimeter, therefore we have:

Density (g. per ccm.) = (numerically) specific gravity.

In the English system, the density of water is 62.4 pounds per cubic foot, therefore in this system we have:

Density (lbs. per cu. ft.) = (numerically) 62.4 × sp. gr.

50. Methods for Finding Density and Specific Gravity
(a) Regular Solids.—Solids of regular shapes such as cubes, spheres, etc., whose volumes may be readily found by measurement, may be weighed. The mass divided by the volume gives the density, or $D = M\mu/v$.

(b) Irregular Solids.—With these the volume cannot be found by measurement but may be obtained by Archimedes' Principle. Weigh the solid first in the air and then in water. The apparent loss of weight equals the weight of the equal volume of water displaced. From this the volume may be found. And then the

density equals $\dfrac{\text{mass}}{\text{volume}}$; the specific gravity =

$$\frac{\text{wt. in air}}{\text{wt. of equal volume of water}} = \frac{\text{wt. in air}}{(\text{wt. in air}) - (\text{wt. in water})}$$

(c) Solids Lighter than Water.—This will require a sinker to hold the body under water. Weigh the solid in air (w). Weigh the sinker in water (s). Attach the sinker to the solid and weigh both in water (w'). The specific gravity equals

$$\frac{\text{wt. of solid in air}}{\text{loss in wt. of solid in water}} \text{ or } \frac{w}{(w + s) - w'}$$

The apparent loss of weight of the solid is equal to the sum of its weight in air plus the weight of the sinker in water, less the combined weight of both in water.

(d) The Density of a Liquid by a Hydrometer.—One may also easily find the density of any liquid by Archimedes' Principle. If one takes the rod described in Art. 46, and places it in water, the number of cubic centimeters of water it displaces indicates its weight in grams. On placing the rod in another liquid in which it floats, it will of course displace its own weight and the height to which the liquid rises on the scale gives the volume. By dividing the *weight* of the rod as shown by its position in *water* by the *volume* of the *liquid* displaced we obtain the density

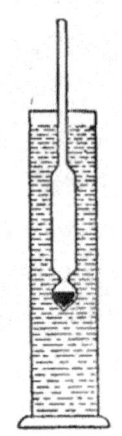

FIG. 28.—A hydrometer used to find the density of a liquid.

of the liquid. Commercial hydrometers for testing the density of milk, alcohol and other liquids are made of glass of the form shown in Fig. 28. The long narrow stem permits small differences in volume to be noticed, hence they are more accurate than the rod described in the preceding paragraph. For convenience this rod contains a paper scale, so that when the height of the liquid on the stem is noted, the density is read at once.

Density of Liquids by Loss of Weight. Weigh a piece of glass in air (W_a), in water (W_w), and in the liquid to be tested (W_1).

Then $(W_a - W_w)$ gives the weight of the water displaced

And $(W_a - W_1)$ gives the weight of the liquid displaced

Hence, $\dfrac{W_a - W_1}{W_a - W_w}$ equals the specific gravity of the liquid.

Important Topics

1. Definitions of density and specific gravity.
2. Methods of finding density: (a) regular solids; (b) irregular solids; (c) solids lighter than water; (d) liquids by hydromter; (e) liquids by loss of weight.

Exercises

Note.—Consider that 1 cu. ft. of water weighs 62.4 lbs. Consider that 1 ccm. of water weighs 1 g.

1. What is meant by the statement that a block of wood has a specific gravity of 0.6?
2. Considering that the density of the human body is the same as that of water, what is the volume of a 125-lb. boy? Of a 250-lb. man? Of a 62.4-lb. boy? What is the volume of your body?
3. How is the weight of large ships found? Give an example.

4. Mention three cases where determinations of density are important.

5. A body weighs 40 g. in air, 15 g. in water, 5 g. in an acid. Find (a) the density of the body; (b) its volume; (c) density of the acid.

6. If the specific gravity of a horse is 1, what is the volume of a horse weighing 500 kg.? Of one weighing 1248 lbs.?

7. A weighted wooden box sinks to a depth of 20 cm. in water and 24 cm. in alcohol, and to a depth of 18 cm. in brine. What is the density of the alcohol and of the brine?

8. A glass stopper weighs in the air 25 g., in water 15 g., in oil 18 g. Find the density and volume of the stopper. Find the density of the oil.

9. What would a cubic foot of wood weigh if the specific gravity were 0.5.?

10. The specific gravity of aluminum is 2.7. Find the weight of a cubic foot of it.

11. A block of wood weighs 40 g. A piece of lead appears to weigh 70 g. in water. Both together appear to weigh 60 g. in water. Find the density of the wood.

12. A stone weighs 30 g. in air, 22 g. in water, and 20 g. in salt water. Find the density of the salt water.

13. Will iron sink in mercury? Why?

14. A submarine boat weighing 200 tons must have what volume in order to float?

15. Find the weight of 2 cu. ft. of copper from its density.

16. What is the weight in water of a mass whose specific gravity is 3.3 and whose weight is 50 kg.?

17. A block of granite weighs 1656 lbs.; its volume is 10 cu. ft., what is its density?

18. If the specific gravity of hard coal is 1.75 how would you determine how many tons of coal a bin would hold?

19. A hollow copper ball weighs 2 kg. What must be its volume to enable it to just float in water?

20. A mass having a volume of 100 ccm. and a specific gravity of 2.67 is fastened to 200 ccm. of wood, specific gravity 0.55. What will the combination weigh in water?

21. A block weighing 4 oz. in air is tied to a sinker which appears to weigh 14 oz. in water. Both together appear to weigh 6 oz. in water. What is the specific gravity of the block?

CHAPTER IV

MECHANICS OF GASES

(1) Weight and Pressure of the Air

51. Weight of Air.—It is said that savages are unaware of the presence of *air*. They feel the *wind* and hear and see it moving the leaves and branches of the trees, but of air itself they have little conception.

To ordinary observers, it seems to have no weight, and to offer little resistance to bodies passing through it. That it has weight may be readily shown as follows: (See Fig. 29.) If a hollow metal sphere, or a glass flask, provided with tube and stopcock, be weighed when the stopcock is open, and then after the air has been exhausted from it by an air pump, a definite loss of weight is noticeable.

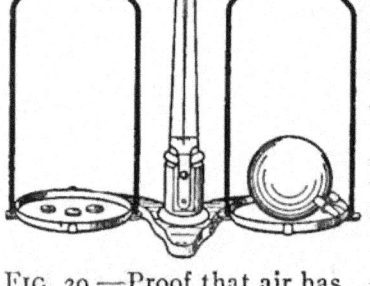

Fig. 29.—Proof that air has weight.

If the volume of the sphere is known and it is well exhausted of air, a fair approximation of the weight of air may be obtained. Under *"standard conditions,"* which means *at the freezing temperature* and a barometric pressure of 76 cm., a liter of air weighs 1.293 g. while 12 cu. ft. of air weigh approximately 1 lb.

52. Pressure of Air.—Since air has weight it may be supposed to exert pressure like a liquid. That it does so may be shown in a variety of ways.

If a plunger fitting tightly in a glass cylinder be drawn upward, while the lower end of the tube is under water, the water will rise in the tube (Fig. 30). The common explanation of this is that the water rises because of "suction." The philosophers of the ancient Greeks explained it by saying that "nature abhors a vacuum," and therefore the water rises. Neither explanation is correct. It was found in 1640 that water would not rise in a pump more than 32 ft. despite the fact that a vacuum was maintained above the water Galileo was applied to for an explanation. He said, "evidently nature's horror of a vacuum does not extend above 32 ft." Galileo began tests upon "the power of a vacuum" but dying left his pupil Torricelli to continue the experiment. Torricelli reasoned that if water would rise 32 ft., then mercury, which is 13.6 times as dense as water, would rise

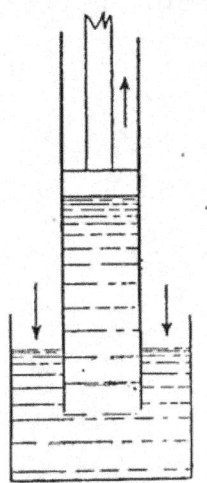

FIG. 30. — Air pressure forces the liquid up the tube.

about $\frac{1}{13}$ as much. To test this, he performed the following famous experiment.

53. Torricelli's Experiment (1643).—Take a glass tube about 3 ft. long, sealed at one end, and fill it with mercury. Close the end with the finger and invert, placing the end closed by the finger under mercury in a dish (Fig. 31). Remove the finger and the mercury sinks until the top of the mercury is about 30 in. above the level of the mercury in the dish. Torricelli concluded that the rise of liquids in exhausted tubes is due to the pressure of the atmosphere acting on the surface of the mercury in the dish.

To test this, place the tube with its mercury upon the plate of an air pump and place a tubulated bell jar over the apparatus so that the tube projects through a tightly fitting stopper. (See Fig. 32.) If the air pressure is the cause of the rise of mercury in the tube, on removing the air from the bell jar the mercury should fall in the tube. This is seen to happen as soon as the pump is started. It is difficult to remove all the air from the receiver so the mercury rarely falls to the same level in the tube as in

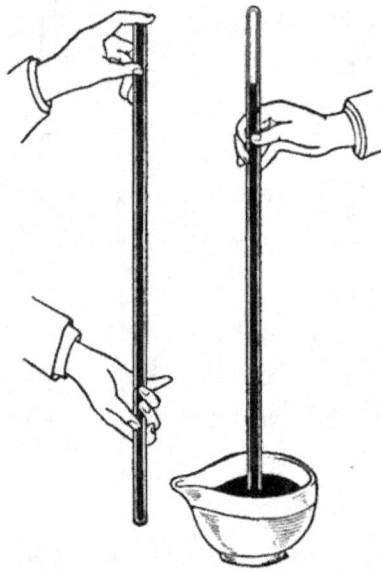
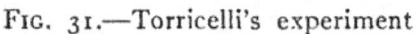

FIG. 31.—Torricelli's experiment.

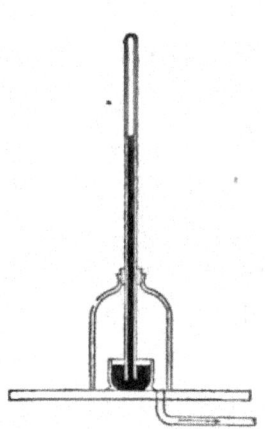

FIG. 32.—The mercury drops as the air is removed.

the dish. A small tube containing mercury is often attached to air pumps to indicate the degree of exhaustion. Such tubes are called *manometers*.

54. The Amount of Atmospheric Pressure.—Torricelli's experiment enables us to compute readily the pressure of the atmosphere, since it is the atmospheric pressure that balances the column of mercury in the tube. By Pascal's Law, the pressure of the atmosphere on the surface of the mercury in the dish is transmitted as an exactly equal pressure on the mercury column in the tube at the same level as the mercury outside.

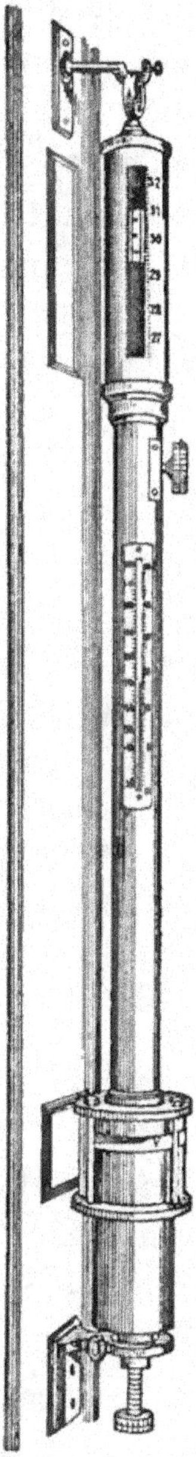

This pressure, due to the air, must balance the weight of the column of mercury in the tube. It therefore equals the weight of the column of mercury of unit cross-section. The average height of the column of mercury at sea-level is 76 cm. Since the weight of 1 cc. of mercury is 13.6 grams, the pressure inside the tube at the level of the surface of the mercury in the dish is equal to 1 × 76 × 13.6 or 1033.6 g. per square centimeter. Therefore the *atmospheric pressure* on the surface of the mercury in the dish is 1033.6 g. per square centimeter, approximately 1 *kg. per square centimeter or 15 lbs. per square inch.*

55. Pascal's Experiment.—Pascal tested in another way the action of atmospheric pressure upon the column of mercury by requesting his brother-in-law, Perrier, who lived near a mountain, to try the experiment on its top. Perrier found that on ascending 1000 meters the mercury fell 8 cm. in the tube. Travelers, surveyors, and aviators frequently determine the altitude above sea-level by reading the barometer, an ascent of 11 meters giving a fall of about 1 mm. in the mercury column, or 0.1 in. for every 90 ft. of ascent.

56. The Barometer.—The modern barometer (Fig. 33), consists of a Torricellian tube properly mounted. Reading a barometer consists in accurately reading the height of the mercury column.

FIG. 33.—A standard barometer.

This height varies from 75 to 76.5 cm. or 29 to 30 in. in localities not far from the sea-level. The atmospheric pressure varies because of disturbances in the atmosphere. It is found that these disturbances of the atmosphere pass across the country from west to east in a somewhat regular manner, hence a series of readings of the barometer may give reliable information of the move-

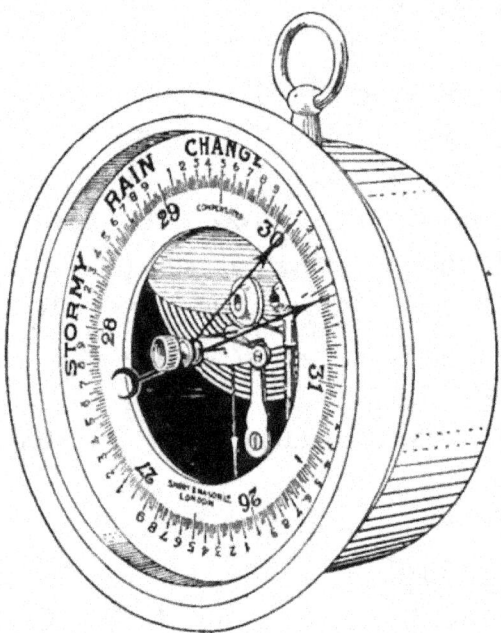

FIG. 34.—An aneroid barometer

ment of these disturbances and so assist in forecasting the weather. The Weather Bureau has observations taken at the same moment at various stations over the country. These observations form the basis for the daily forecast of the weather.

Another form of barometer in common use is the *Aneroid Barometer* (Fig. 34). Its essential parts are a cylindrical air-tight box with an elastic corrugated cover. Inside the box is a partial vacuum. This makes the cover very sensitive to slight changes of pressure. The motion of the top of the box is conveyed by a series of levers to an indicating hand which moves over a dial. This barometer can be made so sensitive as to indicate the change of air pressure from a table top to the floor. It is much used by travelers, explorers, surveying parties and aviators, since the mercurial barometer is inconvenient to carry.

Important Topics

1. *Weight* and *Pressure* of *air* in English and metric units. How shown. Evidences.
2. Work of Galileo, Torricelli, and Perrier.
3. Barometer: construction, action, mercurial, aneroid.

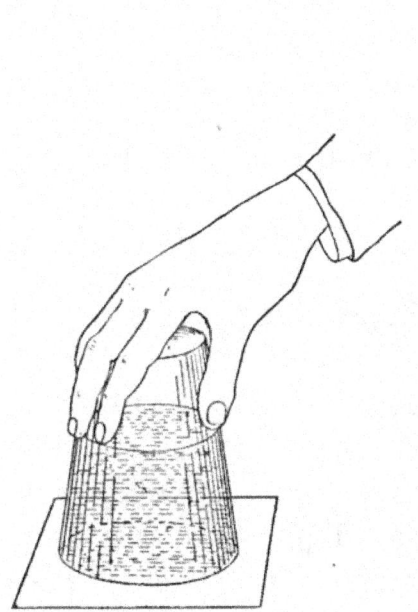

Fig. 35.—Air pressure keeps the water in the tumbler.

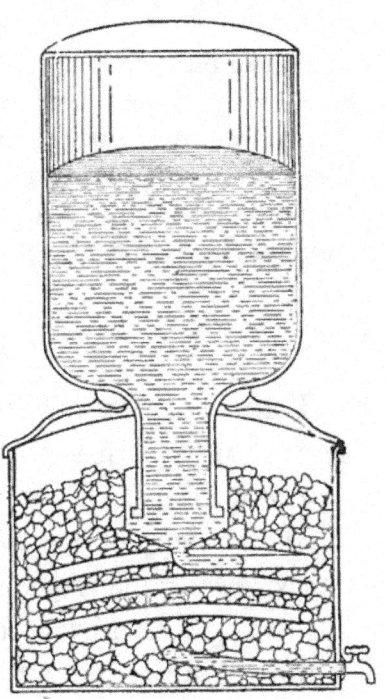

Fig. 36.—Cross-section of a modern drinking fountain.

Exercises

1. Do you think Archimedes' Principle applies to the air? Does Pascal's Law? Why?
2. Find the downward pressure of the mercury in a barometer tube if the cross-section is 1 sq. cm. and the height 75 cm. at the level of the mercury surface in contact with the air. (The density of mercury is 13.6 grams per cc.)

3. What is the weight of the air in a room if it is 10×8×4 meters?
4. What weight of air is in a room 10×15×10 ft.?
5. When smoke rises in a straight line from chimneys, is it an indication of a high or low barometric pressure? Why?
6. Why does a tumbler filled with water and inverted in a dish with its rim under water remain full?
7. If the barometer tube is inclined the mercury remains at the same horizontal level. How can this be explained?
8. When the mercurial barometer stands at 76 cm., how high would a water barometer stand? Explain.
9. Explain why it is possible for one to suck soda water through a tube?
10. Fill a tumbler with water. Place a sheet of paper over the top and invert. The paper clings to the tumbler and prevents the water from escaping. Explain. (See Fig. 35.)
11. Why must a kerosene oil can have two openings in order to allow the oil to flow freely?
12. Explain the action of the modern drinking fountain (Fig. 36).

(2) Compressibility and Expansibility of the Air

57. Effect of Prefssure on Liquids and Gases.—Both classes of fluids, liquids and gases, have many characteristics in common. Both are composed of molecules that move freely; hence both *flow*. At any point within a fluid the *pressure is the same in all directions*. Archimedes' Principle applies, therefore, to both liquids and gases.

We now come to an important *difference* between liquids and gases. *Liquids* are *practically incompressible*. "So much so, that if water is subjected to a pressure of 3000 kg. per sq. cm., its volume is reduced only about one-tenth." Gases show a very different behavior from liquids on being subjected to pressure. They may readily be compressed to a small fraction of their volume as is noticed on inflating a pneumatic tire. A gas has also the *ability to spring back* to a larger volume as soon as the pressure is released, as when a cork is driven from a pop gun. Not only is compressed air able to expand, but air under ordinary conditions will expand if it is released in a space where the pressure is less.

Hollow bodies, animals and plants, are not crushed by atmospheric pressure, because the air and gases contained within exert as much force outward as the air exerts inward.

58. Boyle's Law.—The relation between the volume and pressure of a gas was first investigated by Robert Boyle in the seventeenth century. The experiment by which he first discovered the law or the relation between the volume and the pressure of a gas is briefly described as follows:

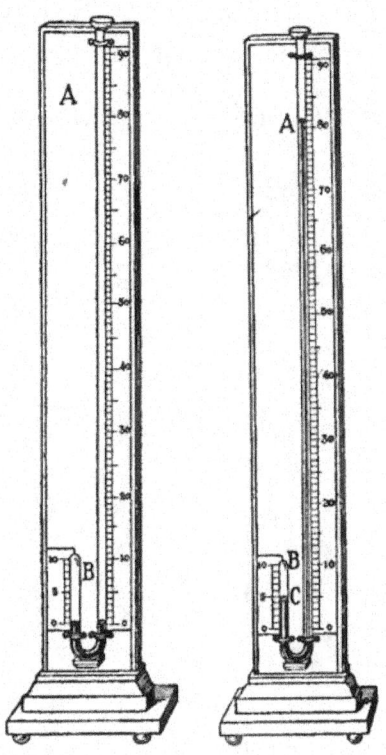

FIG. 37 a.   FIG. 37 b.
FIGS. 37 a AND 37 b.—Boyle's law apparatus.

A glass tube is bent in the form of the capital letter J, the short arm being closed. A little mercury is poured in to cover the bend. (See Fig. 37a.) Since the mercury is at the *same level in both arms*, the pressure in (A) is the same as in (B). Mercury is now poured into (A) until it stands in the long tube at a height above that in (B) which is equal to the height of the mercury column of the barometer. (See Fig. 37b.) The air in (BC) is now under a pressure of two atmospheres (one atmosphere is due to the mercury column). On measurement the air in (BC) will be found to have just one-half of its original volume.

Thus doubling the pressure to which a gas is subjected reduces its volume to one-half. Tripling the pressure, reduces the volume to one-third and so on.

Careful experiments reveal the following law: *The volume of a given mass of gas at constant temperature is inversely proportional to the pressure to which it is subjected.* This law is often expressed mathematically. $\dfrac{P}{P'} = \dfrac{V'}{V}$, or $PV = P'V'$. Since doubling the pressure reduces the volume one-half, it doubles the density. Tripling the pressure triples the density. We therefore have $\dfrac{P}{P'} = \dfrac{D}{D'}$ or the density of a gas directly proportional to its pressure.

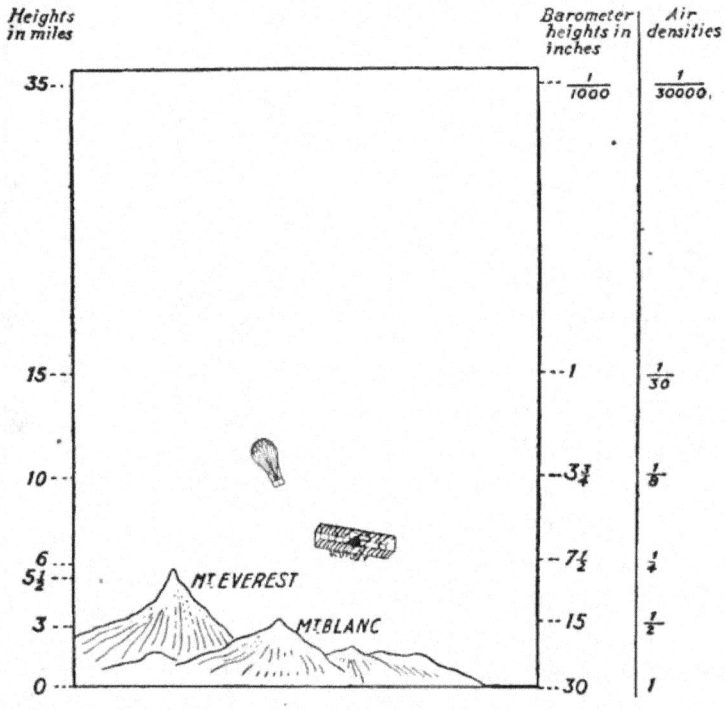

FIG. 38.—Height and density of the air.

59. Height of the Atmosphere.—From its properties of *compression* and *expansion*, the air varies in density and pressure as one ascends in it. At a height of 3 miles the pressure is reduced to about one-half. This is an indication that one-half of the air is below this level. Balloonists have gone to a height of 7 miles, Glaser and

Coxwell in England in 1862 and Berson in France in 1901. The atmosphere has been explored to a height of 30,500 meters (18.95 miles) by sending up self-registering barometers in small balloons which burst at great altitudes. A parachute protects the instruments from breakage from too rapid fall. This height of 30,500 meters was reached by a balloon sent up by William R. Blair, at Huron, South Dakota, September 1, 1910.

At a height of 35 miles, the density is estimated at 1/30,000 of its value at sea-level. (See Fig. 38.) It is believed that some rarefied air exists for a considerable distance above this point, some estimates placing the extent at 100 miles, and others from 200 to 500 miles. Evidences of some air at such heights are shown by: (a) the height at which meteors first appear, (b) the height of the Aurora Borealis, and (c), the distance that the sun is below the horizon when the last traces of color disappear from the sky in the evening.

Although the exact limits of the atmosphere are unknown, the weight of a column of air 1 sq. cm. in cross-section, and extending *upward as high as the atmosphere*, may be accurately computed. For this column of air exactly balances the column of mercury in the tube of the barometer.

Below sea-level, the air increases rapidly in density and it is estimated that at a depth of 35 miles, the density of the air would be a thousand times that at the earth's surface, or more than that of water.

Important Topics

1. Evidence of compressibility of gases and incompressibility of liquids.
2. Boyle's Law. Proof, applications.
3. Extent of the atmosphere—three evidences.

5

Exercises

1. Mention three illustrations of the compressibility and expansibility of air that you know from you own experience.

2. Increasing the pressure increases the amount of a gas that will be absorbed by a liquid? Explain this. Have you ever observed this fact? Where?

3. If a toy balloon containing 2000 ccm. of gas at the earth's surface where the barometer reading is 76 cm., rises to an elevation where the barometer reads 54 cm., the balloon will tend to expand to what volume? Explain. Will it attain this volume?

4. If a gas is compressed, it changes in temperature. How do you explain this?

5. What change in temperature will occur when compressed air is allowed to expand? Explain.

6. Air blowing up a mountain side has its pressure lessened as it approaches the top. How will this affect the temperature? Why? What may result from this change in temperature? Explain.

7. To what pressure must 500 ccm. of air be subjected to compress it to 300 ccm. the barometer reading at first being 75 cm. Explain.

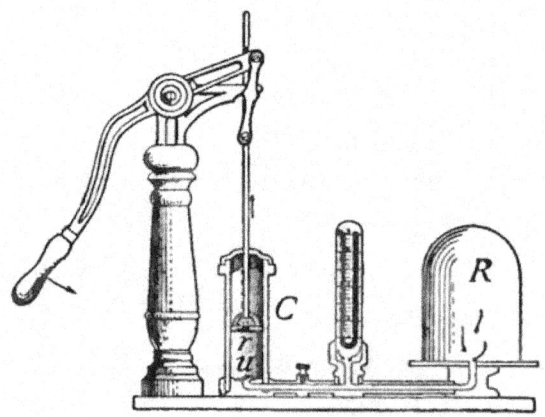

Fig. 39.—The air pump.

(3) Pneumatic Appliances

60. The Air Pump.—The air pump is used to remove air or other gases from a closed vessel. It was invented about 1650 by Otto Von Guericke, burgomaster of Magdeburg, Germany. One form of air pump is shown in Fig. 39. *C* is a cylinder within which slides a tightly fitting piston. *R* is the vessel from which the air is to be exhausted. *r* and *u* are valves opening upward   The action of the pump is as follows:

On pushing the piston down, the air in *C* is compressed. This opens valve *r* allowing the confined air to escape above the piston. The piston is then raised making the space in *C* a partial vacuum. The pressure in *R* now being greater than in *C*, *u* is pushed up and the air from *R* rushes into *C*, until the pressure is equalized. On pushing down the piston again, valve *u* closes and the process is repeated until the pressure in *R* is no longer able to raise the value *u*. Some air pumps are so constructed that the valves are opened and closed automatically by the movement of the piston. With these pumps a higher degree of rarefaction can be obtained.

Air is often partially exhausted from receivers or vessels by the use of a filter pump or *aspirator*. A stream of water flowing through a constriction causes a reduced pressure, draws in air and carries it away, and thus produces a partial vacuum. See Fig. 40 for a section of the device.

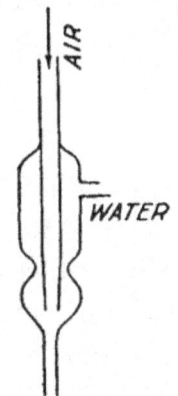

61. The Condensing Pump.—This is like the exhaust pump except that its valves are reversed. It is used in compressing illuminating gases into cylinders

FIG. 40.—An aspirator.

for use in lighting vehicles, stereopticons, Pintsch lights, gas light buoys, etc., and also for compressing air to operate air brakes, pneumatic hammers and drills, and for other uses.

The common condensing pump is the kind used for

inflating tires. (See Fig. 41.) In this, a loosely fitting metal piston is attached to a disc of leather somewhat larger than the cylinder. This device is called a *cup valve*. On raising the piston, air rushes in from the top past the valve, but on pushing the piston down, the valve is pressed tightly against the sides of the cylinder and prevents the escape of any air. The compressed air pushes open a valve on the tire and enters it. This valve closes as soon as the pressure is lessened from outside. It is well to notice in all of these pumps that *two* valves are used. One holds the air already secured while the other opens for a new supply. Both valves are never open at the same time.

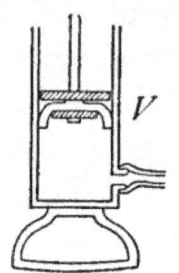

FIG. 41.— Condensing pump used in inflating tires.

62. Water Pumps.—The Common Lift Pump. This, the simplest pump for raising water, consists of a cylinder C (Fig. 42) connected by a pipe R to a supply of water as a cistern or well. A valve opening upward is placed at the bottom of the cylinder over the entrance to the pipe. In the cylinder is a tightly fitting piston connected by a rod to a lever for ease in action. The piston contains a valve opening upward. In operating this pump water is usually first poured into the cylinder to "prime" it. This helps to close the valves and prevents air leaking past them. When the piston is lowered the lower valve closes, the air in the cylinder being compressed pushes the upper valve open and passes above the piston. On raising the piston the upper valve closes. This forms a partial vacuum in the cylinder.

The air pressing on the surface of the water below forces the water and air that may be in the tube upward through the lower valve to fill this partial vacuum.

When the cylinder becomes filled with water, this is lifted out on the up-stroke, whence its name, "lift pump." Since the atmospheric pressure at sea-level can only support a column of water about 34 ft. high, the lower valve must be within this distance of the water surface. In actual practice the limit is about 27 ft. In deeper wells, the cylinder and valves are placed so that they are within 25 or 27 ft. of the surface of the water in the well,

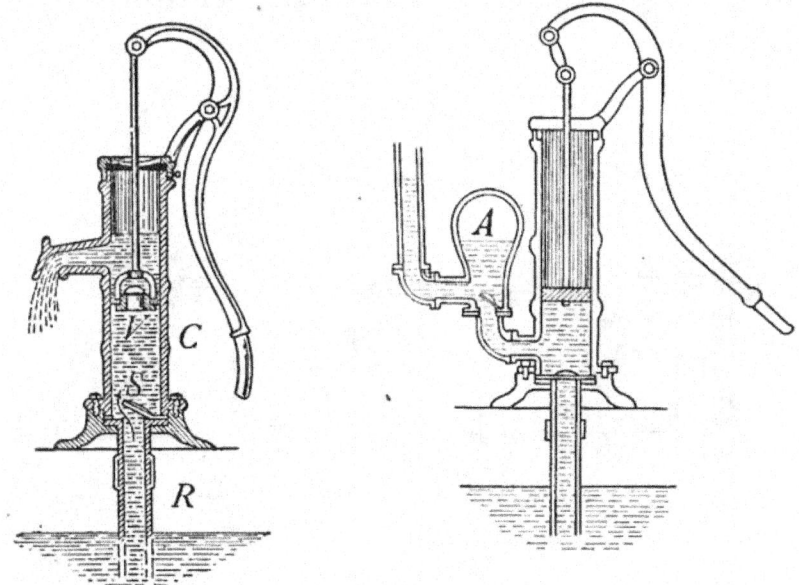

FIG. 42.—The common lift pump.   FIG. 43.—A force pump with an air chamber (*A*)

a long piston rod reaching above the surface of the ground and connected to a pump handle operates the piston. A discharge pipe extends from the cylinder to the surface of the ground above.

63. The Force Pump.—The force pump is used to deliver water under pressure either for spraying or to an elevated reservoir. The piston is solid, the second valve being placed at the entrance of the discharge pipe. (See Fig. 43.) The action is the same as that of the lift pump, with this exception; the piston in its down stroke forces the water out through the discharge pipe, the velocity depending upon the pressure exerted.

A force pump is usually provided with an air chamber which is connected with the discharge pipe. On the down stroke of the piston, water is forced into the air chamber. This compresses the air it contains. The compressed air reacts and exerts pressure on the water forcing it out in a steady stream.

Force pumps are used in deep wells, being placed at the bottom.

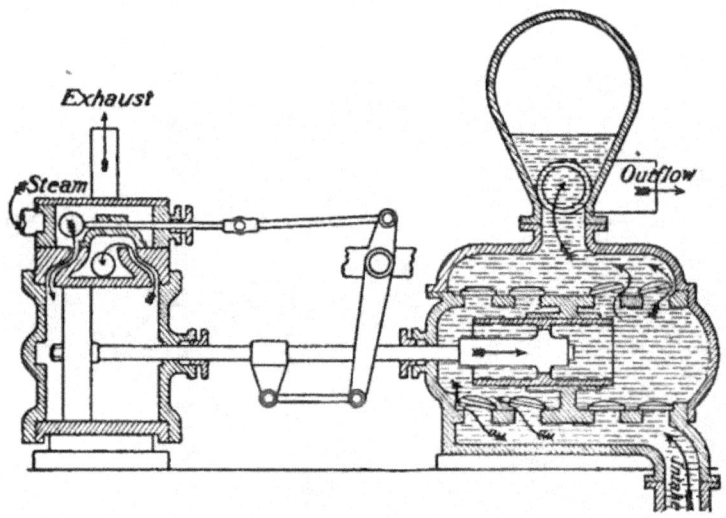

FIG. 44.—A steam pump used on a fire engine.

The pumps used in city water works, fire engines, and all steam pumps, are force pumps. (See Fig. 44.)

64. The Siphon.—The siphon is a tube used to convey a liquid from one level over an elevation to a lower level by atmospheric pressure. It is used to remove liquids from tanks or vessels that have no opening at the bottom.

The siphon cannot be completely understood until one has mastered the laws of the flow of liquids. The following is offered as an incomplete explanation of its behavior. Consider the siphon to be full of water and closed at *d* (Fig. 45). Atmospheric pressure on *a* will hold the siphon full if *ab* does not exceed 34 feet. If *d* is opened the water falls out with a speed equal to that acquired in falling from the level of *a* to that of *d*. This speed is acquired by all the water in the siphon and results in a drop in pressure throughout it. The pressure at *a* inside the siphon becomes less than the pressure at the same level outside as soon as the water starts flowing. The water in the vessel then flows into the siphon and out at *d*. This flow continues as long as there is a fall from the free surface of the water in the vessel to the outlet at *d*.

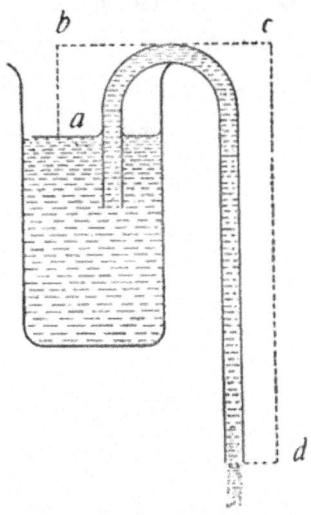

FIG. 45. — Cross-section of a siphon.

65. The Cartesian Diver.—This is a device which illustrates at the same time transmission of pressure by liquids, Archimedes' principle, and compressibility of gases.

FIG. 46.—The Cartesian diver.

It was invented by Des Cartes (1596–1650). As ordinarily made, it is a hollow glass image with a small opening in the foot. It contains air and water in such amounts that the average density of image and contents is slightly less than that of water. It is placed in a tall glass jar filled with water and covered with tightly stretched rubber tissue. (See Fig. 46.) By pressing on the rubber cover the diver may be made to sink, since the air and water transmit the pressure on the cover which compresses the air inside the figure admitting some water to it, thus making the diver more dense than water. By varying the pressure it can be made to sink, rise, or remain stationary at will.[1] A small vial can be used instead of the image.

66. Hydraulic Ram.—The hydraulic ram (see Fig. 47) is an automatic device that is much used for raising water from springs to houses located on higher ground. Water flows through the pipe A through the opening at B. The pressure closes the valve at B. The increased pressure in the pipe due to the closing of B opens the valve C and some of the water flows into the air chamber D. This reduces the pressure against the valve B so

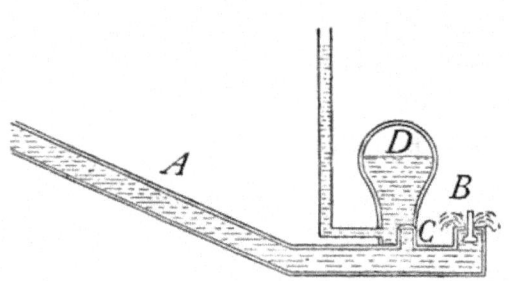

Fig. 47.—Cross-section of a hydraulic ram.

that it drops and allows a little water to escape. Just as this happens, valve C closes. The pressure in the pipe then closes B and forces water past C. This action being continually repeated, the air in D becomes so compressed that it has elastic force enough to raise the water in a steady stream to a height of many feet.

67. The Balloon.—Since air is a fluid, Archimedes' principle applies to it as well as to liquids. Therefore any object in the air is lifted up by a force equal to the weight of the air it displaces. The object will rise, if it weighs less than this displaced air and will continue to rise until both weights are equal.

The Balloon (Fig. 48) rises because it weighs less than the air it displaces, and therefore it is pushed up by the heavier air, the "lifting power" being the difference between its weight and that of the air displaced. The

[1] The position of a submarine in or under water is controlled in a similar manner.

neck at the bottom is left open to allow for expansion of the gas. When the aeronaut wishes to descend, he opens a valve at the top allowing some of the gas to escape.

Hydrogen is the lightest gas, weighing 0.09 kg. per cubic meter, and so gives the greatest lifting power, but as it is

FIG. 48.—Winner of international championship race, Paris, 1913.

expensive to make, coal gas, density 0.75 kg. per cubic meter, is ordinarily employed. Helium has recently been used to fill military balloons because it cannot be set on fire.

The Parachute (Fig. 49) is an umbrella-shaped device for use in descending from a balloon. After falling a few seconds it opens, the large surface exposed to the air caus-

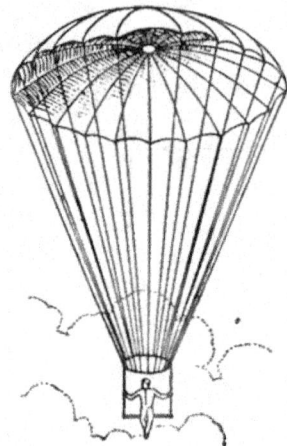

ing it to descend slowly. The hole in the top keeps the parachute upright by allowing the air to escape through it, thus relieving the pressure.

68. The Air Brake.—Compressed air is used to do work in many machines, such as pneumatic drills, hammers, and air brakes. The Westinghouse air brake (Fig. 50) uses air at a pressure of about 70 lbs. to the square inch. The essential

FIG. 49.—A parachute.

parts as shown are a reservoir R, the brake cylinder C and a triple valve V, placed under each car with an air pipe P, leading to the engine. This is connected to R by the triple valve V. When the pressure in P is reduced by the engineer or by accident, the triple valve operates so as to admit air from R into the cylinder C pushing the piston H to the left. H is connected to the brakes by levers which press the brake shoes strongly against the wheels. When the air pressure in P is restored the triple valve acts

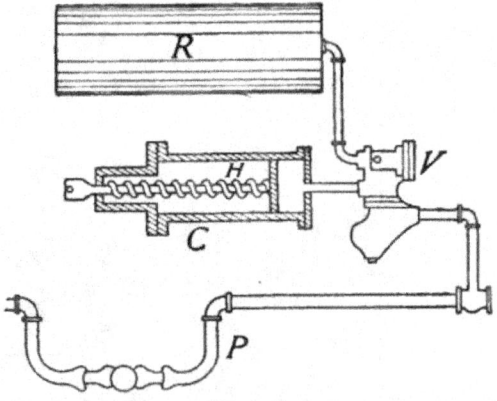

FIG. 50.—Cross-section of a Westinghouse air brake.

so as to permit the air in C to escape while R is filled again from P. The hissing sound heard when a train stops is caused by air escaping from cylinder C. The spring in C keeps the brakes from the wheels except when the "air is on."

69. The Gas Meter.—The gas meter consists of a box divided into two parts by a vertical partition (Fig. 51). Two bellows are at-tached to this partition, one on each side. The valves that regulate the flow of gas to and from the bellows and the chambers A and D are opened and closed by levers connected with the bellows. These levers also operate the hands upon the dials. When the inlet to the bellows B is opened, the outlet of A is also opened. Gas entering B opens the bellows and forces the gas in A out into the house-pipe E. When B is full its inlet valve closes and its outlet valve opens. The inlet of A also opens and its outlet closes. Gas now flows into A, compressing the bel-lows and B, and forcing the gas from it into the house-

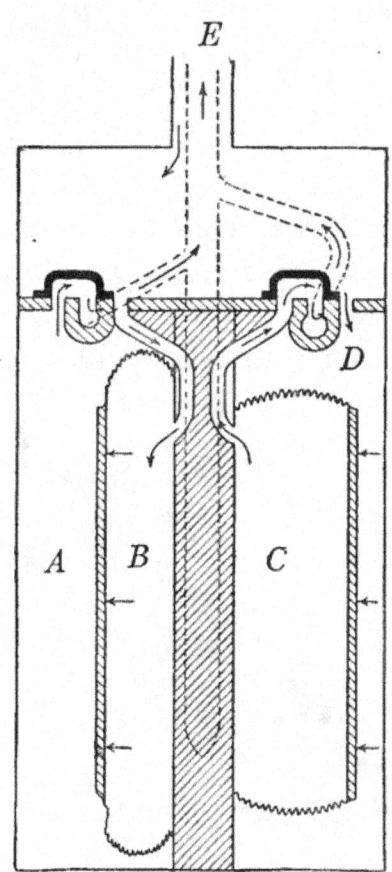

Fig. 51.—Cross-section of a gas meter showing its construc-tion and action.

pipe. At each filling of the bellows B there will be dis-placed from A and forced into the house-pipe as much gas as enters B. It is evident that at each emptying of B an equal amount of gas enters A. Thus we have A and B alternately filling and emptying as long as the gas burner is open. To have a continuous flow of gas in the house-pipes two pipes and two chambers are necessary, one being filled while the other is being emptied.

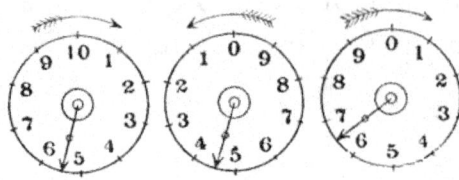

Fig. 52 represents the dials upon a gas meter showing a reading of 54,-600 cu. ft.

FIG. 52.—Dials of a gas meter.

70. Centrifugal Pumps.

Fluids, such as water and air, are often put in motion by devices called *centrifugal pumps* (see Art. 78). These pumps contain a revolving part, like a wheel without a rim, whose spokes are replaced by thin blades. This revolving part resembles the paddle wheel of some steam

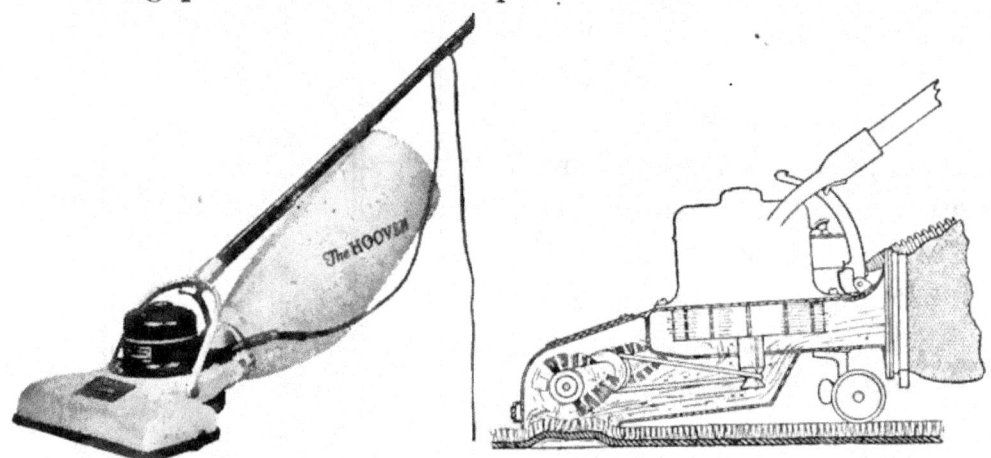

FIG. 53.—A vacuum sweeper. (*Courtesy of the Hoover Suction Sweeper Co.*)

boats and is enclosed in a case or cover having one opening at the rim and another opening on one side about the axle.

When the wheel is rapidly revolved, the fluid is driven out with considerable force through the opening at the rim, while a partial vacuum is produced at the axle causing a rapid flow into the device at this point.

This is the principle of the action of the *vacuum cleaner.*

Fig. 53 is a section of a vacuum sweeper showing the revolving wheel and the current of air passing into the wheel at the lower side and out of the rim of the case at the rear.

Centrifngal water pumps work on the same principle and furnish a continuous flow of water, often large in volume and at considerable pressure.

Important Topics

1. Air pump.
2. Condensing pump.
3. Lift and force pumps.
4. Siphon.
5. Cartesian diver.
6. Hydraulic ram.
7. Balloon.
8. Air brake.
9. Gas meter.
10. Vacuum cleaner.

Exercises

1. Explain why smoke settles to the ground before storms.
2. Why does the water rise in the suction pipe of a pump?
3. Why is it easier to float in water when the lungs are filled with air? than when they are not filled?
4. Why is it easier to swim in salt water than in fresh water?
5. How are submarines made to sink? to rise to the surface?
6. How can a fish rise or sink in water?
7. Explain why a life preserver made of cork will enable a person to float.
8. Hold the open hand out flat with the fingers together. Place underneath the fingers a piece of paper. Blow between the first and second fingers against the paper. As long as you blow hard the paper will not fall but will stick to the hand. Explain.
9. Why does pressing the bulb of an atomizer force out the liquid in a fine spray?
10. Why is air that contains a large amount of water vapor lighter than air that only contains a small amount?
11. How are heights above sea-level ascertained by a barometer?
12. Oil floats on water but sinks in alcohol. Explain.
13. In a balloon the lower end is often open to the air. Why does not the gas escape and prevent the balloon from rising?
14. How long will a balloon continue to rise?

15. If the pressure against the 8-in. piston of an air brake is 70 lbs. per square inch, how much force does the piston exert?

16. The capacity of a balloon is 40,000 cu. ft. The weight of the balloon, car, etc., is 600 lbs.; specific gravity of the gas used is 0.46 that of the air. Find how much weight the balloon can carry.

17. The so-called Magdeburg hemispheres were invented by Otto von Guericke of Magdeburg, Germany. When the hemispheres (see Fig. 54) are placed in contact and the air exhausted it is found very difficult to pull them apart. Explain.

18. Von Guericke's hemispheres had an inside diameter of 22 in. What force would be required to pull them apart if all the air were exhausted from them? (Find the atmospheric force on a circle, 22 in. in diameter.)

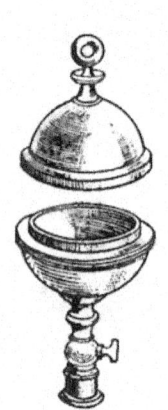

19. Von Guericke made a water barometer whose top extended through the roof of his house. On the top of the water in· the tube was placed a wooden image. In fair weather the image appeared above the roof, but it descended before a storm. Explain.

20. The balloon "Goodyear" (Fig. 48), which won the International championship race at Paris in 1913, has a capacity of 80,000 cu. ft. The gas bag weighs 653 lbs., the net 240 lbs. and the basket 92 lbs. How large a load can it carry when filled with hydrogen specific gravity 0.069 (compared with air).

Fig. 54.—
Magdeburg
hemispheres.

Review Outline: Liquids and Gases

Liquids: Force, pressure, and density. Floating and immersed bodies.

Laws: Liquid force, $F = A.h.d$, Pascal's, Archimedes.

Illustrations and Applications:

Specific gravity, $\dfrac{W_a}{W_a - W_w}$, $\dfrac{W_a - W_1}{W_a - W_w}$, Boyle's, $PV = P'V'$

Devices: Hydraulic press, air cushion, barometer—mercurial and aneroid. Pumps, lift, force, vacuum, compression, centrifugal, balloon, siphon, etc. Construction and action of each.

CHAPTER V

FORCE AND MOTION

(1) FORCE, HOW MEASURED AND REPRESENTED

71. Force.—We have been studying various forces, such as air pressure, pressure in liquids, and the force of elasticity in solids, and have considered them simply as pushes or pulls. A more formal study of forces in general and of devices for representing and measuring them will be helpful at this point of the course.

A force is that which tends to cause a change in the size or shape of a body or in its state of motion. In other words a force is a push or a pull. That is, force tends to produce distortion or change of motion in a body. Force itself is invisible. We measure it by the effect it produces. Forces are usually associated with the objects exerting them. Thus we speak of *muscular force, air pressure, liquid pressure*, the force of a spring, the force of the earth's attraction and so on.

Forces are classified in various ways.

I. With respect to the *duration and steadiness* of the force.

(a) Constant, as the earth's attraction. (b) Impulsive, as the stroke of a bat on a ball. (c) Variable, as the force of the wind.

II. With respect to the *direction* of the force.

(a) Attractive, as the earth's attraction. (b) Repulsive, as air pressure, liquid pressure, etc.

72. Methods of Measuring Force.—Since forces are measured by their effects which are either distortion or change of motion, either of these effects may be used to measure them. For example, the force exerted by a locomotive is sometimes computed by the *speed* it can develop in a train of cars in a given time, or the force of the blow of a baseball bat is estimated by the *distance* the ball goes before it strikes the ground.

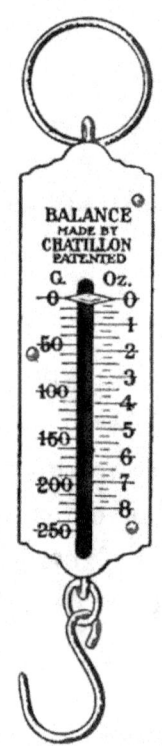

The more common method of measuring force, however, is by *distortion*, that is, by measuring the change of shape of a body caused by the force. In doing this, use is made of Hooke's Law (Art. 32), in which it is stated that "within the limits of perfect elasticity," changes of size or shape are directly proportional to the forces employed. That is, twice as great a force will produce twice as great a change of shape and so on.

A common contrivance using this principle is the spring balance (Fig. 55), with which all are familiar, as ice scales, meat scales, postal scales, etc. The object which changes shape in this device is a coiled spring contained in the case of the instrument. The balance is so constructed that when the spring is pulled out as far as possible it has not reached its limit of

FIG. 55.—A spring balance.

elasticity, since, if the spring were stretched so as to exceed its elastic limit, the index would not return to its first position on removing the load. (See Arts. 30–32.)

73. Graphic Representation of Forces.—A force is said to have three elements. These are (a) *its point* of *application*, (b) *its direction*, and (c) *its magnitude*. For example, if there is hung upon the hook of a spring balance a weight of 5 lbs., then we have: (a) its point of application on the hook of the balance, (b) its downward direction and (c) its magnitude, or 5 lbs. These three elements may be represented by a line. Thus in Fig. 56a, a line *AB* is drawn as shown, five units long; *A* represents the point of application; *B*, the arrow head, shows the direction; and the length of the line (five units) shows the magnitude of the force.

This is called a *graphic representation* since it represents by a line the quantity in question. If another weight of 5 lbs. were hung from the first one, the graphic representation of both forces would be as in Fig. 56b. Here the first force is represented by *AB* as before, *BC* representing the second force applied. The whole line represents the *resultant* of the two forces or the result of their combination. If the two weights were hung one at each end of a short stick *AC* (Fig. 56c), and the latter suspended at its center their combined weight or *resultant* would of course be

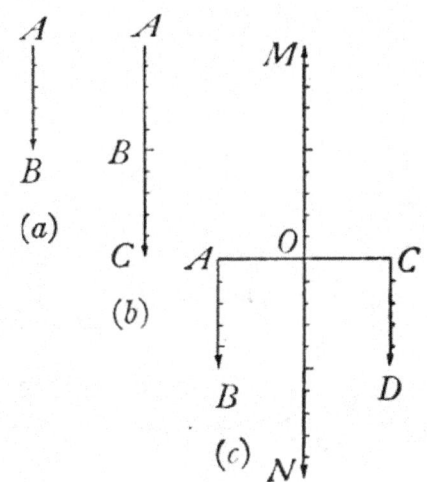

Fig. 56.—Graphic representation of forces acting along the same or parallel lines.

applied at the center. The direction would be the same as that of the two weights. The resultant therefore is represented by *ON*. In order to exactly balance this resultant *ON*, a force of equal magnitude but opposite in direction must be applied at the point of application of *ON*, or *O*. *OM* then represents a force that will just balance or hold in equilibrium the resultant of the two forces *AB* and *CD*. This line *OM* therefore represents the *equilibrant* of the weights *AB* and *CD*. The resultant of two forces at an angle with each other is formed differently, as in Fig. 57 *a*. Here two forces *AB* and *AC* atc at an angle with each other. Lay off at the designated angle the lines *AB* and *AC* of such length as will accurately represent the forces. Lay off *BD* equal to *AC* and *CD* equal to *AB*. The figure *ABCD* is then a parallelogram. Its diagonal *AD* represents the resultant of the forces *AB* and *AC* acting at the angle *BAC*. If *BAC* equals 90

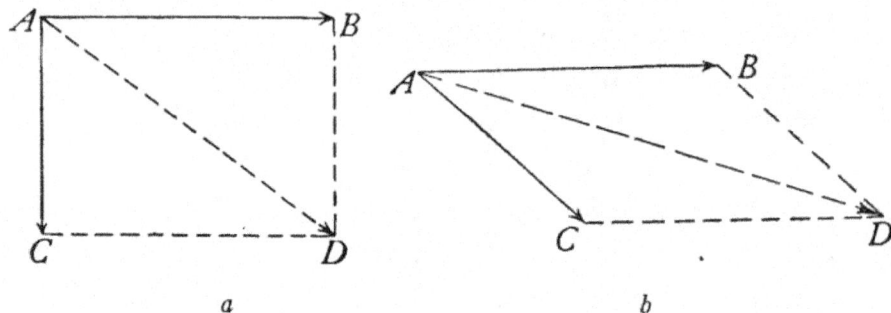

a   *b*

Fig. 57.—Graphic representation of two forces acting (*a*) at a right angle, (*b*) at an acute angle.

degrees or is a right angle, *AD* may be *computed* thus: $AB^2 + BD^2 = AD^2$. Why?

and $$AD = \sqrt{AB^2 + BD^2}.$$

This method of determining the resultant by *computation* may be used when the two forces are at right angles. (In any case, *AD may be measured* using the same scale that is laid off upon *AB* and *AC*, as shown in Fig. 57 *b*.) The three cases of combining forces just given may be classified as follows: The *first* is that of *two forces acting along the same line* in the same or opposite direction, as when two horses are hitched tandem, or in a tug of war. The *second is that of two forces acting along parallel lines*, in the same direction, as when two horses are hitched side by side or abreast. The *third* is that of *two forces acting at the same point at an angle*. It may be represented by the device shown in Fig. 58, consisting of two spring balances suspended from nails at the top of the blackboard at *A* and *B*. A cord is attached to both hooks and is passed through a small ring at *O* from which is suspended a known weight, *W*. Lines are drawn on the blackboard under the stretched cords, from *O* toward *OA*, *OB*, and *OW* and distances measured on each from *O* to correspond to the three forces as read on balance *A* and *B* and the weight *W*. Let a parallelogram be constructed on the lines measured off on *OA* and *OB*. Its diagonal drawn from *O* will be found to be vertical and of the same length as the line measured on *OW*. The diagonal is the *resultant* of the two forces and *OW* is the equilibrant which is equal and opposite to the resultant.

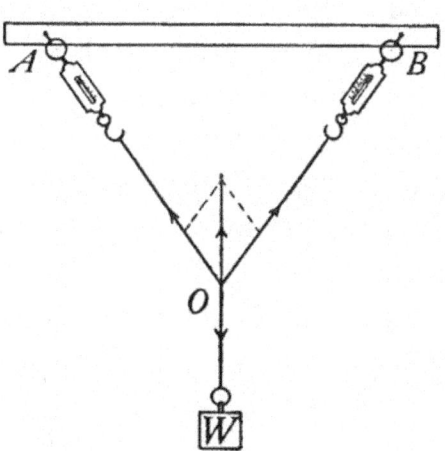

Fig. 58.—Experimental proof of parallelogram of forces.

Again, the *first* case may be represented by a boat moving up or down a stream; the resultant motion being the combined effect of the boat's motion and that of the stream. The *second*, may be represented by two horses attached side by side to the same evener. The resultant force equals the sum of the two component forces. The *third*, may be represented by a boat going across a stream, the resultant motion being represented by the diagonal of the parallelogram formed by using the lines that represent the motion of the stream and of the boat.

74. Units for Measuring Force.—Force is commonly measured in units of weight: in pounds, kilograms, and grams. For example, we speak of 15 lbs. pressure per square inch and 1033.6 g. pressure per square centimeter as representing the air pressure. It should be noted here that the words pound, kilogram, and gram are used not only to represent *weight* or *force* but also the masses of the objects considered. Thus, one may speak of a pound-mass meaning the amount of material in the object.

It will help to avoid confusion if we reserve the simple terms "gram" and "pound" to denote exclusively an amount of matter, that is, a mass, and to use the full expression "gram of force" or "pound of force" whenever we have in mind the pull of the earth upon these masses. Or, one may speak of a *pound-weight* meaning the amount of attraction exerted by the earth upon the object. The same is true of *gram-mass* and *gram-weight*. The mass of a body does not change when the body is transferred to another place. The weight, however, may vary, for on moving a body from the equator toward the poles of the earth the weight is known to increase.

Important Topics

1. Definition of force.
2. Classification of forces. (a) Duration: constant, impulsive, variable. (b) Direction: attractive, repulsive.
3. Methods of measuring force. (a) By distortion. (b) By change of motion.
4. Graphic representation of forces: component, resultant, equilibrant.
5. Three cases of combining forces. (1) Two forces acting on the same line. (2) Two forces acting in parallel lines. (3) Two forces acting at the same point at an angle.
6. Units for measuring force, pound, gram.

Exercises

1. Name five natural forces. Which produce a tension? Which a pressure?
2. How much can you lift? Express in pounds and kilograms.

3. Show graphically the resultatnt of two forces at right angles, one of 12 lbs., the other of 16 lbs. What is the magnitude of this resultant? Then determine the answer, first by measurement and then by computation. Which answer is more accurate? Why?

4. Represent by a parallelogram the two forces that support a person sitting in a hammock and draw the line representing the resultant.

5. Find graphically the resultant of the pull of two forces, one of 500 lbs. east and one of 600 lbs. northwest.

6. Determine the equilibrant of two forces, one of 800 lbs. south and one of 600 lbs. west.

7. Would the fact that weight varies on going from the equator to either pole be shown by a spring balance or a beam balance? Explain.

(2) Motion. Newton's Laws of Motion

75. Motion a Change of Position.—Motion is defined as a continuous *change in the position* of a body. The *position* of a body is usually described as its *distance* and *direction* from some fixed point. Thus a man on a boat may be at rest with respect to the boat and moving with respect to the earth. Or, if he walks toward the stern as fast as the boat moves forward, he may keep directly over a rock on the bottom of the lake and hence not be moving with reference to the rock and yet be in motion with respect to the boat. Motion and rest, therefore, are *relative* terms. The earth itself is in motion in turning on its axis, in moving along its orbit, and in following the sun in its motion through space. Motions are classified in several ways:

(A) Modes of Motion

1. *Translation.*—A body is said to have motion of *translation* when every line in it keeps the same direction.

2. *Rotation.*—A body has motion of *rotation* when it

turns upon a fixed axis within the body, as a wheel upon its axle or the earth upon its axis.

3. *Vibration* or *Oscillation.*—A body is said to have *vibratory* or *oscillatory* motion when it returns to the same point at regular intervals by reversals of motion along a given path, *e.g.*, a pendulum.

(B) Direction of Motion

1. *Rectilinear.*—A body has rectilinear motion when its path is a straight line. Absolute rectilinear motion does not exist, although the motion of a train on a straight stretch of track is nearly rectilinear.

2. *Curvilinear.*—A body has *curvilinear* motion when its path is a curved line, *e.g.*, the path of a thrown ball.

(C) Uniformity of Motion

1. *Uniform.*—A body has uniform motion when its speed and direction of motion do not change. Uniform motion for extended periods is rarely observed. A train may cover, on an average, 40 miles per hour but during each hour its speed may rise and fall.

2. *Variable.*—A body has variable motion when its speed or direction of motion is continually changing. Most bodies have variable motion.

3. *Accelerated.*—A body has accelerated motion when its speed or direction of motion continually changes.

If the speed changes by the same amount each second, *and the direction of motion does not change* the motion is said to be *uniformly* accelerated, *e.g.*, a falling body.

Uniformly accelerated motion will be studied further under the topic of falling bodies.

Velocity is *the rate of motion* of a body in a given direction. For example, a bullet may have a velocity of 1300 ft. a second upwards. *Acceleration* is the *rate of change of velocity* in a given direction, or the change of velocity in a unit of time. A train starting from a station gradually increases its speed. The gain in velocity during one second is its acceleration. When the velocity is decreasing, as when a train is slowing down, the acceleration is opposite in direction to the velocity. A falling body falls faster and faster. It has *downward acceleration*. A ball thrown upward goes more and more slowly. It also has *downward acceleration*.

76. Momentum.—It is a matter of common observation that a heavy body is set in motion with more difficulty than a light one, or if the same force is used for the same length of time upon a light and a heavy body,[1] the light body will be given a greater velocity. This observation has led to the *calculation* of what is called the "quantity of motion" of a body, or its *momentum*. It is computed by multiplying the mass by the velocity. If the C. G. S. system is used we shall have as the momentum of a 12 g. body moving 25 cm. a second a momentum of 12 × 25 or 300 C. G. S. units of momentum. This unit has no name and is therefore expressed as indicated above. The formula for computing momentum is: $M = mv$.

Newton's Laws of Motion

77. Inertia, First Law of Motion.—One often observes when riding in a train that if the train moves forward suddenly the passengers do not get into motion as soon as the train, and apparently are jerked backward. While if the train is stopped suddenly, the passengers tend to keep in motion. This tendency of matter to keep moving when in motion and to remain at rest when at rest is

---

[1] By a light body is meant one of small mass, a heavy body possessing much greater mass.

often referred to as the property of *inertia*. *Newton's first law of motion*, often called the *law of inertia*, describes this property of matter as follows:

Every body continues in a state of rest or of uniform motion in a straight line unless it is compelled to change that state by some external force. This means that if an object like a book is lying on a table it will remain there until removed by some outside force. No inanimate object can move itself or stop itself. ·If a ball is thrown into the air it would move on forever if it were not for the *force* of attraction of the earth and the resistance of the air.

FIG. 59.—The ball remains when the card is driven away.

It takes time to put a mass into motion, a heavy object requiring more time for a change than a light object. As an example of this, note the movements of passengers in a street car when it starts or stops suddenly. Another illustration of the law of inertia is the so-called "penny and card" experiment. Balance a card on the end of a finger. Place on it a coin directly over the finger, snap the card quickly so as to drive the card from beneath the coin. The coin will remain on the finger. (See Fig. 59.)

According to Newton's first law of motion a moving body which could be entirely freed from the action of all external forces would have uniform motion, and would describe a perfectly straight course. The curved path taken by a baseball when thrown shows that it is acted upon by an outside force. This force, the attraction of the earth, is called *gravity*.

78. Curvilinear Motion.—Curvilinear motion occurs when a moving body is pulled or pushed away from a

Sir Isaac Newton (1642–1727) Professor of mathematics at Cambridge university; discovered gravitation; invented calculus; announced the laws of motion; wrote the Principia; made many discoveries in light.

SIR ISAAC NEWTON
"By Permission of the Berlin Photographic Co., New York."

Galileo Galilei (1564–1642). Italian. "Founder of experimental science"; "Originator of modern physics"; made the first thermometer; discovered the laws of falling bodies and the laws of the pendulum; invented Galilean telescope.

GALILEO GALILEI
"By Permission of the Berlin Photographic Co., New York."

straight path. The pull or push is called *centripetal* (center-seeking) force. A moving stone on the end of a string when pulled toward the hand moves in a curve.

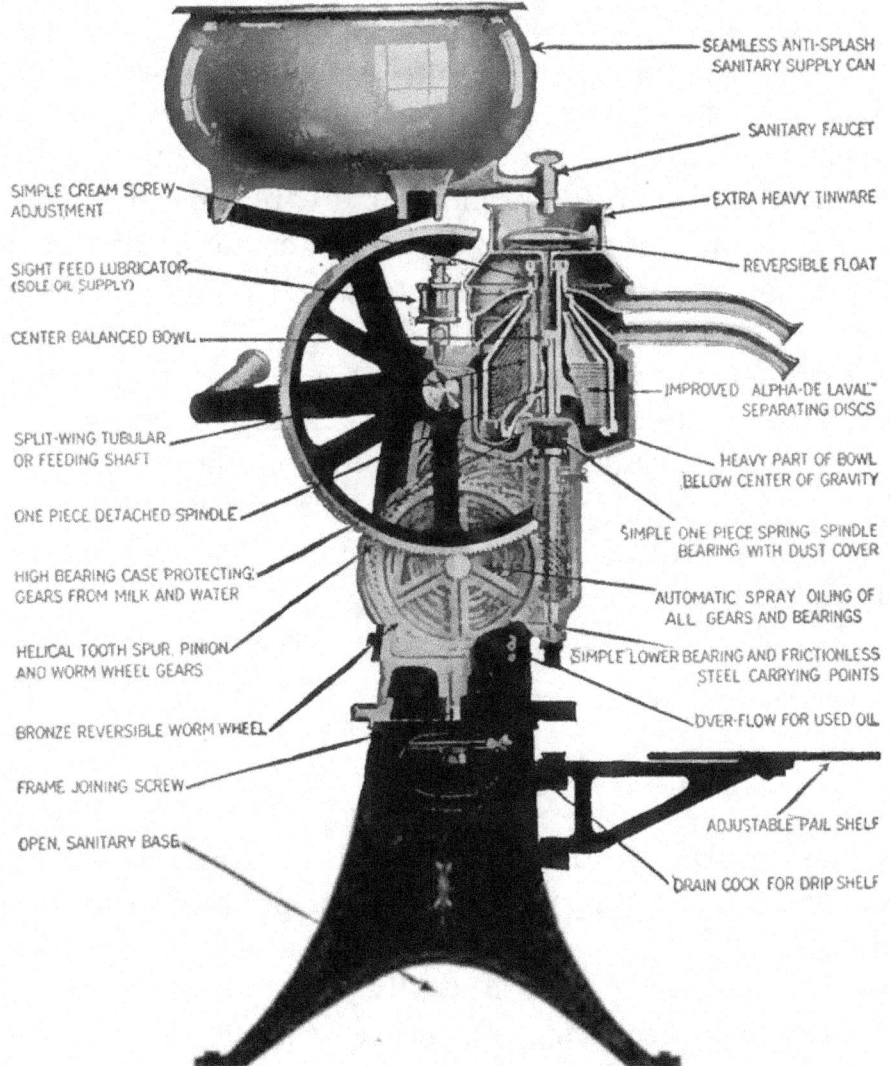

SEAMLESS ANTI-SPLASH SANITARY SUPPLY CAN

SANITARY FAUCET

SIMPLE CREAM SCREW ADJUSTMENT

EXTRA HEAVY TINWARE

SIGHT FEED LUBRICATOR (SOLE OIL SUPPLY)

REVERSIBLE FLOAT

CENTER BALANCED BOWL

IMPROVED ALPHA-DE LAVAL SEPARATING DISCS

SPLIT-WING TUBULAR OR FEEDING SHAFT

HEAVY PART OF BOWL BELOW CENTER OF GRAVITY

ONE PIECE DETACHED SPINDLE

SIMPLE ONE PIECE SPRING SPINDLE BEARING WITH DUST COVER

HIGH BEARING CASE PROTECTING GEARS FROM MILK AND WATER

AUTOMATIC SPRAY OILING OF ALL GEARS AND BEARINGS

HELICAL TOOTH SPUR, PINION AND WORM WHEEL GEARS

SIMPLE LOWER BEARING AND FRICTIONLESS STEEL CARRYING POINTS

BRONZE REVERSIBLE WORM WHEEL

OVER-FLOW FOR USED OIL

FRAME JOINING SCREW

ADJUSTABLE PAIL SHELF

OPEN, SANITARY BASE

DRAIN COCK FOR DRIP SHELF

FIG. 60.—Cross-section of the DeLaval cream separator.

If the string is released the stone moves in a tangent to the curve. The string pulls the hand. This phase of the pull is called *centrifugal* force. The *centripetal* force is the pull on the stone. Centripetal and centrifugal force together cause a tension in the string. Examples of curvilinear motion are very common. The rider and horse in a circus ring lean inward in order to move in a curve. The curve on a running track in a gymnasium is "banked" for the same reason. Mud flying from the wheel of a carriage, the skidding of an automobile when passing rapidly around a corner, and sparks flying from an emery wheel, are illustrations of the First Law of Motion.

Cream is separated from milk by placing the whole milk in a rapidly revolving bowl, the cream being lighter collects in the center and is thrown off at the top. (See Fig. 60.) Clothes in steam laundries are dried by a centrifugal drier. In amusement parks many devices use this principle. (See centrifugal pumps, Art. 70.)

79. The Second Law of Motion, sometimes called the *law of momentum*, leads to the *measurement of force*, by the momentum or the quantity of motion, produced by it. The law is stated as follows:

FIG. 61.—The two balls reach the floor at the same time.

Change of motion, or momentum, is proportional to the acting force and takes place in the direction in which the force acts. In other words, if two or more forces act at the same instant upon a body each produces the same effect that it would if acting alone. If a card be supported on two nails driven horizontally close together into an upright board (see Fig. 61), and two marbles be so placed on the ends as to balance each other, when one marble is snapped horizontally by a blow, the other will fall. Both reach the floor at the same time. The two balls are equally pulled down by the earth's attraction and strike the ground at the same time, though one is shot sidewise, and the other is dropped vertically.

As gravity is a constant force, while the blow was only a momentary force, the actual path or resultant motion will be a curved line.

The constant relation, between the acting force and the change of momentum it produces in a body, has led to the adoption of a convenient C. G. S. unit of force called the *dyne*. *The dyne is that force which can impart to a mass of one gram a change of velocity at the rate of one centimeter per second every second.* This definition assumes that the body acted upon is free to move without hindrance of any kind, so that the acting force has to overcome only the *inertia* of the body. *However*, the *law* applies in every case of application of force, so that each force produces its full effect independently of other forces that may be acting at the same time upon the body.

80. Newton's Third Law.—This law has been experienced by everyone who has jumped from a rowboat near the shore. The muscular action that pushes the body forward from the boat also pushes the boat backward, often with awkward results. The law is stated: *To every action, there is always an opposite and equal reaction, or the mutual actions of any two bodies are always equal and opposite in direction.* Many illustrations of this law are in every one's mind: a stretched rope pulls with the same force in one direction as it does in the opposite direction. If a bat hits a ball, the ball hits the bat with an equal and opposite force. The third law is therefore sometimes called the law of *reaction*. When a weight is hung upon a spring balance the action of the weight pulls down the spring until it has stretched sufficiently (Hooke's Law) to produce an elastic *reaction* that equals and hence supports the weight. When a man stands at the center of a plank supported at its ends, the action of the man's weight bends the plank until the elastic force developed in the plank equals the weight applied. Further, when a train or a wagon is on a bridge the bridge yields until it has developed an elastic reaction equal to the weight applied. If a person stands in the center of a room, the floor beams yield until the third law is satisfied. In fact, whenever a force acts, a contrary equal force always acts.

81. Stress and Strain.—A pair of forces that constitute an action and a reaction is called a *stress*. The two forces are two parts of one *stress*. If the two forces act away from each other, as in the breaking of a string, the stress is called a *tension*, but if they act toward each other as in crushing anything, the stress is called a *pressure*. In order for a body to exert force it must meet with resistance. The force exerted is never greater than the resistance encountered. Thus one can exert but little force upon a feather floating in the air or upon other light objects. A fast moving shot exerts no force unless it encouters some resistance.

Forces, then, are always found in pairs. Thus to break a string, to stretch an elastic band, to squeeze a lemon, one must exert two equal and opposite forces. Such a thing as a single force acting alone is unknown. Usually, however, we give our attention mainly to one of the forces and ignore the other. When a force acts upon a body the change of shape or size resulting is called a *strain*. Hooke's law (Art. 32) is often expressed as follows: "The strain is proportional to the stress," *e.g.*, the stretch of the spring of a spring balance is proportional to the load placed upon it.

Important Topics

1. Motion a change of position. Kinds of motion.
2. Newton's laws of motion.
3. Momentum.
4. Inertia. First law of motion. Curvilinear motion.

5. Second law of motion.
6. Third law of motion. Action and reaction, stress and strain.

Exercises

1 Mention three illustrations of the third law, different from those given.
2. A rifle bullet thrown against a board standing upon edge will knock it down; the same bullet fired at the board will pass through it without disturbing its position. Explain.
3. A hammer is often driven on to its handle by striking the end of the latter. Explain.
4. Consider a train moving 60 miles an hour, with a gun on the rear platform pointing straight backward. If a ball is fired from the gun with a speed of 60 miles an hour, what will happen to the ball?
5. Could one play ball on the deck of an ocean steamer going 25 miles an hour without making allowance for the motion of the ship? Explain.
6. On a railroad curve, one rail is always higher. Which? Why?
7. Why can a small boy when chased by a big boy often escape by dodging?
8. Will a stone dropped from a moving train fall in a straight line? Explain.
9. A blast of fine sand driven against a sheet of glass soon gives it a rough surface. Explain.
10. Explain the use of fly-wheels in steadying the motion of machinery (for example, the sewing machine).
11. Is it easier to walk to the front or rear of a passenger train when it is stopping? Why?
12. Why does lowering the handles of a wheel-barrow on the instant of striking make it easier to go over a bump?
13. Why should a strong side wind interfere with a game of tennis? How can it be allowed for?
14. On which side of a railroad track at a curve is it the safer to walk while a train is passing? Why?
15. Why does a bullet when fired through a window make a clean round hole in the glass, while a small stone thrown against the window shatters the glass?
16. A tallow candle can be fired through a pine board. Why?

17. In cyclones, straws are frequently found driven a little distance into trees; why are the straws not broken and crushed instead of being driven into the tree unbroken?

18. A bullet weighing one-half oz. is fired from a gun weighing 8 lb. The bullet has a velocity of 1800 ft. per second. Find the velocity of the "kick" or recoil of the gun.

19. When football players run into each other which one is thrown the harder? Why?

20. A railroad train weighing 400 tons has a velocity of 60 miles per hour. An ocean steamer weighing 20,000 tons has a velocity of one half mile per hour. How do their momenta compare?

21. Why is a heavy boy preferable to a lighter weight boy for a football team?

22. Why does a blacksmith when he desires to strike a heavy blow, select a heavy sledge hammer and swing it over his head?

23. Why does the catcher on a baseball team wear a padded glove?

(3) Resolution of Forces

82. Resolution of Forces.—We have been studying the effect of forces in producing motion and the results of combining forces in *many* ways; in the *same line*, in *parallel lines*, and in *diverging lines*. Another case of much interest and importance is *the determination of the effectiveness of a force in a direction different from the one in which it acts*. This case which is called *resolution of forces* is frequently used. To illustrate: one needs but to recall that a sailor uses this principle in a practical way whenever he sails his boat in any other direction than the one in which the wind is blowing, *e.g.*, when the wind is blowing, say from the north, the boat may be driven east, west, or to any point south between the east and west and it is even possible to beat back against the wind toward the northeast or northwest. Take a sled drawn by a short rope with the force applied along the line *AB* (see Fig. 62); part of this force tends to lift the front of the sled as *AC* and a part to draw it forward as *AD*. Hence not all of the force applied along

AB is used in drawing the sled forward. Its effectiveness is indicated by the relative size of the component AD compared to AB.

The force of gravity acting upon a sphere that is resting on an *inclined plane* may be readily resolved into two components, one, the *effective* component, as OR, and the other, the *non-effective* as OS. (See Fig. 63.) If the angle ACB is 30 degrees, AB equals

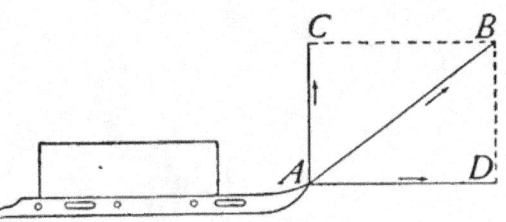

FIG. 62.—AD is the effective component.

½ of AC and OR equals ½ of OG, so that the speed of the sphere down the plane developed in 1 second is less than (about one-half of) the speed of a freely falling body developed in the same time. Why is OS non-effective?

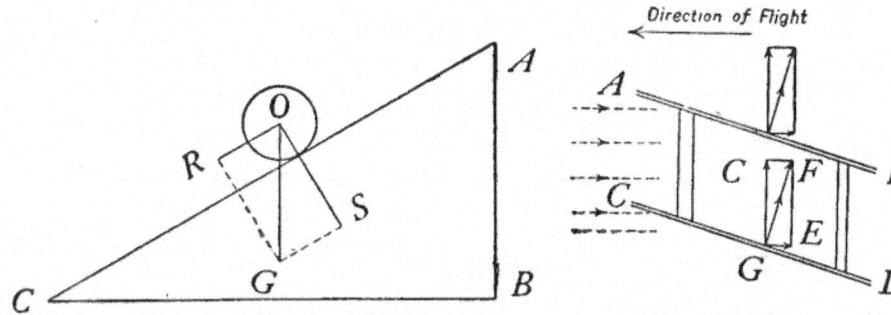

FIG. 63.—The effective component is OR.

FIG. 64.—Resolution of the forces acting on an aeroplane.

83. The Aeroplane.—The aeroplane consists of one or two frames $ABCD$ (see Fig. 64), over which is stretched cloth or thin sheet metal. It is driven through the air by a propeller turned by a powerful gasoline motor. This has the effect of creating a strong breeze coming toward the front of the aeroplane. As in the case of the sailboat a pressure is created at right angles to the plane along GF and this may be resolved into two components

7

as *GC* and *GE*, *GC* acting to lift the aeroplane vertically and *GE* opposing the action of the propeller. Fig. 65 represents the Curtis Flying Boat passing over the Detroit river.

Fig. 65.—The Curtis hydroplane.

Exercises

1. If a wagon weighing 4000 lbs. is upon a hill which rises 1 ft. in 6, what force parallel to the hill will just support the load? (Find the effective component of the weight down the hill.)

2. If a barrel is being rolled up a 16-ft. ladder into a wagon box 3 ft. from the ground, what force will hold the barrel in place on the ladder, if the barrel weighs 240 lbs. Show by diagram.

3. Show graphically the components into which a man's push upon the handle of a lawn mower is resolved.

4. Does a man shooting a flying duck aim at the bird? Explain.

5. What are the three forces that act on a kite when it is "standing" in the air?

6. What relation does the resultant of any two of the forces in problem five have to the third?

7. Into what two forces is the weight of a wagon descending a hill resolved? Explain by use of a diagram.

8. A wind strikes the sail of a boat at an angle of 60 degrees to the perpendicular with a pressue of 3 lbs. per square foot. What is the effective pressure, perpendicular to the sail? What would be the effective pressure when it strikes at 30 degrees?

9. How is the vertical component of the force acting on an aeroplane affected when the front edge of the plane is elevated? Show by diagram.

(4) MOMENT OF FORCE AND PARALLEL FORCES

84. Moment of Force.—In the study of motion we found that the quantity of motion is called *momentum* and is measured by the product of the *mass times the velocity*. In the study of *parallel forces*, especially such as tend to produce *rotation*, we consider a similar quantity. It is called a *moment of force*, which is the term applied to the *effectiveness* of a force in producing change of rotation. It is

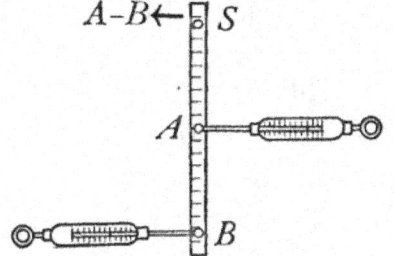

FIG. 66.—The moments about *S* are equal.

also measured by the product of two quantities; *One, the magnitude of the force itself*, and the other, *the perpendicular distance from the axis about which the rotation takes place to the line representing the direction of the force*.

To illustrate: Take a rod, as a meter stick, drill a hole at *S* and place through it a screw fastened at the top of the blackboard. Attach by cords two spring balances and draw to the right and left, *A* and *B* as in Fig. 66. Draw out the balance *B* about half way, hold it steadily, or fasten the cord at the side of the blackboard, and read both balances. Note also the distance *AS* and *BS*. Since the rod is at rest, the tendency to rotate to the right and left must be equal. That is, the moments of the forces at *A* and *B* about *S* are equal. Since these are computed by the product of the *force times the force* arm, multiply *B* by *BS* and *A* by *AS* and see if the computed moments are equal. *Hence a force that tends to turn or rotate a body to the right can be balanced by another of equal moment that acts toward the left.*

85. Parallel Forces.—Objects are frequently supported by two or more upward forces acting at different points and forming in this way a system of parallel forces; as when two boys carry a string of fish on a rod between

FIG. 67.—Law of parallel forces illustrated.

them or when a bridge is supported at its ends. The principle of moments just described aids in determining the magnitude of such forces and of their resultant. To illustrate this take a wooden board 4 in. wide and 4 ft. long of uniform dimensions. (See Fig. 67.) Place several screw hooks on one edge with one set at *O* where the board will hang horizontally when the board is suspended there. Weigh the board by a spring balance hung at *O*. This will be the resultant in the following tests. Now hang the board from two spring balances at *M* and *N* and read both *balances*. Call readings *f* and *f'*. To test the forces consider *M* as a fixed point (see Fig. 67) and the weight of the board to act at *O*. Then the moment of the weight of the board should be equal the moment of the force at *N* since the board does not move, or *w* times *OM* equals *f'* times *NM*. If *N* is considered the fixed point then the moment of the weight of the board and of *f with reference to the point N* should be equal, or *w* times *ON* = *f* times *NM*. Keeping this illustration in mind, the law of parallel forces may be stated at follows: 1. *The resultant of two parallel forces acting in the same direction at different points in a body is equal to their sum and has the same direction as the components.*

The moment of one of the components about the point of application of the other is equal and opposite to the moment of the supported weight about the other.

Problem.—If two boys carry a string of fish weighing 40 lbs. on a rod 8 ft. long between them, what force must each boy exert if the string is 5 ft. from the rear boy?

Solution.—The moment of the force F exerted about the opposite end by the rear boy is $F \times 8$. The moment of the weight about the same point is $40 \times (8 - 5) = 120$. Therefore $F \times 8 = 120$, or $F = 15$, the force exerted by the rear boy. The front boy exerts a force of F

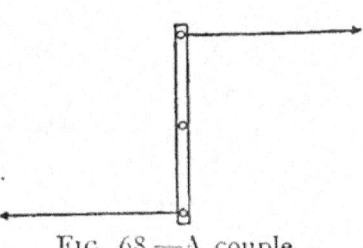

FIG. 68.—A couple.

whose moment about the other end of the rod is $F \times 8$. The moment of the weight about the same point is $40 \times 5 = 200$. Since the moment of F equals this, $200 = F \times 8$, or $F = 25$. Hence the front boy exerts 25 lbs. and the rear boy 15 lbs.

86. The Couple.—If two equal parallel forces act upon a body along different lines in opposite directions, as in Fig. 68, they have no single resultant or there is no one force that will have the same effect as the two components acting together. A combination of forces of this kind is called a *couple*. Its tendency is to produce change of rotation in a body. An example is the action upon a compass needle which is rotated by a force which urges one end toward the north and by an equal force which urges the other end toward the south.

Important Topics

1. Moment of force, how measured.
2. Parallel forces.
3. The two laws of parallel forces.
4. The couple.

Exercises

1. Show by diagram how to arrange a three-horse evener so that each horse must take one-third of the load.
2. Two boys support a 10-ft. pole on their shoulders with a 40-lb. string of fish supported from it 4 ft. from the front boy. What load does each boy carry? Work by principle of moments.
3. If two horses draw a load exerting a combined pull of 300 lbs., what force must each exert if one is 28 in. and the other is 32 in. from the point of attachment of the evener to the load.
4. A weight of 100 lbs. is suspended at the middle of a rope *ACB* 20 ft. long. (See Fig. 69.) The ends of the rope are fastened at points *A* and *B* at the

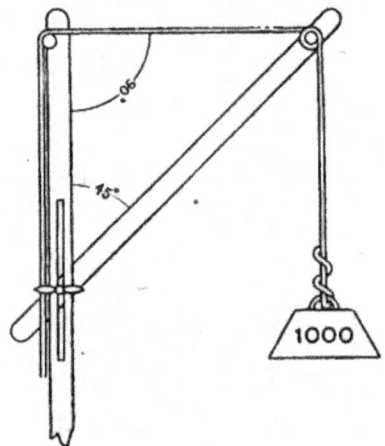

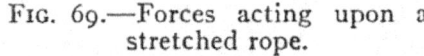

FIG. 69.—Forces acting upon a stretched rope.　　FIG. 70.—A crane with horizontal tie.

same height. Consider *D* as the center of the line *AB*. What is the tension of the rope when *CD* is 3 ft.? When *CD* is 1 ft.? When *CD* is 1 in.?

5. A crane is set up with the tie horizontal. (See Fig. 70.) If 1000 lbs. is to be lifted, find the tie stress and the boom stress if the boom angle is 30 degrees? If 45 degrees? 60 degrees?
6. A ball is placed on a plane inclined at an angle of 30 degrees to the horizontal. What fraction of its weight tends to cause motion down the plane? What effect does the other component of the weight have? Why?
7. A person weighing 150 lbs. is lying in a hammock. The distance between the supports is 15 ft. The hammock sags 4 ft. What is the tension in the supports at each end? What is the tension when the sag is only 1 ft.?

8. A ladder 30 ft. long and weighing 80 lbs. leans against the side of a building so that it makes an angle of 30 degrees with the building. Find the direction and magnitude of the component forces on the ground and at the building.

9. A traveling crane 50 ft. long weighing 10 tons moves from one end of a shop to the other, at the same time a load of 4000 lbs. moves from end to end of the crane. Find the pressure of the trucks of the crane on the track when the load is at a distance of 5, 10, 15, and 25 ft. from either end.

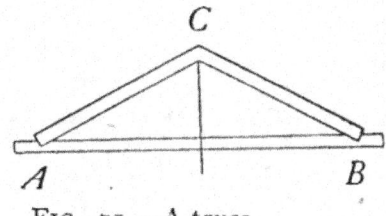

FIG. 71.—A truss.

10. Resolve a force of 500 lbs. into two components at right angles to each other, one of which shall be four times the other.

11. A truss (see Fig. 71), carries a load of 1000 lbs. at C. Find the forces acting along AC, BC, and AB. If AC and BC are each 12 ft. and AB 20 ft., which of these forces are tensions and which are pressures?

(5) GRAVITATION AND GRAVITY

87. Gravitation.—Gravitation is the force of attraction that exists between all bodies of matter at all distances. This attraction exists not only between the heavenly bodies, the stars and planets, etc., but is also found between bodies on the earth. A book attracts all objects in a room and outside of a room as well, since its weight shows that it is attracted by the earth itself. The gravitational attraction between ordinary bodies is so slight that it requires careful experiments to detect it. In fact, it is only when one of the attracting bodies is large, as for example the earth, that the force becomes considerable. Careful studies of the motions of the heavenly bodies, especially of that of the moon in its orbit about the earth, led Sir Isaac Newton to the statement of the *law of gravitation* which is well expressed in the following statement:

88. Law of Gravitation.—*Every particle of matter in the universe attracts every other particle with a force that is directly proportional to the product of their masses and inversely proportional to the square of the distance between them.*

The law may be separated into two parts, one referring to the masses of the bodies concerned, the other to the effect of the distance between them. The first part is easily understood since we all know that two quarts of milk will weigh just twice as much as one quart. To illustrate the second part of the law, suppose that the moon were removed to *twice* its present distance from the earth, then the attraction between the earth and the moon would be *one-fourth* its present attraction. If removed to *three* times its present distance, the attraction would be *one-ninth*, etc.

The attraction of the earth for other bodies on or near it is called *gravity*. The *weight* of a body is the measure of the earth's attraction for it; or it is the force of gravity acting upon it. Newton's third law of motion states that every action is accompanied by an equal and opposite reaction (Art. 80). Hence, the attraction of the earth for a book or any other object is accompanied by an equal attraction of the book for the earth.

89. Weight.—In advanced physics it is proved that a sphere attracts as if it were concentrated at its center. Thus if the earth's radius be considered as 4000 miles, then a body 4000 miles above the earth's surface would be 8000 miles above the earth's center, or twice as far from the center of the earth as is a body upon the earth's surface. A body then 4000 miles above the earth's surface will weigh then but one-fourth as much as it will at the surface of the earth.

Since the earth is flattened at the poles, the surface at

the equator is farther from the center of the earth than at points north or south. Thus a body weighing 1 lb. at the equator weighs 1.002 lb. at Chicago, or about 1/500 more. The rotation of the earth also affects the weight of a body upon it so that at the equator the weight of a body is 1/289 less than at the pole. Both effects, that of flattening and of rotation, tend to diminish the weight of bodies at the equator, so that a body at the latter place weighs about 1/192 less than at the poles.

In studying the effect of the earth's gravity, the following illustration will be helpful: Imagine an open shaft a mile square extending through the earth. What would happen to a stone thrown into the shaft? At first it would have the attraction of the whole earth drawing it and continually increasing its speed downward. As it descends from the surface, the pull toward the center grows less and less. Halfway to the center the body has lost half its weight. When the stone reaches the center, the pull in all directions is the same, or in other words, *it has no weight*. It would, however, continue moving rapidly on account of its inertia, and as it continues on from the center, the greater part of the earth being left behind, the attraction pulling toward the center will gradually stop it. It will then fall again toward the center and be stopped again after passing it, and after repeatedly moving up and down will finally come to rest at the center of the earth. At this point it will be found to be a body without weight since it is pulled equally in all directions by the material of the earth. What force brings the body to rest?

90. Center of Gravity.—A body is composed of a great many particles each of which is pulled toward the center of the earth by the force of gravity. A single force that would exactly equal the combined effect of the pull of the earth for all the particles of a body would be their resultant. The *magnitude* of this resultant is the weight of the body. The *direction* of this resultant is in a line passing toward the earth's center, while the *point of application* of this resultant is called the *center of gravity* of the body. The center of gravity of a body may also be briefly defined as *the point about which it may be balanced*. As the location of this point depends upon the distribution of matter in the body, the center of gravity is also sometimes called the *center of mass* of the body.

The earth's attraction for a body is considered for the

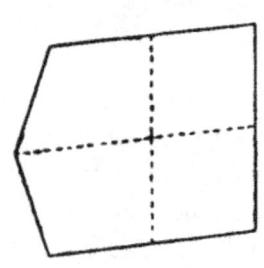

sake of simplicity, not as a multitude of little forces, but as a single force applied at its center of gravity. To find the center of gravity of a body find two intersecting lines along which it balances, see Fig. 72, and the center of gravity will be at the intersection. A vertical line through this point is sometimes called the *line of direction of the weight*.

FIG. 72.—The center of gravity is at the intersection of the lines of direction.

91. Equilibrium of Bodies.—Equilibrium means equally balanced. A body at rest or in uniform motion is then in equilibrium. An object is in equilibrium under gravity when a vertical line through its center of gravity passes through the point of support. A trunk is an example of a body in equilibrium since a vertical line from its center of gravity falls within the base formed by the area upon which it rests. Work will be necessary to tip the trunk from its position. The amount of work required will depend upon the weight of the body and the location of the center of gravity.

92. Kinds of Equilibrium.—(a) **Stable.**—A body is in stable equilibrium under gravity if its center of gravity is raised whenever the body is displaced. It will return to its first position if allowed to fall after being slightly displaced. In Fig. 73, *a* and *b* if slightly tipped will return to their first position. They are in stable equilibrium.

Other examples are a rocking chair, and the combination shown in Fig. 74.

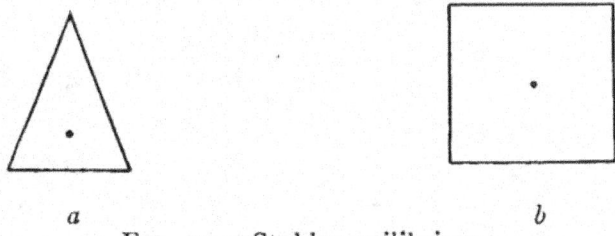

a   *b*

FIG. 73.—Stable equilibrium.

(b) Unstable.—A body is in unstable equilibrium under gravity if its center of gravity is lowered whenever the body is slightly displaced. It will fall farther from its first position. A pencil balanced on its point or a broom balanced on the end of the handle are in unstable equilibrium. The slightest disturbance will make the line of direction of the weight fall outside of (away from) the point of support (Fig. 75 *a*).

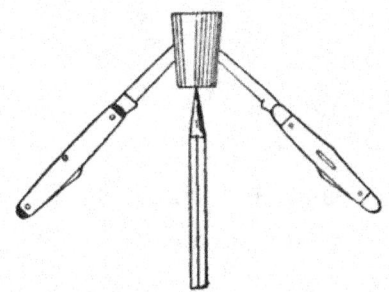

FIG. 74.—An example of stable equilibrium. Why?

(c) Neutral.—A body is in neutral equilibrium if its center of gravity is neither raised nor lowered whenever the body is moved. Familiar examples are a ball lying

a   *b*

FIG. 75.—Unstable equilibrium *a*, neutral equilibrium *b*.

on a table (Fig. 75 *b*) and a wagon moving on a level street (referring to its forward motion).

93. Stability.—When a body is in stable equilibrium, effort must be exerted to overturn it, and the degree of stability is measured by the effort required to overturn it. To overturn a body, it must be moved so that the vertical line through its center of gravity will pass outside of its supporting base. This movement in stable bodies necessitates a raising of the center of gravity. The higher this center of gravity must be raised in overturning the body,

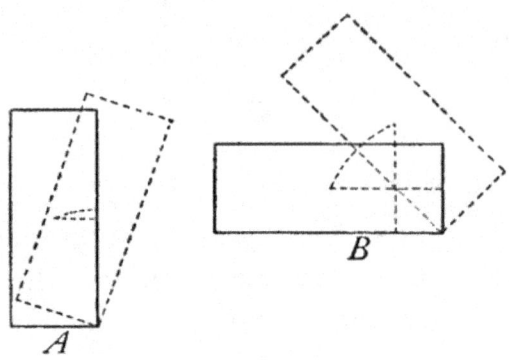

FIG. 76.—*B* is more stable than *A*.

the more stable it is, *e.g.*, see Fig. 76. Thus a wagon on a hillside will not overturn until its weight falls outside of its base, as in Fig. 77 *B* The stability of a body depends upon the position of its center of gravity and the area of its base. *The lower the center of gravity and the larger the base*, the more stable the body. What means are employed to give stability to bodies, in every-day use (such as clocks, ink-stands, pitchers, vases, chairs, lamps, etc).?

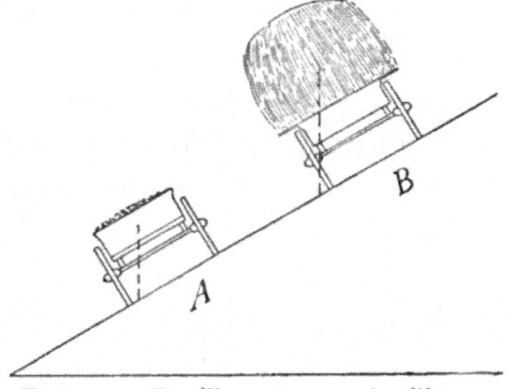

FIG. 77.—*B* will overturn; *A* will not.

Important Topics

1. Gravitation; law of gravitation, gravity, weight.
2. Center of gravity.
3. The three states of equilibrium.   Stability.

Exercises

1. Why is a plumb-line useful in building houses?
2. What is the center of gravity of a body?
3. Explain the action of a rocking chair that has been tipped forward.
4. Is the stability of a box greater when empty or when filled with sand? Explain.
5. How can you start yourself swinging, in a swing, without touching the ground?
6. Is the center of gravity of the beam of a balance above, below, or at the point of a support? How did you find it out?
7. Why are some ink bottles cone shaped with thick bottoms?
8. Would an electric fan in motion on the rear of a light boat move it? Would it move the boat if revolving under water? Explain.
9. What turns a rotary lawn sprinkler?
10. Why, when you are standing erect against a wall and a coin is placed between your feet, can you not stoop and pick it up unless you shift your feet or fall over?
11. What would become of a ball dropped into a large hole bored through the center of the earth?
12. When an apple falls to the ground, does the earth rise to meet it?
13. How far from the earth does the force of gravity extend?
14. Why in walking up a flight of stairs does the body bend forward?
15. In walking down a steep hill why do people frequently bend backward?
16. Why is it so difficult for a child to learn to walk, while a kitten or a puppy has no such difficulty?
17. Explain why the use of a cane by old people makes it easier for them to walk?

(6) Falling Bodies

94. Falling Bodies.—One of the earliest physical facts learned by a child is that a body unsupported falls toward the earth. When a child lets go of a toy, he soon learns to look for it on the floor. It is also of common observation that light objects, as feathers and paper, fall much slower than a stone. The information, therefore, that all bodies actually fall at the same rate in a vacuum or when removed from the retarding influence of the air is received with surprise.

This fact may be shown by using what is called a coin and feather tube. On exhausting the air from this tube, the feather and coin within are seen to fall at the same rate. (See Fig. 78.) When air is again admitted, the feather flutters along behind.

95. Galileo's Experiment.—The fact that bodies of different weight tend to fall at the same rate was first experimentally shown by Galileo by dropping a 1-lb. and a 100-lb. ball from the

FIG. 78.—Bodies fall alike in a vacuum.

FIG. 79.—Leaning tower of Pisa.

top of the leaning tower of Pisa in Italy (represented in Fig. 79). Both starting at the same time struck the ground together. Galileo inferred from this that feathers and other light objects would fall at the same rate as iron or lead were it not for the resistance of the air. After the invention of the air pump this supposition was verified as just explained.

96. Acceleration Due to Gravity.—If a body falls freely, that is without meeting a resistance or a retarding influence, its motion will continually increase. The *increase* in motion is found to be constant or uniform during each second. This uniform increase in motion or in velocity of a falling body gives one of the best illustrations that we have of uniformly accelerated motion. (Art. 75.) On the other hand, a body thrown upward has uniformly retarded motion, that is, its acceleration is downward. The velocity acquired by a falling body in unit time is called its *acceleration*, or the *acceleration due to gravity*, and is equal to 32.16 ft. (980 cm.) per second, downward, each second of time. In one second, therefore, a falling body gains a velocity of 32.16 ft. (980 cm.) per second, downward. In two seconds it gains twice this, and so on.

In formulas, the acceleration of gravity is represented by "g" and the number of seconds by t, therefore the formula for finding the velocity, V,[1] of a falling body starting from rest is $V = gt$. In studying gravity (Art. 89) we learned that its force varies as one moves toward or away from the equator. (How?) In latitude 38° the acceleration of gravity is 980 cm. per second each second of time.

97. Experimental Study of Falling Bodies.—To study falling bodies experimentally by observing the fall of unobstructed bodies is a difficult matter. Many devices have been used to reduce the motion so that the action of a falling body may be observed within the limits of a laboratory or lecture room. The simplest of these, and in some respects the most satisfactory, was used by Galileo. It consists of an inclined plane which reduces the effective component of the force of gravity so that the motion of a body rolling down the plane may be observed

---

[1] V represents the velocity of a falling body at the end of t seconds.

for several seconds. For illustrating this principle a steel piano wire has been selected as being the simplest and the most easily understood. This wire is stretched taut across a room by a turn-buckle so that its slope is about one in sixteen. (See Fig. 80.) Down this wire a weighted pulley is allowed to run and the distance it travels in 1, 2, 3, and 4 seconds is observed. From these observations we can compute the distance covered each second and the velocity at the end of each second.

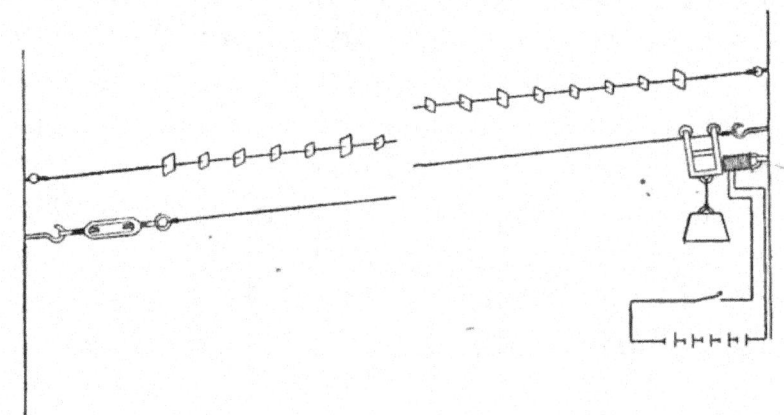

FIG. 80.—Apparatus to illustrate uniformly accelerated motion.

In Fig. 63, if *OG* represents the weight of the body or the pull of gravity, then the line *OR* will represent the effective component along the wire, and *OS* the non-effective component against the wire. Since the ratio of the height of the plane to its length is as one to sixteen, then the motion along the wire in Fig. 80 will be one-sixteenth that of a falling body.

98. Summary of Results.—The following table gives the results that have been obtained with an apparatus arranged as shown above.

In this table, column 2 is the one which contains the results directly observed by the use of the apparatus. Columns, 3, 4, and 5 are computed from preceding columns.

(1) No. of seconds	(2) Total distance moved	(3) Distance each second	(4) Velocity at end of second	(5) Acceleration each second
			Per second	Per second
1	30 cm.	30 cm.	60 cm.	60 cm.
2	120 cm.	90 cm.	120 cm.	60 cm.
3	270 cm.	150 cm.	180 cm.	60 cm.
4	480 cm.	210 cm.	240 cm.	60 cm.

Column 5 shows that the acceleration is uniform, or the same each second. Column 4 shows that the velocity increases with the number of seconds or that $V = at$. Column 3 shows that the increase in motion from 1 second to the next is just equal to the acceleration or 60 cm. This is represented by the following formula: $s = \frac{1}{2} a(2t - 1)$.

The results of the second column, it may be seen, increase as 1:4:9:16, while the number of seconds vary as 1:2:3:4. That is, the *total distance covered is proportional to the square of the number of seconds.*

This fact expressed as a formula gives: $S = \frac{1}{2} at^2$.

Substituting g, the symbol for the acceleration of gravity, for a in the above formulas, we have:

(1) $V = gt$, (2) $S = \frac{1}{2} gt^2$, (3) $s = \frac{1}{2} g(2t - 1)$.

99. Laws of Falling Bodies.—These formulas may be stated as follows for a body which falls from rest:

1. The velocity of a freely falling body at the end of any second is equal to 32.16 ft. per sec. or 980 cm. per second multiplied by the number of the second.

2. The distance passed through by a freely falling body during any number of seconds is equal to the square of the number of seconds multiplied by 16.08 ft. or 490 cm.

3. The distance passed through by a freely falling body during any second is equal to 16.08 feet or 490 cm. multiplied by one less than twice the number of the second.

8

Important Topics

1. Falling bodies.
2. Galileo's experiment.
3. Acceleration due to gravity.
4. Laws of falling bodies.

Exercises

1. How far does a body fall during the first second? Account for the fact that this distance is numerically equal to half the acceleration.

2. (a) What is the velocity of a falling body at the end of the first second? (b) How far does it fall during the second second? (c) Account for the difference between these numbers.

3. What is the velocity of a falling body at the end of the fifth second?

4. How far does a body fall (a) in 5 seconds (b) in 6 seconds (c) during the sixth second?

5. (a) What is the difference between the average velocity during the sixth second and the velocity at the beginning of that second?

   (b) Is this difference equal to that found in the second problem? Why?

6. A stone dropped from a cliff strikes the foot of it in 5 seconds. What is the height of the cliff?

7. Why is it that the increased weight of a body when taken to higher latitudes causes it to fall faster, while at the same place a heavy body falls no faster than a light one?

8. When a train is leaving a station its acceleration gradually decreases to zero, although the engine continues to pull. Explain.

9. Would you expect the motion of equally smooth and perfect spheres of different weight and material to be equally accelerated on the same inclined plane? Give reason for your answer. Try the experiment.

10. A body is thrown upward with the velocity of 64.32 ft. per sec. How many seconds will it rise? How far will it rise? How many seconds will it stay in the air before striking the ground?

11. 32.16 feet = how many centimeters?

12. The acceleration of a freely falling body is constant at any one place. What does this show about the pull which the earth exerts on the body?

(7) THE PENDULUM

100. The Simple Pendulum.—Any body suspended so as to swing freely to and fro is a pendulum, as in Fig. 81. A simple pendulum is defined as a single particle of matter suspended by a cord without weight. It is of course impossible to construct such a pendulum. A small metal ball suspended by a thread is approximately a simple pendulum. When allowed to swing its vibrations are made in equal times. This feature of the motion of a pendulum was first noticed by Galileo while watching the slow oscillations of a bronze chandelier suspended in the Cathedral in Pisa.

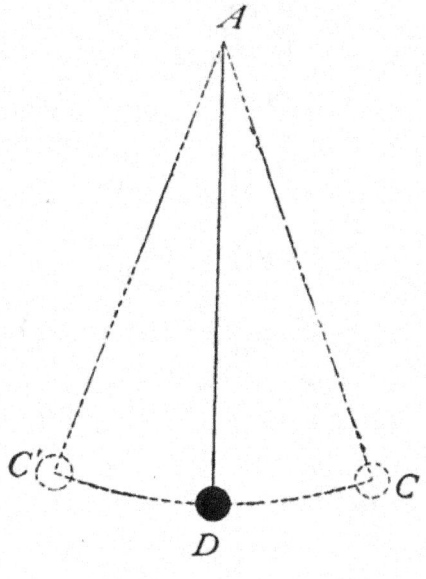

FIG. 81.—A simple pendulum.

101. Definition of Terms. *The center of suspension* is the point about which the pendulum swings. A *single vibration* is one swing across the arc. A *complete* or *double* vibration is the swing across the arc and back again. The time required for a double vibration is called the *period*. The *length* of a simple pendulum is approximately the distance from the point of support to the center of the bob.

A *seconds pendulum* is one making a single vibration per second. Its length at sea-level, at New York is 99.31 cm. or 39.1 in., at the equator 39.01 in., at the poles 39.22 in.

A *compound pendulum* is one having an appreciable portion of its mass elsewhere than in the small compact body or sphere called a bob. The ordinary clock pendulum or a meter stick suspended by one end are examples of compound pendulums.

The *amplitude* of a vibration is one-half the arc through which it swings, for example, the arc *DC* or the angle *DAC* in Fig. 81.

102. Laws of the Pendulum.—The following laws may be stated:

1. The period of a pendulum is not affected by its mass or the material of which the pendulum is made.

2. For small amplitudes, the period is not affected by the length of the arc through which it swings.

3. The period is directly proportional to the square root of the length. Expressed mathematically, $\dfrac{t}{t'} = \dfrac{\sqrt{l}}{\sqrt{l'}}$.

103. Uses of the Pendulum.—The chief use of the pendulum is to regulate motion in clocks. The wheels are kept in motion by a spring or a weight and the regulation is effected by an escapement (Fig. 82). At each vibration of the pendulum one tooth of the wheel *D* slips past the prong at one end of the escapement *C*, at the same time giving a slight push to the escapement. This push transmitted to the pendulum keeps it in motion. In this way, the motion of the wheel work and the hands is controlled. Another use of the pendulum is in finding the acceleration of gravity, by using the formula, $t = \pi\sqrt{\dfrac{l}{g}}$, in which t is the time in seconds of a single vibration and l the length of the pendulum. If, for example, the length of the seconds pendulum is 99.31 cm., then $1 = \pi\sqrt{\dfrac{99.31}{g}}$; squaring both sides of the equation, we have $1^2 = \pi^2\dfrac{99.31}{g}$, or $g = \dfrac{\pi^2 \times 99.31}{1^2} = 980.1$ cm. per sec., per sec. From this it follows that, since the force of gravity depends upon the distance from the center of the earth, the pendulum may be used to determine the elevation of a place above sea level and also the shape of the earth.

Important Topics

1. Simple pendulum.
2. Definitions of terms used.
3. Laws of the pendulum.
4. Uses of the pendulum.

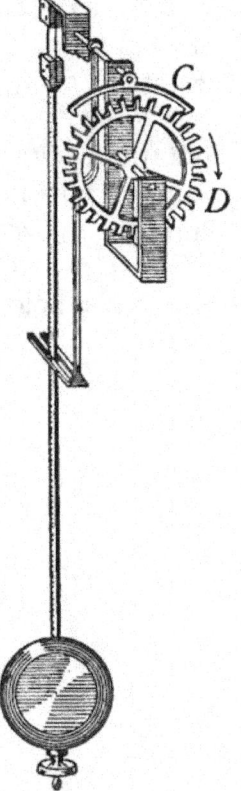

FIG. 82.—Escapement and pendulum of a clock.

Exercises

1. What is the usual shape of the bob of a clock pendulum? Why is this shape used instead of a sphere?

2. Removing the bob from a clock pendulum has what effect on its motion? Also on the motion of the hands?

3. How does the expansion of the rod of a pendulum in summer and its contraction in winter affect the keeping of time by a clock? How can this be corrected?

4. Master clocks that control the time of a railway system have a cup of mercury for a bob. This automatically keeps the same rate of vibration through any changes of temperature. How?

5. How will the length of a seconds pendulum at Denver, 1 mile above sea-level, compare with one at New York? Why?

6. What is the period of a pendulum 9 in. long? *Note.* In problems involving the use of the third law, use the length of a seconds pendulum for *l*, and call its period 1.

7. A swing is 20 ft. high, find the time required for one swing across the arc.

8. A pendulum is 60 cm. long. What is its period?

9. If in a gymnasium a pupil takes 3 sec. to swing once across while hanging from a ring, how long a pendulum is formed?

10. A clock pendulum makes four vibrations a second, what is its length?

Review Outline: Force and Motion

Force; definition, elements, how measured, units, dyne.

Graphic Representation; typical examples of finding a component, a resultant, or an equilibrant.

Motion; Laws of motion (3), inertia, curvilinear motion, centrifugal force, momentum, ($M = mv$), reaction, stress and strain.

Moment of Force; parallel forces, couple, effective and non-effective component.

Gravitation; law; gravity, center of; weight.   Equilibrium 3 forms; stability, how increased.

Falling Bodies; velocity, acceleration, " g," Laws; $V = gt, S = \frac{1}{2}gt^2$ — $s = \frac{1}{2}g(2t - 1)$.

Pendulum; simple, seconds, laws (3), $t = \pi\sqrt{l/g}$.

CHAPTER VI

WORK AND ENERGY

104. Work.—"Whenever a force moves a body upon which it acts, it is said to do work upon that body." For example, if a man pushes a wheelbarrow along a path, he is doing work on it as long as the wheelbarrow moves, but if the wheelbarrow strikes a stone and the man continues to push and no motion results, from a scientific point of view he is then doing no work on it.

"Work signifies the overcoming of resistance," and unless the resistance is overcome no work is done. Lifting a weight is doing work on it, supporting a weight is not, although the latter may be nearly as tiresome as the former. Work as used in science is a technical term. Do not attach to it meanings which it has in everyday speech.

105. Measurement of Work.—Work is measured by the product of the force by the displacement caused in the direction of the force, that is $W = fs$. Therefore if a unit of force acts through a unit of space, a unit of work will be done. There are naturally several units of work depending upon the units of force and space employed.

English Work Unit.—If the force of one *pound* acts through the distance of one *foot*, a *foot-pound* of work is done. A foot-pound is defined as the work done when 1 lb. is lifted 1 ft. against the force of gravity.

Metric Work Unit.—If the force is one kilogram and the distance one meter, *one kilogram-meter* of work is done.

Absolute Work Unit.—If the force of one *dyne* acts through the distance of one *centimeter* a *dyne-centimeter* of work is done. This usually is called an *erg*. Other work units are sometimes used depending upon the force and distance units employed. One, the *joule*, is equal to 10,000,000 ergs or 10^7 ergs.

Problem.—If a load is drawn 2 miles by a team exerting 500 lbs. force, how much work is done?

Solution.—Since the force employed is 500 lbs., and the distance is 2×5280 ft., the work done is $500 \times 2 \times 5280$ or 5,280,000 ft.-lbs.

106. Energy.—In the various cases suggested in the paragraphs upon work, an agent, a man, an animal or a machine, was mentioned as putting forth an effort in order to do the work. It is also true that in order to perform work an agent must employ *energy, or the energy of a body is its capacity for doing work.* Where an agent does work upon a body, as in winding up a spring or in lifting a weight, the body upon which the work has been done may acquire energy by having work done upon it. That is, it may become able to do work itself upon some other body. For instance, a lifted weight in falling back to its first position may turn wheels, or drive a post into the ground against resistance; a coiled spring may run clock work, strike a blow, or close a door. Hence the energy, or the capacity for doing work, is often acquired by a body because work has first been done upon that body.

107. Potential Energy.—The wound up spring may do work because work has first been done upon it. The lifted weight may also do work because work has first been done in raising it to its elevated position since in falling it may grind an object to powder, lift another weight or do some other kind of work. *The energy that a body possesses on account of its position or shape and a stress to which it is subjected is called potential energy.* The potential energy of a body is measured by the work done in lifting it, changing its shape, or by bringing about the conditions by which it can do work. Thus if a block of iron weighing 2000 lbs. is lifted 20 ft., it possesses 40,000 ft.-lbs. of potential energy. It is therefore able to do 40,000 ft.-lbs. of work in falling back to its first position. If the block just mentioned should fall from its elevated position upon a post, it could drive the post into the ground because its motion at the instant of striking enables it to do work. To compute potential energy you compute the work done upon the body. That is, $P.E. = w \times h$ or $f \times s$.

108. Kinetic Energy.—*The energy due to the motion of a body is called kinetic energy.* The amount of kinetic energy in a body may be measured by the amount of work done to put it in motion. It is usually computed, however, by using its mass and velocity on striking. To illustrate, a 100-lb. ball is lifted 16 ft. The work done upon it, and hence its potential energy, is 1600 ft.-lbs. On falling to the ground again, this will be changed into kinetic energy, or there will be 1600 ft.-lbs. of kinetic energy on striking. It will be noted that since energy is measured by the work it can do, work units are always used in measuring energy. To compute the kinetic energy of a falling body by simply using its mass and velocity one proceeds as follows, in solving the above problem:

First, find the velocity of the falling body which has fallen 16 ft. A body falls 16 ft. in *one* second. In this time it gains a velocity of 32 ft. per second. Now using the formula for kinetic energy $K.E. = \dfrac{wv^2}{2g}$, we have $K.E. = \dfrac{100 \times 32 \times 32}{2 \times 32} = 1600$ ft.-lbs. as before.

The formula, $K.E. = \dfrac{wv^2}{2g}$, may be derived in the following manner:

The kinetic energy of a falling body equals the work done in giving it its motion, that is, $K.E. = w \times S$, in which, $w =$ the weight of the body and $S =$ the distance the body must fall freely in order

to acquire its velocity.   The distance fallen by a freely falling body,
$S, = \frac{1}{2}gt^2 = \frac{g^2t^2}{2g}$ (Art. 98, p. 111).   Now, $v = gt$ and $v^2 = g^2t^2$.
Substituting for g^2t^2, its equal v^2, we have $S = \frac{v^2}{2g}$.   Substituting
this value of S in the equation $K.E. = w \times S$, we have $K.E$
$= \frac{wv^2}{2g}$.

Since the kinetic energy of a moving body depends upon its mass
and velocity and not upon the *direction* of motion, this formula may
be used to find the kinetic energy of any moving body.   Mass and
weight in such problems may be considered numerically equal.

Important Topics

1. Work defined.
2. Work units, foot-pound, kilogram-meter, erg.
3. Energy defined.
4. Kinds of energy, potential and kinetic.

Problems

1. How much work will a 120-lb. boy do climbing a mountain 3000
   ft. high?   Should the vertical or slant height be used?   Why?
2. In a mine 4000 kg. of coal are lifted 223 meters: how much work
   is done upon the coal?   What is the kind and amount of energy
   possessed by the coal?
3. A pile driver weighs 450 lbs.   It is lifted 16 ft.   How much work
   has been done upon it?   What kind and amount of energy will
   it have after falling 16 ft. to the pile?
4. A train weighing 400 tons is moving 30 miles per hour.   Compute its kinetic energy.   (Change its weight to pounds and
   velocity to feet per second.)
5. What would be the kinetic energy of the train in problem 4 if it
   were going 60 miles per hour?   If it were going 90 miles per
   hour?   How does doubling or trebling the speed of an object
   affect its kinetic energy?   How does it affect its momentum?
6. What is the kinetic energy of a 1600-lb. cannon ball moving
   2000 ft. per second?
7. Mention as many kinds of mechanical work as you can and show
   how each satisfies the definition of work.
8. A pile driver weighing 3000 lbs. is lifted 10 ft.   How much work
   is done upon it?

9. If the pile driver in problem 8 is dropped upon the head of a pile which meets an average resistance of 30,000 lbs., how far will one blow drive it?

10. A 40 kg. stone is placed upon the top of a chimney 50 meters high. Compute the work done in kilogram-meters and foot-pounds.

(2) POWER AND ENERGY

109. Horse-power.—In computing work, no account is taken of the time required to accomplish it. But since the time needed to perform an undertaking is of much importance, the rate of work, or the *power or activity* of an agent is an important factor. Thus if one machine can do a piece of work in one-fifth the time required by another machine, it is said to have five times the power of the other. Therefore the power of a machine is *the rate at which it can do work.* James Watt (1736–1819), the inventor of the steam-engine, in *expressing* the power of his engine, used as a unit a *horse-power.* He considered that a horse could do 33,000 ft.-lbs. of work a minute. This is equal to 550 ft.-lbs. per second or 76.05 kg.-m. per second. This is too high a value but it has been used ever since his time. Steam engines usually have their power rated in horse-power. That is, locomotives produce from 500 to 1500 horse-power. Some stationary and marine engines develop as high as 25,000 horse-power. The power of an average horse is about $\frac{3}{4}$ horse-power and of a man about $\frac{1}{7}$ horse-power when working continuously for several hours.

110. The Watt.—In the metric system, the erg as a unit of work would give as a unit of power 1 erg per second. This amount is so small, however, that a larger unit is usually employed, the practical unit being 10,000,000 ergs a second, that is, one joule per second. (See Art. 105.) This practical unit is called a *Watt* after James Watt.

The power of dynamos is usually expressed in kilowatts, a kilowatt representing 1000 watts. Steam-engines in modern practice are often rated in kilowatts instead of horse-power. A horse-power is equivalent to 746 watts, or is nearly 3/4 of a kilowatt.

111. Energy. Its Transference and Transformation. We have considered energy as the capacity for doing work, and noted the two kinds, potential and kinetic, and the facility with which one may change into another. In fact, the transference of energy from one body to another, and its transformation from one form to another is one of the most common processes in nature. Take a pendulum in motion, at the *end* of a swing, its energy being entirely due to its elevated position is all *potential;* at the *lowest* point in its path its energy being entirely due to its motion is all *kinetic.* The change goes on automatically as long as the pendulum swings. A motor attached by a belt to a washing machine is started running. The energy of the motor is transferred by the belt to the washer where it is used in rubbing and moving the clothes.

The heat used in warming a house is usually obtained by burning coal or wood. Coal is believed to be formed from the remains of plants that grew in former geologic times. These plants grew through the help of the radiant energy of the sun. The following are transformations of energy that have occurred: The radiant energy of sunlight was transformed into the *chemical* energy of the plants. This remained as chemical energy while the plants were being converted into coal, was mined, brought to the stove or furnace and burned. The burning transformed the chemical energy into heat energy in which form we use it for warming rooms. Take the energy used in running a street car whose electrical energy comes from a waterfall. The energy of the car itself is mechanical.

Its motor, however, receives electrical energy and transforms it into mechanical. This electrical energy comes along a wire from a dynamo at the waterfall, where waterwheels and generators transform into electrical energy the mechanical energy of the falling water. The water obtained its energy of position by being evaporated by the heat of the radiant energy of the sun. The vapor rising into the air is condensed into clouds and rain, and falling on the mountain side, has, from its elevated position, potential energy. The order of transformation, therefore, is in this case, radiant, heat, mechanical, electriaal, and mechanical. Can you trace the energy from the sun step by step to the energy you are using in reading this page?

112. Forms of Energy.—A steam-engine attached to a train of cars employs its energy in setting the cars in motion, *i.e.*, in giving them kinetic energy and in overcoming resistance to motion. But what is the source of the energy of the engine? It is found in the coal which it carries in its tender. But of what kind? Surely not kinetic, as no motion is seen. It is therefore potential. What is the source of the energy of the coal? This question leads us back to the time of the formation of coal beds, when plants grew in the sunlight and stored up the energy of the sun's heat and light as *chemical* energy. The sun's light brings to the earth the energy of the sun, that central storehouse of energy, which has supplied nearly all the available energy upon the earth. Five *forms* of energy are known, viz., mechanical, heat, electrical, radiant, and chemical.

113. Energy Recognized by its Effects.—Like force, energy is invisible and we are aware of the forms only by the effects produced by it.

We recognize *heat* by *warming*, by expansion, by pressure.

We recognize *light* by *warming*, by its affecting vision.

We recognize *electrical* energy by its heat, light, motion, or magnetic effect. We recognize *mechanical* energy by the *motion* that it produces. We recognize *chemical* energy by knowing that the source of energy does not belong to any of the foregoing.

A boy or girl is able to do considerable work. They therefore possess energy. In what form does the energy of the body mainly occur? One can determine this for himself by applying questions to each form of energy in turn as in Art. 114.

114. Source of the Energy of the Human Body.—Is the energy of the human body mostly heat? No, since we are not very warm. Is it light or electrical? Evidently not since we are neither luminous nor electrical. Is it mechanical? No, since we have our energy even when at rest. Is it chemical? It must be since it is none of the others. Chemical energy is contained within the molecule.

It is a form of potential energy and it is believed to be due to the position of the atoms within the molecule. As a tightly coiled watch spring may have much energy within it, which is set free on allowing the spring to uncoil, so the chemical energy is released on starting the chemical *reaction*. Gunpowder and dynamite are examples of substances containing chemical energy. On exploding these, heat, light, and motion are produced. Gasoline, kerosene, and illuminating gas are purchased because of the potential energy they contain. This energy is set free by burning or exploding them.

The source of the energy of our bodies is of course the food we eat. The energy contained in the food is also chemical. Vegetables obtain their energy from the sunlight (radiant energy).   This is why plants will not grow in the dark.   The available energy is mostly contained in the form of starch, sugar and oil.   Digestion is employed principally to dissolve these substances so that the blood may absorb them and carry them to the tissues of the body where they are needed.   The energy is set free by oxidation (burning), the oxygen needed for this being supplied by breathing.   Breathing also removes the carbon dioxide, which results from the combustion.   It is for its energy that our food is mostly required.

115. Conservation of Energy.—In the study of matter we learned that it is indestructible.   Energy is also believed to be indestructible.   This principle stated concisely teaches that *despite the innumerable changes which energy undergoes the amount in the universe is unchangeable*, and while energy may leave the earth and be lost as far as we are concerned, that it exists somewhere in some form.   The principle which teaches this is called the "Conservation of Energy."   The form into which energy is finally transformed is believed to be heat.

Important Topics

1. Power defined.   Units.   Horse-power.   Watt.
2. Transference and transformations of energy.
3. Forms of energy; heat, electrical, mechanical, radiant, chemical.
4. Effects of the several forms of energy.
5. Energy of the human body.
6. Conservation of energy.

Exercises

1. A boy weighing 110 lbs. ran up a stairs 10 ft. high, in 4 seconds.   How much work was done?   What was his *rate* of work (foot-pounds per second)?   Express also in horse-power.
2. A locomotive drawing a train exerts a draw bar pull of 11,000 lbs.   How much work does it do in moving 3 miles?   What

is its *rate* of work if it moves 3 miles in 5 minutes? Express in horse-power.[1]

3. If 400 kg. are lifted 35 meters in 5 seconds what work is done? What is the rate of work? Express in horse-power, watts and kilowatts.

4. Trace the energy of a moving railway train back to its source in the sun.

5. Why does turning the propeller of a motor boat cause the boat to move?

6. Does it require more power to go up a flight of stairs in 5 seconds than in 10 seconds? Explain. Is more work done in one case than in the other? Why?

7. Can 1 man carrying bricks up to a certain elevation for 120 days do as much work as 120 men carrying up bricks for 1 day?

8. If the 1 man and 120 men of problem 7 do the same amount of work have they the same power? Explain.

9. If 160 cu. ft. of water flow each second over a dam 15 ft. high what is the available power?

10. What power must an engine have to fill a tank 11 × 8 × 5 ft. with water 120 ft. above the supply, in 5 minutes?

11. A hod carrier weighing 150 lbs. carries a load of bricks weighing 100 lbs. up a ladder 30 ft. high. How much work does he do?

12. How much work can a 4-horse-power engine do in 5 minutes?

13. Find the horse-power of a windmill that pumps 6 tons of water from a well 90 ft. deep in 30 minutes.

14. How many horse-power are there in a waterfall 20 ft. high over which 500 cu. ft. of water pass in a minute?

15. The Chicago drainage canal has a flow of about 6000 cu. ft. a second. If at the controlling works there is an available fall of 34 ft. how many horse-power can be developed?

16. How long will it take a 10-horse-power pump to fill a tank of 4000 gallons capacity, standing 300 ft. above the pump?

17. A boy weighing 162 lbs. climbs a stairway a vertical height of 14 ft. in 14.6 seconds. How much power does he exert?

18. The same boy does the same work a second time in 4.2 seconds. How much power does he exert this time? What causes the difference?

19. What is a horse-power-hour? a kilowatt-hour?

---

[1] The following formula is of assistance in computing *horse-power* in problems: H. p. $= \dfrac{\text{lbs.} \times \text{ft.}}{550 \times \text{sec.}}$.

(3) SIMPLE MACHINES AND THE LEVER

116. Machines and Their Uses.—A man, while standing on the ground, can draw a flag to the top of a pole, by using a rope passing over a pulley.

A boy can unscrew a tightly fitting nut that he cannot move with his fingers, by using a wrench.

A woman can sew a long seam by using a sewing machine in much less time than by hand.

A girl can button her shoes much quicker and easier with a button-hook than with her fingers.

These illustrations show some of the reasons why machines are used. In fact it is almost impossible to do any kind of work efficiently without using one or more machines.

117. Advantages of Machines.—(a) Many machines make possible an *increased speed* as in a sewing machine or a bicycle.

(b) Other machines exert an *increased force*. A rope and a set of pulleys may enable a man to lift a heavy object such as a safe or a piano. By the use of a bar a man can more easily move a large rock. (See Fig. 83.)

FIG. 83.—The rock is easily moved.

(c) The *direction* of a force may be changed thus enabling work to be done that could not be readily accomplished otherwise. As, e.g., the use of a pulley in raising a flag to the top of a flag pole, or in raising a bucket of ore from a mine by using a horse attached to a rope passing over two or more pulleys. (See Fig. 84.)

(d) *Other agents* than man or animals *can be used* such as electricity, water power, the wind, steam, etc. Fig. 85 represents a windmill often used in pumping water.

A machine is a device for transferring or transforming energy. It is usually therefore an instrument for doing work. An electric motor is a machine since it *transforms* the energy of the electric current into motion or mechanical energy, and *transfers* the energy from the wire to the driving pulley.

118. A Machine Cannot Create Energy.—Whatever does work upon a machine (a man, moving water, wind, etc.) *loses energy* which is employed in doing the work of the

FIG. 84.—The horse lifts the bucket of ore.

machine. A pair of shears is a machine since it transfers energy from the hand to the edges that do the cutting. Our own bodies are often considered as machines since they both transfer and transform energy.

We must keep in mind that a *machine cannot create energy.* The principle of "Conservation of Energy" is just as explicit on one side as the other. Just as energy, cannot be destroyed, so energy cannot be created. A machine can give out no more energy than is given to it. It acts simply as an agent in transferring energy from one body to another. Many efforts have been made to construct machines that when once started will run themselves, giving out more energy than they receive. Such efforts, called seeking for *perpetual motion*, have never succeeded. This fact is strong evidence in favor of the principle of the conservation of energy.

119. Law of Machines.— When a body receives energy, work is done upon it. Therefore work is done upon a machine when it receives energy and the machine does work upon the body to which it gives the energy. In the operation of a machine, therefore, two quantities of work are to be considered and by the principle of the conservation of energy, these two must be equal. *The work done by a machine equals the work done upon it, or the energy given out by a machine equals the energy received by it.* These two

FIG. 85.—A windmill.

quantities of work must each be composed of a *force* factor and a *space* factor. Therefore two forces and two spaces are to be considered in the operation of a machine. The force factor of the work done on the machine is called the *force* or *effort*. It is the force applied to the machine. The force factor of the work done by a machine is called the *weight or resistance*. It is the force exerted by the machine in overcoming the resistance and equals the resistance overcome.

If f represents the force or effort, and D_f the space it acts through, and w represents the weight or resistance, and D_w the space it acts through, then the law of machines may be expressed by an equation, $f \times D_f = w \times D_w$. That is, *the effort times the distance the effort acts equals the resistance times the distance the resistance is moved or overcome.* When the product of two numbers equals the product of two other numbers either pair may be made the means and the other the extremes of a proportion. The equation given above may therefore be expressed $w : f = D_f : D_w$. Or the resistance is to the effort as the effort distance is to the resistance distance. The law of machines may therefore be expressed in several ways. One should keep in mind, however, that the same law of machines is expresed even though the form be different. What two ways of expressing the law are given?

120. The Simple Machines.—There are but six *simple machines.* All the varieties of machines known are simply modifications and combinations of the six simple machines. The six simple machines are more easily remembered if we separate them into two groups of three each. The first or *lever* group consists of those machines in which a part revolves about a fixed axis. It contains the *lever, pulley* and *wheel and axle.* The second or *inclined plane group* includes those having a sloping surface. It contains the *inclined plane,* the *wedge,* and the *screw.*

121. The Lever.—The *lever* is one of the simple machines most frequently used, being seen in scissors, broom, coal shovel, whip, wheelbarrow, tongs, etc. *The lever consists of a rigid bar capable of turning about a fixed axis called the fulcrum.* In studying a lever, one wishes to know what weight or resistance it can overcome when a certain force is applied to it. Diagrams of levers, therefore, contain the letters w and f. In addition to these, O

stands for the fulcrum on which it turns. By referring to Fig. 86, *a*, *b*, *c*, one may notice that each of these may occupy the middle position between the other two. The two forces (other than the one exerted by the fulcrum) acting on a lever always oppose each other in the matter of changing rotation. They may be considered as a pair of parallel forces acting on a body, each tending to produce rotation.

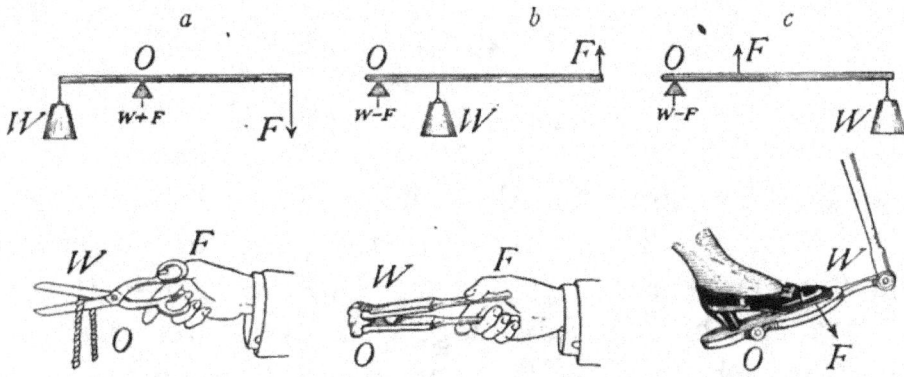

Fig. 86.—The three classes of levers.

122. Moment of Force.—The *effectiveness* of each force may therefore be determined by computing its *moment* about the fixed axis (see Art. 84), that is, by multiplying each force by its distance to the fulcrum or axis of rotation. Let a meter stick have a small hole bored through it at the 50 cm. mark near one edge, and let it be mounted on a nail driven into a vertical support and balanced by sliding a bent wire along it. Suspend by a fine wire or thread a 100 g. weight, 15 cm. from the nail and a 50 g. weight 30 cm. from the nail, on the other side of the support. These two weights will be found to balance. When viewed from this side *A* (Fig. 87) tends to turn the lever in a clockwise direction (down at right), *B* in the counter-clockwise direction (down at left). Since the lever balances, the forces have equal and opposite effects in changing its rotation as may also be computed by determining the moment of each force by multiplying each by its distance from the fulcrum. Therefore the *effectiveness* of a force in changing rotation depends upon the distance from it to the axis as well as upon the magnitude of the force.

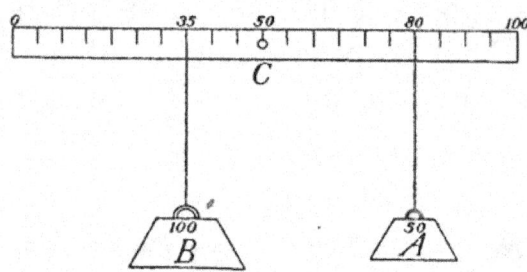

FIG. 87.—The two moments are equal about C. $100 \times 15 = 50 \times 30$.

From the experiment just described, the moment of the acting force equals the moment of the weight or $f \times D_f = w \times D_w$, or the effort times the effort arm equals the weight times the weight arm. This equation is called the law of the lever. It corresponds to the general law of machines and may also be written $w:f = D_f:D_w$.

123. Mechanical Advantage.—A lever often gives an advantage because by its use one may lift a stone or weight which the unaided strength of man could not move. If the lever is used in lifting a stone weighing 500 lbs., the force available being only 100 lbs., then its *mechanical advantage* would be 5, the ratio of $w:f$. In a similar way, the mechanical advantage of any machine is found by finding the ratio of the resistance or weight to the effort. What must be the relative lengths of the effort arm and resistance or weight arm in the example just mentioned? Since the effort times the effort arm equals the weight times the weight arm, if $f \times D_f = w \times D_w$, then D_f is five times D_w. Hence the mechanical advantage of a lever is easily found by finding the ratio of the effort arm to the weight arm.

Important Topics

1. Advantage of machines.
2. Machines cannot create energy.
3. Law of machines.

4. Six simple machines.
5. Lever and principle of moments.
6. Mechanical advantage of a machine.

Exercises

1. Give six examples of levers you use.
2. Fig. 88a represents a pair of paper shears, 88b a pair of tinner's

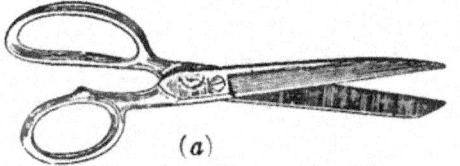 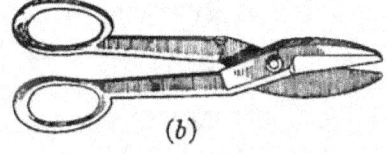

(a) (b)

FIG. 88.—(a) Paper shears. (b) Tinner's shears.

shears. Which has the greater mechanical advantage? Why? Explain why each has the most effective shape for its particular work.

3. Find examples of levers in a sewing machine.
4. What would result if, in Art. 122, the 100 g. weight were put 25 cm. from O and the 50 g. weight 45 cm. from O? Why? Explain using principle of moments.
5. How is the lever principle applied in rowing a boat?
6. When you cut cardboard with shears, why do you open them wide and cut near the pivot?
7. In carrying a load on a stick over the shoulder should the pack be carried near the shoulder or out on the stick? Why?
8. How can two boys on a see-saw start it without touching the ground?
9. In lifting a shovel full of sand do you lift up with one hand as hard as you push down with the other? Why?

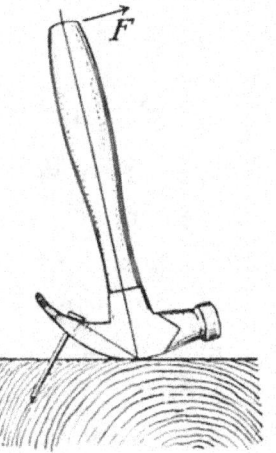

FIG. 89.—The hammer is a bent lever. What is its mechanical advantage?

10. Why must the hinges of a gate 3 ft. high and 16 ft. wide be stronger than the hinges of a gate 16 ft. high and 3 ft. wide?
11. When one sweeps with a broom do the hands do equal amounts of work? Explain.

12. A bar 6 ft. long is used as a lever to lift a weight of 500 lbs. If the fulcrum is placed 6 in. from the weight, what will be the effort required? Note: two arrangements of weight, fulcrum and effort are possible.

13. The handle of a hammer is 12 in. long and the claw that is used in drawing a nail is 2.5 in. long. (See Fig. 89.) A force of 25 lbs. is required to draw the nail. What is the resistance of the nail?

14. The effective length of the head of a hammer is 2 in. The handle is 15 in. long and the nail holds in the wood with a force of 500 lbs. Only 60 lbs. of force is available at the end of the handle. What will be the result?

15. If an effort of 50 lbs. acting on a machine moves 10 ft., how far can it lift a weight of 1000 lbs.?

16. A bar 10 ft. long is to be used as a lever. The weight is kept 2 ft. from the fulcrum. What different levers can it represent?

17. The effort arm of a lever is 6 ft., the weight arm 6 in. How long will the lever be? Give all possible answers.

18. Two boys carry a weight of 100 lbs. on a pole 5 ft. long between them. Where should the weight be placed in order that one boy may carry one and one-fourth times as much as the other?

(4) The Wheel and Axle and the Pulley

124. The Wheel and Axle.—1. One of the simple machines most commonly applied in compound machines is the *wheel* and *axle*. It consists of a wheel H mounted on a cylinder Y so fastened together that both turn on the same axis. In Fig. 90, ropes are shown attached to the circumferences of the wheel and axle. Sometimes a hand wheel is used as on the brake of a freight or street car, or simply a crank and handle is used, as in Fig. 91. The *capstan* is used in moving buildings. Sometimes two or three wheels and axles are geared together as on a derrick or crane as in Fig. 92.

Fig. 93 is a diagram showing that the wheel and axle acts like a lever. The axis D is the fulcrum, the effort is applied at F, at the extremity of a radius of the wheel

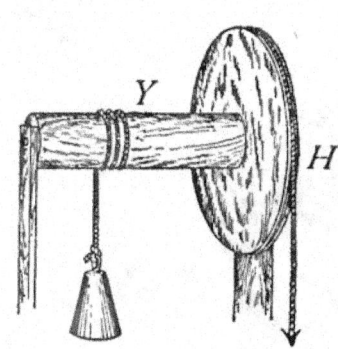

FIG. 90.—The wheel and axle

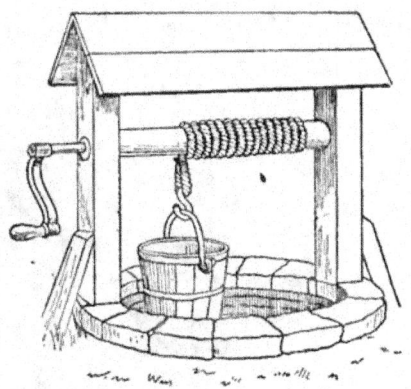

FIG. 91.—Windlass used in drawing water from a well.

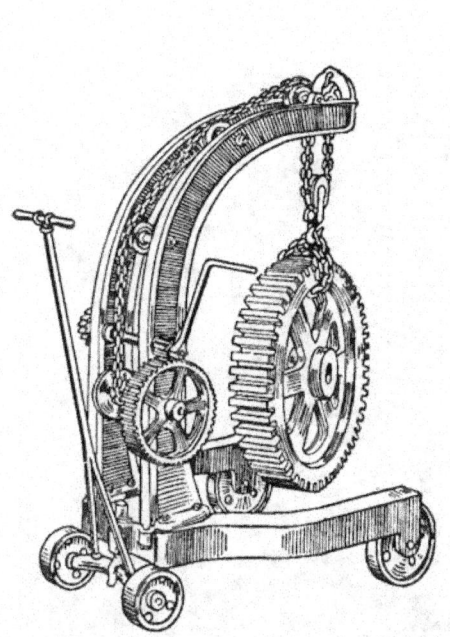

FIG. 92.—A portable crane.

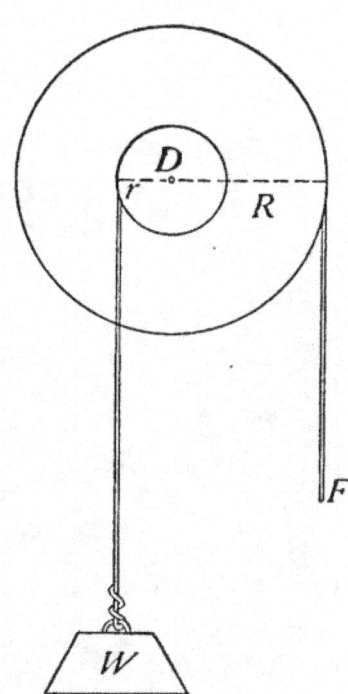

FIG. 93.—The wheel and axle considered as a lever.

and the resisting weight W at the extremity of a radius of the axle. Hence, if D_f, the effort distance, is three

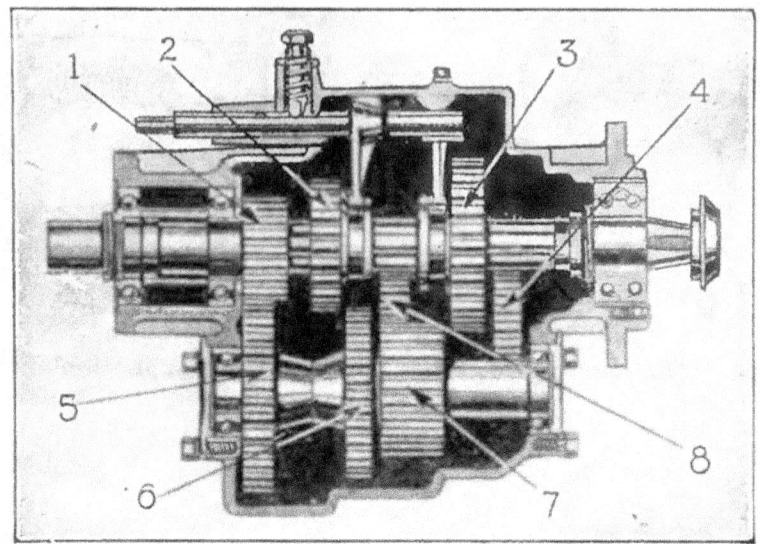

FIG. 94.—View of transmission gears in an automobile. 1, Drive gear; 2, High and intermediate gear; 3, Low and reverse gear; 4, 8, Reverse idler gears; 5, 6, 7, Countershaft gears. (*Courtesy of the Automobile Journal.*)

FIG. 95.—Reducing gear of a steam turbine.

times D_w, the weight distance, the weight that can be supported is three times the effort. Here as in the lever,

$f \times D_f = w \times D_w$, or $w:f = D_f:D_w$, or *the ratio of the weight to the effort equals the ratio of the radius of the wheel to the radius of the axle.* This is therefore the mechanical advantage of the wheel and axle. Since the diameters or circumferences are in the same ratio as the radii these can be used instead of the radii. Sometimes, when *increased speed* instead of increased force is desired, the radius of the wheel or part to which power is applied is less than that of the axle. This is seen in the bicycle, buzzsaw, and blower. Sometimes geared wheels using the principle of the wheel and axle are used to reduce speed, as in the *transmission* of an automobile (see Fig. 94), or the reducing gear of a steam turbine. (See Figs. 95 and 293.)

A *bevel gear* is frequently used to change the direction of the force. (See Fig. 94.)

125. The Pulley.—The *pulley* consists of a wheel turning on an axis in a frame. The wheel is called a sheave and the frame a block. The rim may be smooth or grooved. The grooved rim is used to hold a cord or rope. One use of the pulley is to change the *direction* of the acting force as in Fig. 84, where pulley B changes a horizontal pull at H to a downward force and pulley A changes this into an upward force lifting the weight W. These pulleys are fixed and simply change the direction. Without considering the loss by friction, the pull at W will equal that at F. Sometimes, a pulley is attached to the weight and is lifted with it. It is then called a *movable pulley*. In Fig. 96 the *movable pulley* is at P, a fixed pulley is at F. When *fixed pulleys* are used, a single cord runs through from the weight to the effort, so that if a force of 100 lbs. is applied by the effort the same force is received at the weight. But with movable pulleys several sections of cord may extend upward from the weight each with the force of the effort upon it. By this arrangement, a weight several times larger than the effort can be lifted. Fig. 97 represents what is called a *block and tackle*. If a force of 50 lbs. is exerted at F, each section of the rope will have the same tension and hence the six sections of the rope will support 300 lbs. weight. The *mechanical advantage of the pulley* or the *ratio of the weight* to the effort, therefore, *equals the number of sections of cord supporting the weight*. The fixed pulley represents a lever, see Fig. 98, where the effort and weight are equal. In the movable pulley,

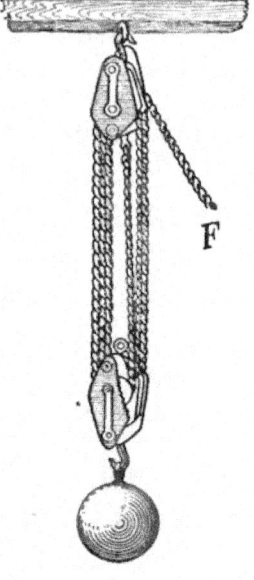

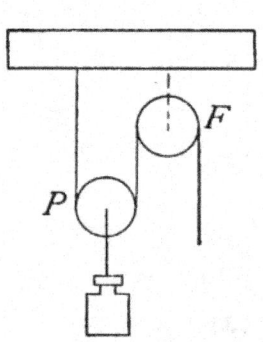

FIG. 96.—A single movable pulley.

FIG. 97.—Block and tackle.

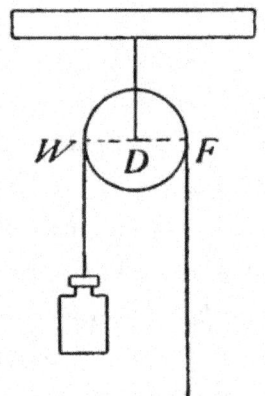

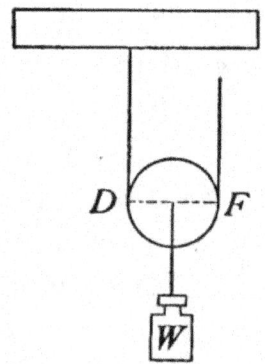

FIG. 98.—The fixed pulley considered as a lever.

FIG. 99.—The movable pulley considered as a lever.

the fulcrum (see Fig. 99) is at D; the weight, W, is applied at the center of the pulley and the effort at F. The weight distance, D_w, is the radius, and the effort distance, D_f, is the diameter of the pulley. Since $W/F = D_f/D_w = 2$ in a movable pulley, the weight is twice the effort, or its mechanical advantage is 2.

Important Topics

1. Wheel and Axle, Law of Wheel and Axle.
2. Pulley, Fixed and Movable, Block and Tackle, Law of Pulley.

Exercises

1. Why do door knobs make it easier to unlatch doors? What simple machine do they represent? Explain.

2. What combination of pulleys will enable a 160-lb. man to raise a 900-lb. piano?

3. When you pull a nail with an ordinary claw hammer, what is the effort arm? the resistance arm?

4. How much work is done by the machine in problem 2 in lifting the piano 20 ft.? How much work must be done upon the machine to do this work?

5. The pilot wheel of a boat has a diameter of 60 in.; the diameter of the axle is 6 in. If the resistance is 175 lbs., what force must be applied to the wheel?

6. Four men raise an anchor weighing 1½ tons, with a capstan (see Fig. 110) having a barrel 9 in. in diameter. The circle described by the hand-spikes is 13½ ft. in diameter. How much force must each man exert?

FIG. 100.—The capstan.

7. A bicycle has a 28-in. wheel. The rear sprocket is 3 in. in diameter,[1] the radius of the pedal crank is 7 in.; 24 lbs. applied to the pedal gives what force on the rim of the wheel? What will be the speed of the rim when the pedal makes one revolution a second?

8. Measure the diameters of the large and small pulleys on the sewing-machine at your home. What mechanical advantage

[1] Consider the diameter of the front sprocket as 6 inches.

in number of revolutions does it give?. Verify your computation by turning the wheel and counting the revolutions.

9. What force is required with a single fixed pulley to raise a weight of 200 lbs.? How far will the effort move in raising the weight 10 ft.? What is the mechanical advantage?

10. In the above problem substitute a single movable pulley for the fixed pulley and answer the same questions.

11. What is the smallest number of pulleys required to lift a weight of 600 lbs. with a force of 120 lbs.? How should they be arranged?

12. A derrick in lifting a safe weighing 2 tons uses a system of pulleys employing 3 sections of rope. What is the force required?

13. Name three instances where pulleys are used to do work that otherwise would be difficult to do.

14. Draw a diagram for a set of pulleys by means of which 100 lbs. can lift 400 lbs.

(5) The Inclined Plane. Efficiency

126. Efficiency.—The general law of machines which states that the work done by a machine equals the work put into it requires a modification, when we apply the law in a practical way, for the reason that in using any machine there is developed more or less friction due to parts of the machine rubbing on each other and to the resistance of the air as the parts move through it. Hence the statement of the law that accords with actual working conditions runs somewhat as follows: *The work put into a machine equals the useful work done by the machine plus the wasted work done by it.* The *efficiency* of a machine is the ratio of the *useful* work done by it to the *total* work done on the machine. If there were no friction or wasted work, the efficiency would be perfect, or, as it is usually expressed, would be 100 per cent. Consider a system of pulleys into which are put 600 ft.-lbs. of work. With 450 ft.-lbs. of useful work resulting, the efficiency would be 450 ÷ 600 = ¾, or 75 per cent. In this case 25 per cent. of the

work done on the machine is wasted.   In a simple lever
the friction is slight so that nearly 100 per cent. efficiency
is often secured.

Some forms of the wheel and axle have high efficiencies
as in bicycles with gear wheels.   Other forms in which
ropes are employed have more friction.   Pulleys have
sometimes efficiencies as low as 40 per cent. when heavy
ropes are used.

127. Inclined Plane.—We now come to a type of *simple
machine of lower efficiency* than those previously mentioned.   These belong to the inclined plane group, which

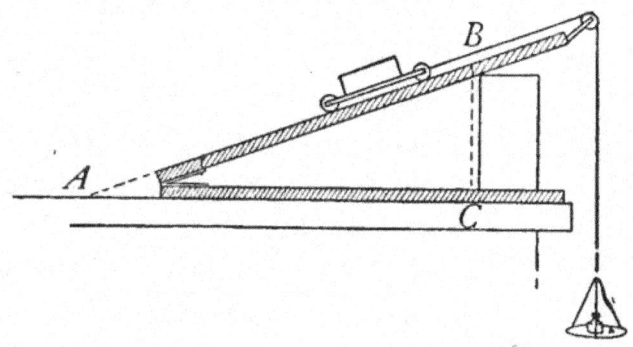

FIG. 101.—An inclined plane.

includes the inclined plane (see Fig. 101), the wedge and
the screw.   They are extensively used, however, notwith-
standing their low efficiency, on account of often giving a
high mechanical advantage.   The *relation between these
machines may be easily shown*, as the *wedge* is obviously
a double inclined plane.   In Art. 82 it is shown that the
effort required to hold a weight upon an inclined plane is
to the *weight* supported as the *height* of the plane is to its
length.

Or while the weight is being lifted the vertical height
BC, the effort has to move the length of the plane *AC*.
Since by the law of machines the effort times its distance
equals the weight times its distance, or the weight is to

the effort as the effort distance is to the weight distance, therefore the mechanical advantage of the inclined plane is the ratio of the length to the height of the inclined plane.

Inclined planes are used to raise heavy objects short distances, as barrels into a wagon, and iron safes into a building. Stairways are inclined planes with steps cut into them.

128. The Wedge.—Wedges are used to separate objects, as in splitting wood (see Fig. 102), cutting wood, and where great force is to be exerted for short distances. An axe is a wedge, so is a knife. A fork consists of several round wedges set in a handle. The edge of any cutting tool is either an inclined plane or a wedge. Our front teeth are wedges. Numerous examples of inclined planes may be seen about us.

FIG. 102.—One use of the wedge.

No definite statement as to the mechanical advantage of the wedge can be given as the work done depends largely on friction. The force used is generally applied by blows on the thick end. In general, the longer the wedge for a given thickness the greater the mechanical advantage.

129. The Screw.—The screw is a cylinder around whose circumference winds a spiral groove. (See Fig. 103.) The raised part between the two adjacent grooves is the *thread* of the screw. The screw turns in a block called a *nut*, within which is a spiral groove and thread exactly corresponding to those of the screw. The distance between two consecutive threads measured parallel to the axis is called the *pitch* of the screw. (See Fig. 104.) If the thread winds around the cylinder ten times in the space of 1 in., the screw is said to have ten threads to the inch, the pitch being $\frac{1}{10}$ in. The screw usually is turned by a lever or wheel with the effort applied at the end of the lever, or at the circumference of the wheel. While the effort moves once about the circumference of the wheel the weight is pushed forward a distance equal to the distance between two threads (the pitch of the screw). The work done by the effort therefore equals $F \times 2\pi r$, r being the radius of the wheel, and the work done on the weight equals $W \times s$, s being the pitch of the screw. By the law of machines $F \times 2\pi r = W \times s$ or $W/F = 2\pi r/s$. Therefore the mechanical advantage of the screw equals $2\pi r/s$. Since the distance the weight moves is small compared to that the power travels, there is a great gain in force. The

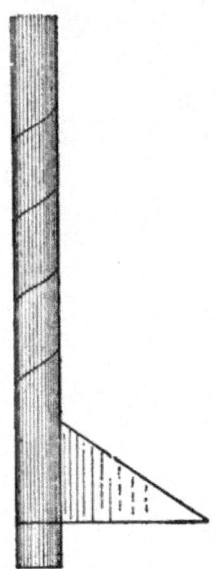

FIG. 103.—The screw is a spiral inclined plane.

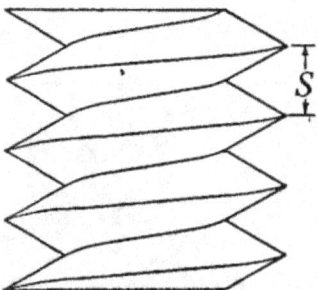

FIG. 104.—The pitch is S.

screw is usually employed where *great force* is to be exerted through small distances as in the vise (Fig. 105) the jack screw (Fig. 106), screw clamps, to accurately measure small distances as in the micrometer (Fig. 107) and spherometer, and to lessen the motion in speed-reducing devices. The worm gear (Fig. 108) is a modification of the screw that is sometimes used where a considerable amount of speed reduction is required.

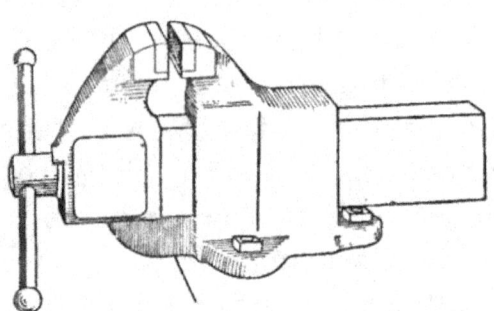

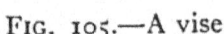

FIG. 105.—A vise.

FIG 106.—A jack screw.

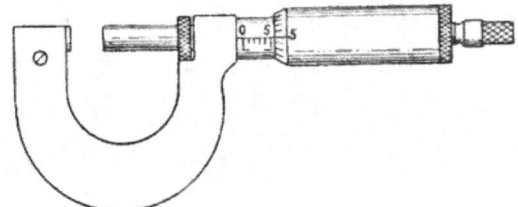

FIG. 107.—A micrometer screw.

FIG. 108.—This large worm-wheel is a part of the hoisting mechanism employed for the lock gates of the Sault Ste. Marie Canal.

Important Topics

1. Efficiency of machines.
2. The inclined plane, wedge and screw.   Applications.

Exercises

1. A plank 12 ft. long is used to roll a barrel weighing 200 lbs. into a wagon 3 ft. high.   Find the force required parallel to the incline.
2. How long a plank will be needed to roll an iron safe weighing $1\frac{1}{2}$ tons into a wagon 3 ft. high using a pull of 600 lbs. parallel to the incline.
3. An effort of 50 lbs. acting parallel to the plane prevents a 200-lb. barrel from rolling down an inclined plane.   What is the ratio of the length to the height of the plane?
4. A man can push with a force of 150 lbs. and wishes to raise a box weighing 1200 lbs. into a cart 3 ft. high.   How long a plank must he use?
5. The radius of the wheel of a letter press is 6 in., the pitch of its screw is $\frac{1}{4}$ in.   What pressure is produced by a force of 40 lbs.?
6. The pitch of a screw of a vice is $\frac{1}{4}$ in., the handle is 1 ft. long. What pressure can be expected if the force used is 100 lbs.?
7. A jackscrew is used to raise a weight of 2 tons.   The bar of the jackscrew extends 2 ft. from the center of the screw.   There are two threads to the inch.   Find the force required.

(6) Friction, Its Uses and Laws

130. Friction.—Although often inconvenient and expensive, requiring persistent and elaborate efforts to reduce it to a minimum, friction has its uses, and advantages. Were it not for friction between our shoes and the floor or sidewalk, we could not keep our footing. *Friction is the resistance that must be overcome when one body moves over another.*   It is of two kinds, *sliding* and *rolling*.   If one draws a block and then a car of equal weight along a board, the force employed in each case being measured by a spring balance, a large difference in the force required will be noticed, showing how much less rolling friction is than sliding friction.

131. Ways of Reducing Friction.—(a) Friction is often caused by the minute projections of one surface sinking into the depressions of the other surface as one moves over the other. It follows, therefore, that if these projections could be made as small as possible that friction would be lessened. Consequently *polishing* is one of the best means for reducing friction. In machines all moving surfaces are made as smooth as possible. In different kinds of materials these little ridges and depressions are differently arranged.

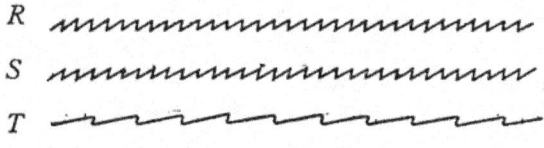

Fig. 109.—The friction between *R* and *S* is greater than beween *R* and *T*.

(b) In Fig. 109 the friction between *R* and *S* would be greater than between *R* and *T*. In *R* and *S* the surfaces will fit closer together than in *R* and *T*. The *use of different materials will reduce friction.* The iron axles of car wheels revolve in bearings of brass. Jewels are used in watches for the same reason. (c) Another very common method of reducing friction is by the use of *lubricants.* The oil or grease used fills up the irregularities of the bearing surfaces and separates them. *Rolling friction* is frequently substituted for sliding friction by the use of ball and roller bearings. These are used in many machines as in bicycles, automobiles, sewing machines, etc. (See Fig. 110.)

132. Value of Friction.—*Friction always hinders motion* and whenever one body moves over or through another the energy used in overcoming the friction is transformed into heat which is taken up by surrounding bodies and usually lost. Friction is therefore the great obstacle to perfect efficiency in machines. Friction, however, like most afflictions *has its uses*. We would find it hard to get along without it. Without friction we could neither walk nor run; no machines could be run by belts; railroad trains, street cars, in fact all ordinary means of travel would be impossible, since these depend upon friction between the moving power and the road for propulsion.

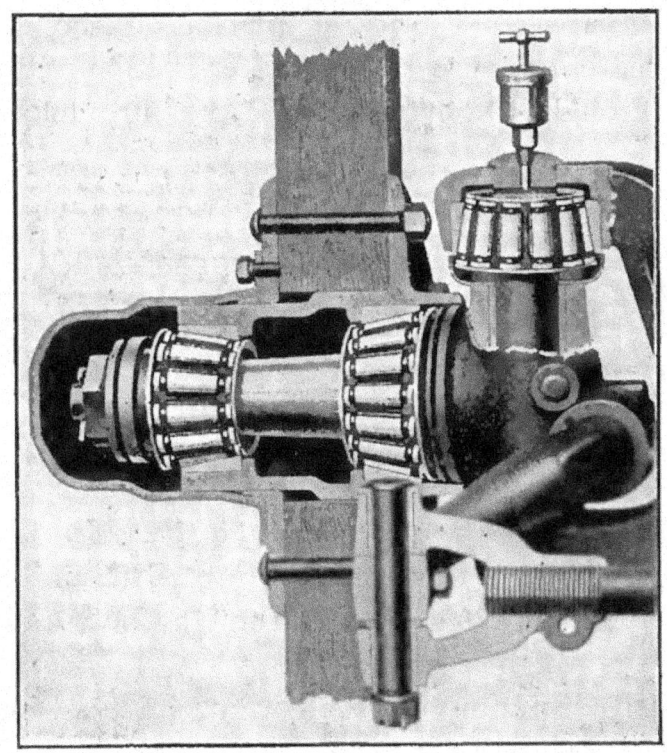

FIG. 110.—Timken roller bearings. As used in the front wheel of an automobile.

133. Coefficient of Friction.—The ratio between the friction when motion is just starting and the force pushing the surfaces together is called the *coefficient of friction*.

If the block in Fig. 111 is drawn along the board with uniform motion, the reading of the spring balances indicates the amount of friction. Suppose the friction is found to be 500 g., and the weight of the block to be 2000 g.

Then the coefficient of friction for these two substances

will be $\dfrac{500}{2000} = \frac{1}{4}$, or 25 per cent.

134. Laws of Friction, Law I.—*The friction when motion is occurring between two surfaces is proportional to the force holding them together.* Thus if one measures the friction when a brick is drawn along a board, he will find that it is doubled if a second brick is placed on the first. On brakes greater pressure causes greater friction. If a rope is drawn through the hands more pressure makes more friction.

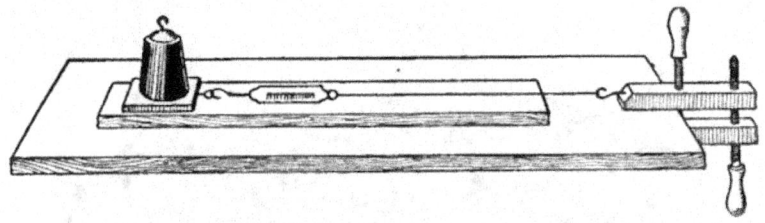

FIG. 111.—A method for testing the friction between surfaces

Law II.—*Friction is independent of the extent of surface in contact.* Thus a brick has the same friction drawn on its side as on its edge, since, although the surface is increased, the weight is unchanged.

Law III.—Friction is greatest at starting, but after starting is practically the same for all speeds.

135. Fluid Friction.—When a solid moves through a fluid, as when a ship moves through the water or railroad trains through the air, the resistance encountered is not the same as with solids but increases with the square of the velocity for slow speeds and for high speeds at a higher rate. This is the reason why it costs so much to increase the speed of a fast train, since the resistance of the air becomes the prominent factor at high speeds. The resistance to the motion of a ship at high speed is usually considered to increase as the cube of the velocity so that to double the speed of a boat its driving force must be eight times as great.

Important Topics

1. Friction: two kinds; sliding and rolling.
2. Four ways of reducing friction.
3. Uses of friction.
4. Coefficient of friction. Three laws of friction.
5. Fluid friction.

Exercises

1. How long must an inclined plane be which is 10 meters high to enable a car weighing 2000 kg. to be pushed up its length by a force of 100 kg. parallel to the incline?

2. State how and where friction is of use in the operation of the inclined plane, the wedge, the screw, the wheel and axle.

3. A wheelbarrow has handles 6 ft. long. If a load of 300 lbs. is placed 18 in. from the axis of the wheel, what force placed at the end of the handles will be required to lift it?

4. A jackscrew has 3 threads to the inch, and the lever used to turn it is 4 ft. long. If the efficiency of the screw is 60 per cent. what force must be applied to raise a load of 5 tons?

5. In problem 4 how far must the force move in raising the weight 3 in. Compute the work done upon the weight, the work done by the power and the efficiency of the machine from these two amounts of work.

6. What simple machines are represented in a jackknife, a sewing-machine, a screw-driver, a plane, a saw, a table fork?

7. A laborer carries 1500 lbs. of brick to a platform 40 ft. high. How much useful work does he do?

8. If he weighs 150 lbs. and his hod weighs 10 lbs., how much useless work does he do in taking 30 trips to carry up the bricks of problem 7? What is his efficiency?

9. If the laborer hoists the brick of problem 7 in a bucket weighing 50 lbs., using a fixed pulley and rope, what is the useless work done if it takes 12 trips to carry up the brick? What is the efficiency of the device?

10. The efficiency of a set of pulleys is 70 per cent. How much force should be applied if acting through 100 ft. it is to raise a load of 400 lbs. 20 ft.?

11. The spokes of the pilot wheel of a motor-boat are 1 ft. long, the axle around which the rudder ropes are wound is 3 in. in diameter. What effort must be applied if the tension in the ropes is 50 lbs.?

12. Why are the elevated railway stations frequently placed at the top of an incline, the tracks sloping gently away in both directions?

13. The screw of a press has 4 threads to the inch and is worked by a lever of such length that an effort of 25 lbs. produces a force of 2 tons. What is the length of the lever?

14. It takes a horizontal force of 10 lbs. to draw a sled weighing 50 lbs. along a horizontal surface. What is the coefficient of friction?

15. The coefficient of rolling friction of a railroad train on a track is 0.009. What pull would an engine have to exert to haul a train weighing 1000 tons along a level track?

16. How heavy a cake of ice can be dragged over a floor by a horizontal force of 20 lbs., if the coefficient of friction is 0.06?

17. The coefficient of friction of iron on iron is 0.2. What force can a switch engine weighing 20 tons exert before slipping?

18. Using a system of pulleys with a double movable block a man weighing 200 lbs. is just able to lift 600 lbs. What is the efficiency of the system?

19. What is the horse-power of a pump that can pump out a cellar full of water 40 ft. × 20 ft. by 10 ft. deep, in 30 minutes?

20. How many tons of coal can a 5 horse-power hoisting engine raise in 30 minutes from a barge to the coal pockets, a height of 50 ft.?

(7) Water Power

136. Energy of Falling Water.—The energy of falling and running water has been used from the earliest times for developing power and running machinery. The energy is derived from the action of the moving water in striking and turning some form of *water-wheel*, several varieties of which are described below.

The Overshot Wheel.—The overshot wheel (Fig. 112) is turned by the weight of the water in the buckets. It was formerly much used in the hilly and mountainous

sections of this country for running sawmills and grist
mills as it is very easily made and requires only a small
amount of water. Its efficiency is high, being from 80
to 90 per cent., the loss being due to friction and spilling
of water from the buckets. To secure this high efficiency
the overshot wheel must have
a diameter equal to the height
of the fall which may be as
much as 80 or 90 ft.

FIG. 112.—Overshot water
wheel.

The Undershot Wheel.—
The old style undershot
wheel (Fig. 113) is used in
level countries, where there is
little fall, often to raise water
for irrigation. Its efficiency is
very low, seldom rising more
than 25 per cent. The principle of the undershot wheel,
however, is extensively used
in the water motor and the Pelton wheel (Fig. 114).
In these the water is delivered from a nozzle in a jet
against the lower buckets of the wheel. They have

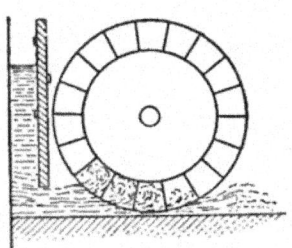

FIG. 113.—Undershot
water wheel.

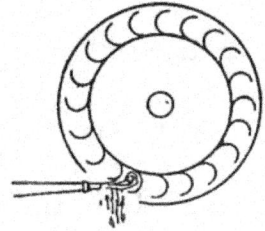

FIG. 114.—Diagram illustrating the principle of the Pelton wheel.

an efficiency of about 80 per cent. and are much used in
cities for running small machines, washing machines, pipe

organ blowers, etc., and in mountainous districts where the head is great.

137. The Turbine.—The turbine is now used more than any other form of water-wheel. It was invented in 1827 by De Fourneyron in France. It can be used with a small or large amount of water, the power depending on the head (the height of the water, in the reservoir above

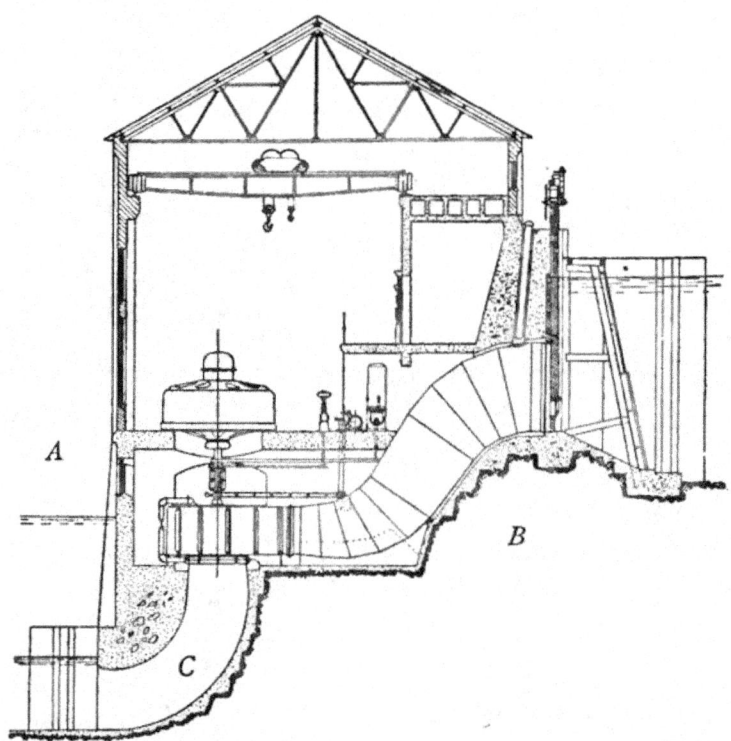

Fig. 115.—Diagram of a hydro-electric power house showing a vertical turbine *A* with penstock *B* and tail race *C*.

the wheel). It is the most efficient type of water-wheel, efficiencies of 90 per cent. often being obtained. The wheel is entirely under water (Fig. 115). It is enclosed in an outer case (Fig. 116) which is connected with the reservoir by a penstock or pipe and is always kept full of water. The wheel itself is made in two parts, a rotating part called the runner (see Fig. 118) and an inner case

FIG. 116.—The outer case of a turbine showing the mechanism for controlling the gates.

FIG. 117.—Inner case of a turbine showing the gates and the lower end of the runner within.

(Fig. 117) with gates that regulate the amount of water entering the wheel. This case has blades curved so that the water can strike the curved blades of the rotating

FIG. 118.—The runner of a turbine.

FIG. 119.—Turbine and generator of the Tacoma hydro-electric power plant.

part (Fig. 118) at the angle that is best adapted to use the energy of the water. The water then drops through the central opening into the tail race below (see Fig. 115).

The energy available is the product of the weight of the water and the head. The turbine is extensively used to furnish power for generating electricity at places where there is a sufficient fall of water. The electrical energy thus developed is transmitted from 50 to 200 miles to cities where it is used in running street cars, electric lighting, etc. Turbines can be made to revolve about either vertical or horizontal axes. Fig. 119 represents a *horizontal* water turbine connected to a dynamo. Compare this with the *vertical* turbine in Fig. 115.

Exercises

1. Does a person do more work when he goes up a flight of stairs in 5 seconds than when he goes up in 15 seconds? Explain.
2. A motorcycle has a 4 horse-power motor and can go at a rate of 50 miles per hour. Why cannot 4 horses draw it as fast?
3. What is the efficiency of a motor that is running fast but doing no useful work?
4. What horse-power can be had from a waterfall, 12 ft. high, if 20 cu. ft. of water pass over it each second?
5. What is the horse-power of a fire engine if it can throw 600 gallons of water a minute to a height of 100 ft.?
6. Why are undershot wheels less efficient than the overshot wheel or turbine?
7. A revolving electric fan is placed on the stern of a boat. Does the boat move? Why? Place the fan under water. Does the boat now move? Why?
8. Why does an electric fan produce a breeze?
9. Explain the action of the bellows in an organ.
10. At Niagara Falls the turbines are 136 ft. below the surface of the river. Their average horse-power is 5000 each. 430 cu. ft. of water each second pass through each turbine. Find the efficiency.
11. At Laxey on the Isle of Man is the largest overshot wheel now in use. It has a horse-power of 150, a diameter of 72.5 ft., a width of 10 ft., and an efficiency of 85 per cent. How many cubic feet of water pass over it each second?

12. The power plant at the Pikes Peak Hydro-electric Company utilizes a head of 2150 ft., which is equal to a pressure of 935 lbs. per square inch, to run a Pelton wheel. If the area of the nozzle is 1 sq. in. and the jet has a velocity of 22,300 ft. per minute, what is the horse-power developed if the efficiency is 80 per cent.?

13. A test made in 1909 of the turbines at the Centerville power house of the California Gas and Electric Corporation showed a maximum horse-power of 9700, speed 400 r.p.m. under a head of 550 ft. The efficiency was 86.25 per cent. How many cubic feet of water passed through the turbines each second?

14. The turbine in the City of Tacoma Power Plant (see Fig. 120) uses a head of 415 ft. 145 cu. ft. a second pass through the turbine. Calculate the horse-power.

15. In problem 14, what is the water pressure per square inch at the turbine?

16. The power plant mentioned in problem 13 develops 6000 kw. What is the efficiency?

Review Outline: Work and Energy

Work; how measured, units, foot-pound, kilogram meter, erg.

Energy; how measured, units, potential, $P.\ E. = w \times h$, or $f \times s$.

Kinetic $= \dfrac{wv^2}{2g}$.

Power; how measured, units horse power, watt, 5 forms of energy, conservation. $H.p. = \dfrac{lbs. \times ft.}{550 \times sec.}$.

Machines; 6 simple forms, 2 groups, advantages, uses, Law: $W \times D_w = F \times D_f$.

Lever; moments, mechanical advantage, uses and applications.

Wheel and Axle and Pulley; common applications, mechanical advantage.

Inclined Plane, Wedge, and Screw; mechanical advantage and efficiency.

Friction; uses, how reduced, coefficient of, laws (3).

Water Wheels; types, efficiency, uses.

CHAPTER VII

HEAT, ITS PRODUCTION AND TRANSMISSION

(1) Sources and Effects of Heat

138. Importance of the Study of Heat.—Heat is brought to our attention through the sensations of heat and cold. In winter, we warm our houses and prevent the escape of heat from them as much as possible. In summer we endeavor to keep our living rooms cool and our bodies from being overheated.

A clear understanding of the several *sources, effects,* and *modes of transferring* heat is of importance to everyone living in our complex civilization, especially when we consider the multitudes of objects that have as their principal use the *production, transfer or utilization* of heat.

139. Principal Sources of Heat.—*First* and most important is the *Sun,* which is continually sending to us *radiant energy* in the form of light and heat waves. These warm the earth, make plants grow, evaporate water, besides producing many other important effects.

Second, chemical energy is often transformed into heat. One has but to think of the heat produced by burning coal, wood, oil, and gas, to recognize the importance of this source. Chemical energy is also the source of the heat produced within our bodies. The action of quicklime and water upon each other produces much heat. This action is sometimes employed during balloon trips as a means of warming things.

Third, Electrical Energy.—In many cities electric cars are heated by the electric current. We have all heard of

electric toasters and other devices for heating by electricity. *Electric* light is pro-
duced by the heating of some
material to incandescence by
an electric current. The
electric furnace has a wide
application in the preparation
and refining of metals.

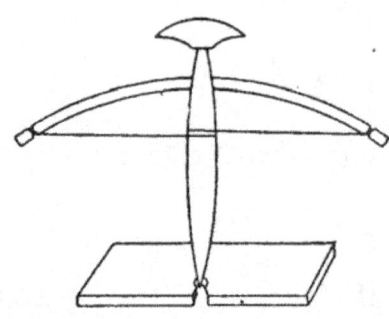

FIG. 120.—Boy-scout method of
making fire by friction.

Fourth, heat is also produced whenever *mechanical
energy* of motion is overcome,
whether it be by *friction, concussion,* or *compression.*
Friction *always* results in the production of heat, as
when we warm our hands by rubbing them, together. When friction is excessive, such as in the
case of a heavy bearing not properly oiled, the
bearing may get very hot. This is the cause of
the "hot box" on a railway car. Friction may
produce heat enough to set wood on fire. Some
fires in mills are believed to be due to this cause.
Every *boy scout* must learn how to produce fire by
friction. (See .Fig. 120.) *Concussion* may be
illustrated by the heating of a piece of metal by
hammering it, while the compression of a gas
always makes it warmer, as those who have used
a bicycle pump have observed. The production
of heat by compressing a gas is illustrated by
the "fire syringe" (Fig. 121). This consists of a
glass tube with a tightly fitted piston. A sudden
compression of the air contained may ignite a
trace of carbon bisulfid vapor.

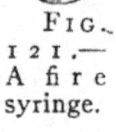

FIG.
121.—
A fire
syringe.

The *interior of the earth* is hot, but its heat
seldom gets to the surface except at *hot springs* and
volcanoes.

140. The Effects of Heat.—There are five important changes produced by heat: (a) change of *size*, (b) change of *temperature*, (c) change of *state*, as the melting of ice or evaporating of water, (d) *chemical* change, as the charring of sugar when it is overheated, and (e) *electrical* change. This is illustrated by the production of an electric current, by the heating of the junction of two different metals.   A thermo-electric generator (see Fig. 122) has been constructed upon this principle and works successfully.

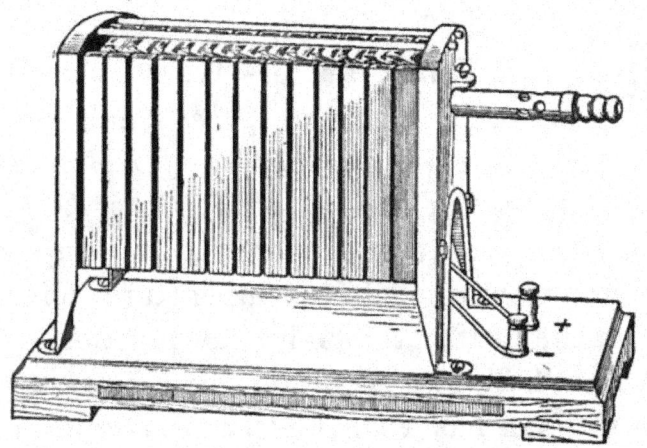

FIG. 122.—A thermo-electric generator.

Important Topics

1. Importance of a study of heat.
2. Four sources of heat.
3. Five effects of heat.
4. Examples of each.
5. Illustrations of transformation of energy which involve *thea*.

Exercises

1. Write a list of the *sources* of heat in the order of their importance to you.   State why each is important to you.
2. Which *three* of the *effects* of heat do *you* make most use of?   Explain what use you make of each of these effects.
3. Which of the forms of energy can be transformed into heat? How in each case?

11

4. Into what other forms of energy may heat be transformed? Name the device or process used in each case.

5. What five different commodities are purchased by people in your neighborhood for the production of heat? Which of these costs least for the amount of heat furnished? Which is most expensive? How do you determine these answers?

6. Why do many people buy heat in an expensive form, as in using an electric toaster, when they can obtain it in a cheaper form by burning gas or coal?

7. How many of the five effects of heat have you observed outside of school?

(2) Temperature and Expansion

141. Heat and Temperature.—We should now clearly distinguish between the terms, *heat* and *temperature*. Heat is *a form of energy consisting of molecular motion.* The temperature of a body is its *degree of hotness.* The *amount of heat* present in a body and its *temperature* are very different things. The temperature refers to the intensity of the heat in the body. A quart of water and a red hot iron ball may contain *equal amounts* of heat, although the ball has a *much higher temperature* than the water. A cup of boiling water will have the same temperature as a tank full of boiling water, but the tank will contain more heat. Every one knows that it will take longer to boil a kettle full of water than a cupful. A hot-water bag, holding 2 quarts of water will give off heat longer than a 1-quart bag, both being filled with water at the same temperature. To put it in another way, more work is done in heating a large amount of water, than a small amount through the same change of temperature.

142. Units of Heat and Temperature.—There are two common units for measuring heat: the *Calorie* and the *British thermal unit.* The *calorie is the amount of heat required to raise the temperature of a gram of water one centigrade degree.* The British thermal unit is *the* *amount of heat required to raise the temperature of one pound of water one Fahrenheit degree.* One of the units plainly belongs to the metric system, the other to the English.

An instrument for measuring temperature is called a *thermometer.* Various scales are placed upon thermometers. The two thermometer scales most commonly used in this country are the *Centigrade* and the *Fahrenheit.* The *Fahrenheit thermometer scale* has the temperature of melting ice marked 32°. The boiling point or steam temperature of pure water under standard conditions of atmospheric pressure is marked 212° and the space between these two fixed points is divided into 180 parts.

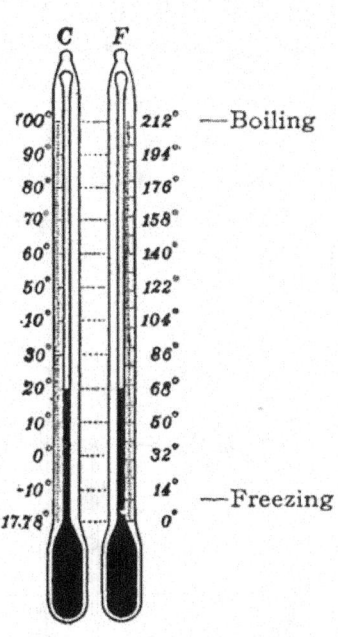

FIG. 123.—Comparison of centigrade and Fahrenheit scales.

The centigrade thermometer scale has the same fixed points marked 0 and 100 and the space between divided into 100 parts. (See Fig. 123.) The centigrade scale is the one used by scientists everywhere.

143. Comparison of Thermometer Scales.—It is often necessary to express in centigrade degrees a temperature for which the Fahrenheit reading is given or *vice versa.* Since there are 180 Fahrenheit degrees between the "fixed points" and 100 centigrade degrees, the Fahrenheit degrees are smaller than the centigrade, or 1°F. = 5⁄9°C. and 1°C. = 9/5°F. One must also take into account the fact that the melting point of ice on the Fahrenheit scale is marked 32°. Hence the following rule: To change a Fahrenheit reading to centigrade subtract 32 and take ⅝ of the remainder, while to change centigrade to Fahrenheit multiply the centigrade by ⅑ and add 32 to the product. These two rules are expressed by the following formulas.

$$\frac{(F.° - 32)5}{9} = C.°, \quad \frac{9C.°}{5} + 32° = F.°$$

Another method of changing from one thermometric scale to another is as follows:

A temperature of −40°F. is also *represented* by −40°C., therefore to change a Fahrenheit reading into centigrade, we add 40 to the given reading, then divide by 1.8 after which subtract 40. To change from a centigrade to Fahrenheit reading the only difference in this method is to multiply by 1.8 or

$$C. = \frac{(F. + 40)}{1.8} - 40 \text{ and } F. = 1.8(C. + 40) - 40.$$

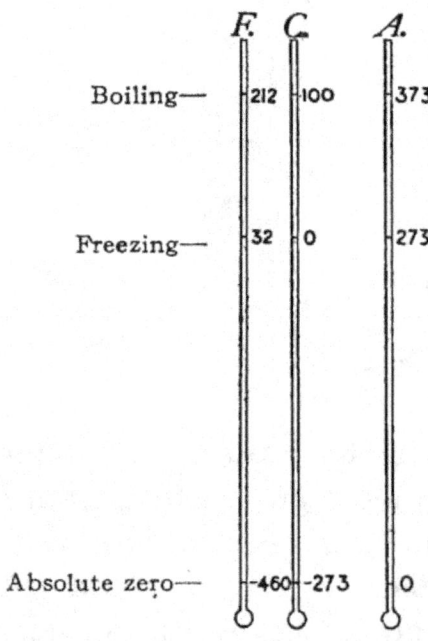

FIG. 124.—Comparison of absolute, centigrade and Fahrenheit scales.

144. The Absolute Scale of Temperature.—One often hears the statement "as cold as ice." This expresses the incorrect idea that ice cannot become colder than its freezing temperature. The fact is that ice *may be cooled* below freezing down to the temperature of its surroundings. If a piece of ice is placed where the temperature is below the melting point, the ice, like any other solid, cools to the temperature of the surrounding space. For example, a piece of ice out of doors is at 10°F. when the air is at this temperature. It follows then, that when ice has been cooled below the freezing temperature that heat is required to warm the ice up to its melting point; or in other words that ice at its melting temperature possesses some heat. The temperature at which absolutely no heat exists is called *absolute zero*. There has been devised an *absolute scale of* temperature. This scale is based upon the centigrade scale, *i.e.*, with 100° between the two fixed points; the scale, however, extends down, below the centigrade zero, 273°, to what is called *absolute zero*. It follows therefore that upon the absolute scale, the melting point of ice, and the boiling point of water are 273° and 373° respectively. (See Fig. 124.)

The means employed to find the location of absolute zero are of much interest. It has been observed that when heated a gas tends to expand. If a measured volume of air at 0°C. is cooled or heated 1°C., it changes its volume $\frac{1}{273}$, the pressure remaining the same. If it is cooled 10° it loses $\frac{10}{273}$, if cooled 100° it loses $\frac{100}{273}$ and so on. No matter how far it is cooled the same rate of reduction continues as long as it remains in the gaseous state. From these facts it is concluded that if the cooling could be carried down 273° that the volume would be reduced $\frac{273}{273}$ or that the volume of the gas would be reduced to nothing. This is believed to mean that the molecular motion constituting heat would cease rather than that the matter composing the gas would disappear. Scientists have been able to obtain temperatures of extreme cold far down on the absolute scale. Liquid air has a temperature of −292°F., or −180°C. or 93°A. The lowest temperature thus far reported is 1.7°A. or −271.3°C., obtained in 1911, by evaporating liquid helium.

145. The Law of Charles.—The facts given in the last paragraph mean that if 273 ccm. of a gas at 0°C. or 273° A. are cooled 100°, or to −100°C., or 173°A., then it will lose $100\frac{9}{273}$ of its volume or have a volume of 173 ccm. If warmed 100°, or up to 100°C., or 373°A., it will have a volume of 373 ccm. It follows then that in every case the volume will correspond to its absolute temperature, providing the pressure remains unchanged. The expression of this fact in scientific language is called the law of *Charles*. *At a constant pressure the volume of a given mass of gas is proportional to its absolute temperature.*

Expressed mathematically, we have $\dfrac{V_1}{V_2} = \dfrac{T_1}{T_2}$. Compare the statement and mathematical expression of the laws of Charles and Boyle.

The formulas for the laws of Boyle and Charles are sometimes combined into one expression as follows:

$$\frac{PV}{T} = \frac{P'V'}{T'}$$

or the product of the volume and pressure of a constant mass of gas is proportional to its absolute temperature.

Important Topics

1. Heat units; calorie, British thermal unit.
2. Three thermometer scales, fixed points on each.
3. Absolute zero, how determined. Its value on each scale.
4. Law of Charles, its meaning. Combination of laws of Boyle and Charles.

Exercises

1. Does ice melt at the same temperature at which water freezes? Express the temperature of freezing water on the three thermometer scales.
2. A comfortable room temperature is 68° F. What is this temperature on the centigrade and absolute scales?
3. Change a temperature of 15°C. to F.; 15°F. to C.; −4°C. to F.; −20°F. to C.

4. The temperature of the human body is 98.6°F. What is this temperature on the absolute and centigrade scales?

5. The temperature of liquid air is—180°C. What is it on the Fahrenheit scale?

6. Mercury is a solid at −40°F. What is this on the centigrade scale?

7. How much heat will be required to raise the temperature of 8 lbs. of water 32°F.; 5 lbs. 10°F.?

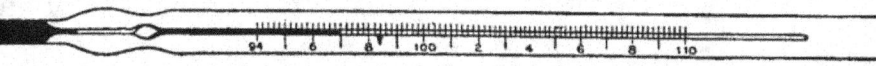

FIG. 125.—A clinical thermometer used to take the temperature of the body.

8. How much heat will be required to raise the temperature of 30 g. of water 43°C.; 20 g., 50°C.?

9. Compute the temperature of absolute zero on the Fahrenheit scale.

10. Take three basins of water, one hot, one cold, and one lukewarm. If one hand be placed in the hot water while the other is placed in the cold and after a few minutes both are placed in the lukewarm water, this water will feel cool to one hand and warm to the other. Explain.

11. If 200 ccm. of air at 200° absolute is heated to 300°A. under constant pressure, what volume will the air occupy at the latter temperature?

12. How does one change a reading on the centigrade scale to a corresponding reading on the absolute scale?

(3) EXPANSION OF LIQUIDS AND SOLIDS

146. Expansion of Gases.—The law of Charles is found to apply to all gases. That is, all gases change in volume in proportion to the change of temperature provided the pressure remains constant. It is for this reason that we have the *gas thermometer* (see Fig. 126) which gives in skillful hands more accurate temperature readings than the best mercurial thermometer. Galileo devised and used the first *air thermometer* which consisted of a hollow bulb blown on a glass tube and inverted in a dish of water. (See Fig. 1.) The *water thermometer* consists of a glass bulb filled with water which rises into a tube attached to the bulb. One disadvantage of the water thermometer is its limited range since it cannot be used below 0° or above 100°. Why?

147. Expansion of Liquids.—The expansion of liquids differs from that of gases in several important respects:

(a) Liquids have a smaller rate of expansion than gases. The *rate* of expansion per degree is called the *Coefficient of Expansion.* For example, the coefficient of expansion of a gas under constant pressure at 0°C. is $\frac{1}{273}$ of its volume per degree centigrade.

(b) Different liquids expand at wholly different rates, that is, their coefficients of expansion differ widely. For example, the coefficient of expansion of mercury is 0.00018 per degree centigrade, of glycerine 0.0005 per degree centigrade, of petroleum 0.0009 per degree centigrade.

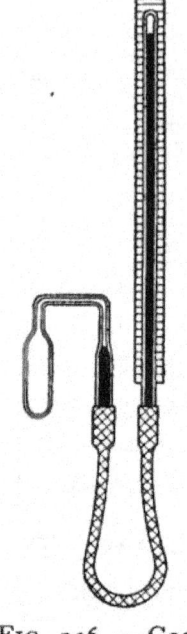

FIG. 126.—Gas thermometer.

(c) The same liquid often has different coefficients of expansion at different temperatures. Water between 5°C. and 6°C. has a coefficient expansion of 0.00002 per degree centigrade, between 8° and 50° of 0.0006, between 99° and 100° of 0.00076. The coefficient of expansion of mercury, however, is constant for a wide range of temperature and, therefore, it is well adapted for use in thermometers.

148. Peculiarity in the Expansion of Water.—Water has a peculiar rate of expansion. This is illustrated by the following experiment:

A test-tube filled with cold water is closed by a stopper containing a small glass tube, the water extending up into the small tube. (See Fig. 127.) The test-tube is placed in a freezing mixture of salt and ice contained in a tumbler. As the water cools, the level of the water in the small tube at first *sinks*. But before the water freezes it *rises* again, showing that after the water cools to a certain temperature that *expansion of the water occurs with further cooling.*

Careful tests show that the water on cooling contracts until it reaches 4°C. On cooling below this temperature it expands. For this reason, when the water of a lake or river freezes, the coldest water is at the surface. On account of this the ice forms at the top instead of at the bottom. If water contracted as it cooled to the freezing temperature the coldest water would be at the bottom. Freezing would begin at the bottom instead of at the surface. Lakes and rivers would freeze solid. In the summer only in shallow waters would all the ice melt. The result would be that fish and other aquatic life would be killed. Climate would be so changed that the earth might become uninhabitable. Since water is densest at 4°C. all the water in a lake or river, when it is covered with ice, is at 4°C. except that near the surface.

Fig. 127. — Apparatus used in testing the expansion of water.

149. The Expansion of Solids.—Most solids when heated expand less than liquids and gases. Careful experiments show that expansion is:

(a) Proportional to the change in temperature.

(b) Different in different solids.

Here are a few coefficients of linear (length) expansion.

Brass...	0.000018 per degree C.
Glass...	0.000009 per degree C.
Ice...	0.000052 per degree C.
Iron...	0.000012 per degree C.
Platinum...	0.000009 per degree C.
Zinc...	0.000027 per degree C.

The coefficient of linear expansion is the fraction of its length that a body expands when heated one degree.

The coefficient of cubical expansion is the fraction of its volume that a body expands when heated one degree.

The expansion of solids is used or allowed for in many cases:

a. Joints between the rails on a railroad allow for the expansion of the rails in summer.

b. One end of a steel truss bridge is usually supported on rollers so that it can expand and contract with changing temperatures. (See Fig. 128.)

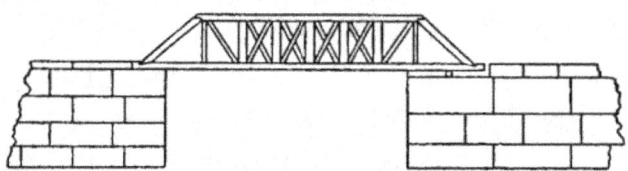

FIG. 128.—Truss bridge showing roller support at one end.

c. Suspension bridges have expansion joints where the ends of the iron girders can move in or out of an expansion joint thus making the bridge longer or shorter according to the temperature.

d. Iron tires are heated, slipped on to wagon wheels and then cooled, the contraction on cooling setting them tightly in place

e. Metallic thermometers depend upon the movement due to the expansion of a coiled strip of metal which turns a pointer on the dial of the instrument. (See Fig. 129.)

f. The wires that are fused into glass in incandescent light bulbs must have the same coefficient of expansion as the glass. Platinum has therefore been used for this purpose. (See table above.)

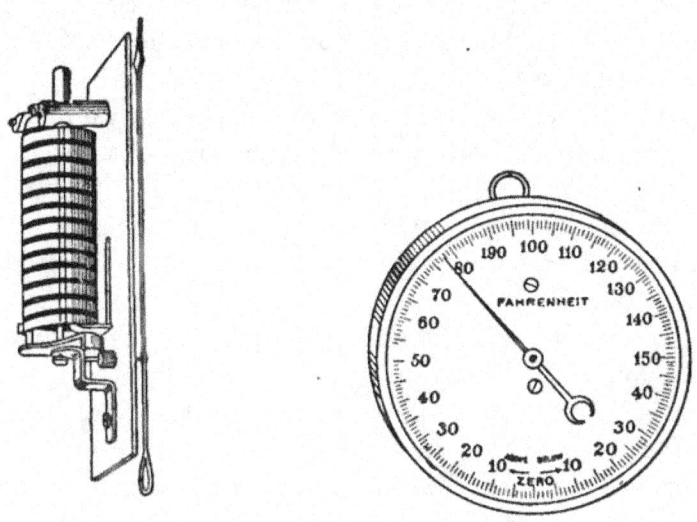

FIG. 129.—Metallic thermometer.

Important Topics

1. Expansion of Liquids; peculiarities. Anomalous expansion of water and its results.
2. Expansion of solids; peculiarities, applications.
3. Coefficient of linear expansion.
4. Coefficient of cubical expanison.

Exercises

1. The gas within a partly inflated balloon has a volume of 1000 cu. ft. at a pressure of 74 cm., and a temperature of 15°C. What will be the volume of the gas when its pressure is 37 cm. and the temperature is −17°C.?
2. A man taking a full breath on the top of a mountain fourteen thousand feet high inhales 4 liters of air, the pressure being 40 cm. What volume would this same mass of air have in a place 600 ft. above sea-level when the barometer reads 75 cm. and the temperature is the same as on the mountain top?
3. If the coefficient of linear expansion of iron is 0.000012 per

degree C., how much will an iron bridge 1000 ft. long change in length in warming from −20°C. on a winter day to 30°C. upon a summer day.

4. What are some of the results that would follow in freezing weather if water continually contracted on being cooled to zero instead of beginning to expand when cooled below 4°C.?

5. Mention two instances that you have noticed of expansion occurring when a body is heated?

6. Compare the density of air at 30°C. with that at 10°C. at the same pressure. If both are present in a room, where will each be found? Why?

7. Compare the density of water at 40°C. with that at 10°C. If water at the two temperatures are in a tank, where will each be found? Why?

8. If water at 0°C. and at 4°C. are both in a tank, where will each be found? Why?

9. How much heat will be required to raise the temperature of a cubic foot of water 10°F.?

10. How much heat will be required to raise the temperature of 4 liters of water 25°C.?

11. How much longer would the cables of the Brooklyn suspension bridge be on a summer's day when the temperature is 30°C. than in winter at −20°C., the length of cable between the supports being about 1600 ft.

12. If 25 liters of air at −23°C. is warmed to 77°C. under constant pressure, what will be the resulting volume of air? Explain.

13. White pig iron melts at about 2000°F. Express this temperature upon the centigrade and absolute scales.

14. If 200 ccm. of air at 76 cm. pressure and 27°C. temperature be heated to 127°C. at a pressure of 38 cm. what will be the resulting volume?

15. A balloon contains 10,000 cu. ft. of gas at 75.2 cm. pressure and 24°C. It ascends until the pressure is 18 cm. and the temperature is −10°C. What is the volume of gas it then contains.

16. A gas holder contains 50 "cu. ft." of gas at a pressure of one atmosphere and 62°F. How much gas will it hold at 10 atmospheres and 32°F.

17. One thousand "cubic feet" of illuminating gas has what volume with 75 lbs. pressure and temperature of 10°C.

18. Define a "cubic foot" of illuminating gas.

150. Methods of Transmitting Heat.—One of the most practical benefits of the study of heat is clearer understanding of the different methods by which heat is transferred from one place to another and an intelligent idea of the means employed to prevent the transfer of heat.

It should be definitely understood at the beginning that *cold signifies the absence of heat*, just as darkness implies the absence of light, so when one speaks of cold getting into a house what is really meant is either the entrance of cold air by some opening or else the escape of the heat.

There are three distinct methods by which heat energy is transferred from one place to another, depending upon the medium or substance that transfers the heat.

a. A solid transmits heat by the method called *conduction*.

b. A fluid, either a liquid or a gas, transmits heat mainly by the method called *convection*.

c. Space transmits the energy of hot objects by the method called *radiation*.

151. Conduction.— To illustrate conduction, place in a gas flame the ends of same metal wires supported as in Fig. 130. In a short time the other ends of the wires become hot enough to burn one's hand. This may be explained as follows: The hot gas flame contains molecules in violent vibration and those striking the wire set its molecules rapidly vibrating. Since, in a solid, the molecules are held in the same relative positions, when one end of a wire is heated the rapidly vibrating molecules at the hot end set their neighbors vibrating and these the next in turn and so on until the

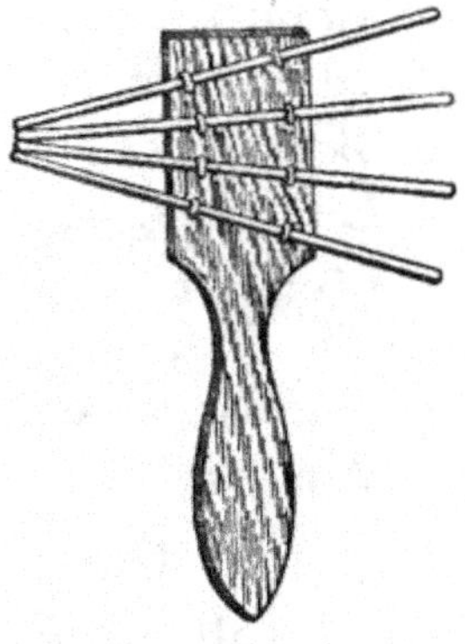

Fig. 130.—Solids conduct heat.

whole wire is hot. It is a fortunate circumstance that different substances have different rates of conductivity for heat. To realize this, suppose that our clothing were as good a conductor as iron, clothing would then be very uncomfortable both in hot and in cold weather. The best conductors for heat are metals. It is interesting to note that, as a rule good conductors of heat are also good conductors of electricity, while poor conductors of heat are also poor electric conductors. Careful experiments in testing the rate that heat will be conducted through different substances show the following rates of conductivity.

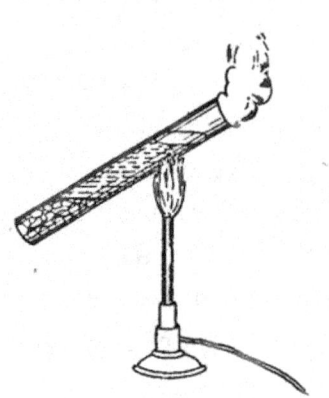

FIG. 131.—Water is a poor conductor of heat.

These figures are averages taken mainly from the Smithsonian Physical Tables:

Silver............	100	Granite.........	.53
Copper..........	74	Limestone.......	.52
Aluminum.......	35	Ice.............	.5
Brass...........	27	Glass...........	.2
Zinc............	26	Water..........	.124
Iron............	15	Pine, with grain.	.03
Tin.............	14.7	Pine, across grain	.01
German silver.....	8.4	Felt...........	.008
Mercury.........	1.7	Air............	.005

To test the conductivity of *liquids*, take a test-tube nearly full of cold water, hold the lower end in the hand while the tube is inclined so that the upper end is heated by a gas flame until the water boils. The lower end will be found to remain cold. (See Fig. 131.) Careful measurements of the conductivity of water show that heat is transmitted through it only $\frac{1}{800}$ as rapidly as in silver, while air conducts but $\frac{1}{25}$ as rapidly as water.

152. Non-conductors and Their Uses.—Many solids, however, are poor conductors, as leather, fur, felt, and woolen cloth. These substances owe their non-conductivity mainly to the fact that they are porous. The air which fills the minute spaces of these substances is one of the poorest conductors known and hinders the transfer of

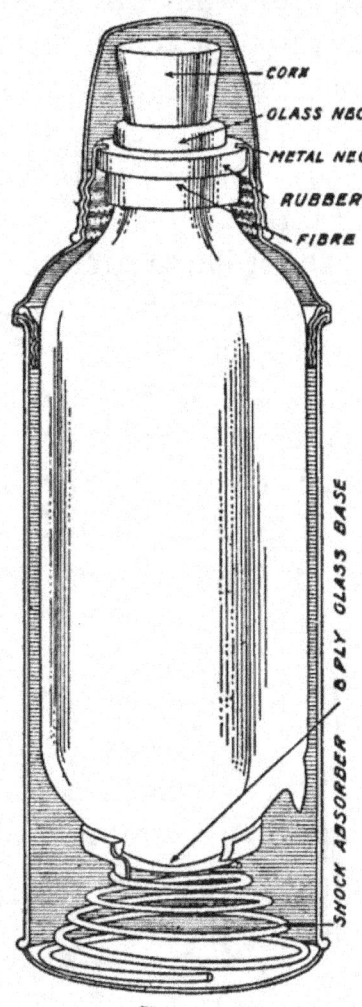

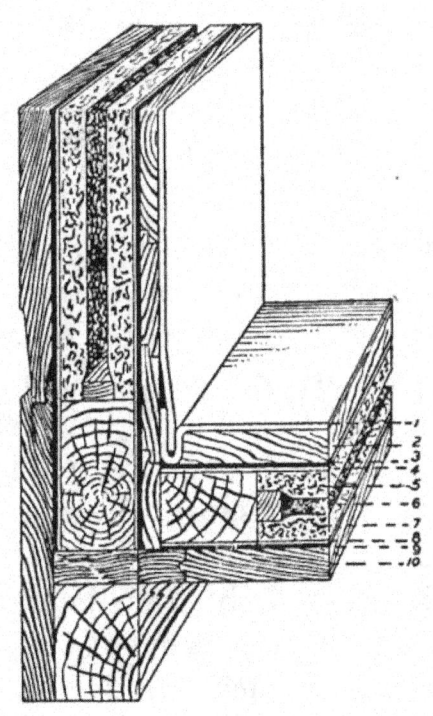

FIG. 132.

FIG. 133.

FIG. 132.—Wall construction of a refrigerator. 1, Porcelain enamel lining lock joint; 2, inside wood lining; 3, 3-ply red rope waterproof paper; 4, wool felt deafening paper; 5, flaxlinum insulation; 6, dead air space; 7, flaxlinum insulation; 8, wool felt deafening paper; 9, 3-ply red rope waterproof paper; 10, outside wood case.

FIG. 133.—Sectional view of a Thermos bottle.

heat through these solids. For the same reason loosely packed snow is a protection to vegetation covered by it during a period of severe cold in winter. The efficiency of storm sash or double windows, and of the double and triple walls of ice-houses and refrigerators (see Fig. 132) in preventing the conduction of heat is also largely due to the poor conductivity of the air confined in the spaces between the walls. To prevent the circulation of the air, sawdust, charcoal, and other porous material is often loosely packed into the space between the walls of such structure.

Other illustrations of effective non-conductors will occur to every one; such as *woolen* clothing, *wooden* handles for hot objects, and the *packing* used in fireless cookers. A *Thermos* bottle is effective as a non-conductor of heat because the space between the double walls has the air exhausted from it (Figs. 133 and 134).

Of several objects in a cold room, some feel much colder to the touch than others, thus iron, marble, oil cloth, and

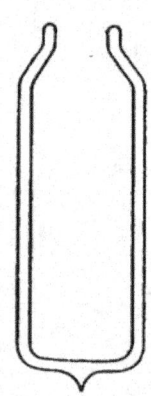

FIG. 134.— Cross-section of the vacuum flask in a Thermos bottle.

earthenware will feel colder than woolen cloth, carpet, feathers, or paper. The first four objects feel cold because they are conductors, and conduct the heat away from the hand rapidly. The other substances named are non-conductors and hence remove heat from the hand less rapidly, and therefore do not feel so cold. In a similar way, if several hot objects are touched by the hand, the good conductors are the ones which will burn one most quickly by conducting heat rapidly to the hand. The non-conductors, however, will rarely burn one. Why are the handles of hot utensils often made of non-conducting materials such as wood, cloth, asbestos, etc.?

153. Radiation is the method by which heat comes to us from the sun across space containing no tangible matter. It is also the method by which heat gets to us when we stand near a fire. Everyone has noticed that this heat is cut off by holding an object between the person and the fire. This fact indicates that radiant heat travels in *straight* lines.

The radiation of heat is believed to be accomplished by means of waves in a medium called *ether*, which is invisible and yet pervades everything. Three of the most important characteristics of radiation are *first, heat is transferred by radiation with the speed of light*, or 186,000 miles per second. This fact is shown by the cutting off of both the sun's heat and light at the same instant during an eclipse of the sun. *Second, radiant heat*[1] *travels in straight lines*, while other modes of transferring heat may follow irregular paths. The straight line motion of radiant heat is shown by its being cut off where a screen is placed between the source of heat and the object sheltered. *Third, radiant heat may pass through an object without heating it.* This is shown by the coldness of the upper layers of the atmosphere and also by the fact that a pane of glass may not be heated appreciably by the heat and light from the sun which passes through it.

When radiant energy falls upon any object it may be (a) *reflected* at the surface of the object, (b) *transmitted* through the substance, (c), absorbed. All three of these effects occur in different degrees with different portions of the radiation. *Well-polished surfaces are good reflectors.* Rough and blackened surfaces are *good absorbers*. Transparent objects are those which transmit light well, but even they absorb some of the energy.

154. The Radiometer.—Radiant heat may be detected by means of the radiometer (Fig. 135). This consists of a glass bulb from which the air has been nearly exhausted.

---

[1] Radiant heat is really *radiant energy* and becomes heat when it is absorbed by a body.

Within it is a wheel with four vanes of mica or of aluminum mounted on a vertical axis. One side of each vane is covered with lampblack, the other being highly polished. When exposed to radiant heat from any source the vanes revolve with the bright side in advance.

The bulb is so nearly exhausted of air that a single molecule remaining may travel from the walls of the bulb to the vanes without coming in contact with another molecule.

The blackened sides absorb more heat than the highly polished sides. The air molecules striking these blackened sides receive more heat and so rebound with greater velocity than from the other side, thus exerting greater pressure. The blackened sides therefore are driven backward.

FIG. 135.—A If the air were not so rarified the air
radiometer.
molecules would hit each other so frequently as to equalize the pressure and there would be no motion.

Sun's Radiation.—Accurate tests of the amount of the sun's radiation received upon a square centimeter of the earth's surface perpendicular to the sun's rays were made at Mt. Wilson in 1913. The average of 690 observations gave a value of 1.933 calories per minute. These results indicate that the sun's radiation per square centimeter is sufficient to warm 1 g. of water 1.933°C. each minute. Although the *nature* of *radiation* is not discussed until Art. 408–411 in light, it should be said here that all bodies are radiating heat waves at all temperatures, the heat waves from cool bodies being much longer than those from hot bodies. Glass allows the short luminous waves to pass through freely but the longer heat waves from objects at the room temperature pass through with difficulty. This is the reason why glass is used in the covering of greenhouses and hot beds. Water also absorbs many of the longer heat waves. It is therefore used in stereopticons to prevent delicate lantern slides from being injured by overheating.

Important Topics

1. Conduction in solids, liquids, gases.
2. Non-conductors; uses, best non-conductors.
3. Radiation, three characteristics.
4. The sun's radiation, amount. The radiometer.

Exercises

1. Does clothing ever afford us heat in winter? How then does it keep us warm?
2. Why are plants often covered with paper on a night when frost is expected?
3. Will frost form in the fall of the year sooner on a wooden or a cement sidewalk? Why? On which does ice remain longer? Why?
4. Why in freezing ice-cream do we put the ice in a wooden pail and the cream in a tin one?
5. Is iron better than brick or porcelain as a material for stoves? Explain.
6. Which is better, a good or a poor conductor for keeping a body warm? for keeping a body cool?
7. Should the bottom of a teakettle be polished? Explain.
8. How are safes made fireproof?
9. Explain the principle of the Thermos bottle.
10. Explain why the coiled wire handles of some objects as stove-lid lifters, oven doors, etc., do not get hot.

(5) Transmission of Heat in Fluids. Heating and Ventilation

155. Convection.—While fluids are poor conductors, they may transmit heat more effectively than solids by the mode called *convection*. To illustrate: if heat is

applied at the *top* of a test-tube of water, the hot water
being lighter is found at the top, while at the bottom the
water remains cold.   On the other hand, if heat is applied
at the *bottom* of the vessel, as soon as the water at the bottom is warmed (above 4°C.) it expands, becomes lighter
and is pushed up to the top by the colder, denser water

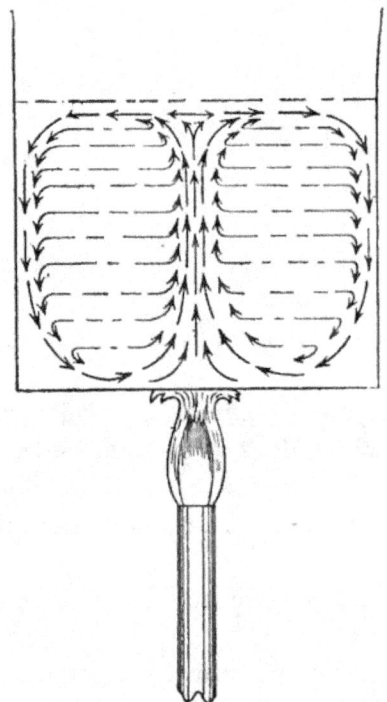

FIG.   136.—Convection   in   a
liquid.

about it.   This circulation of
water continues as long as heat
is applied below, until all of the
water is brought to the boiling
temperature.   (See Fig. 136.)

When a liquid or a gas is
heated in the manner just
described, the heat is said to
be transferred by *convection*.
Thus the air in the lower part
of a room may receive heat by
conduction from a stove or
radiator.   As it expands on
being warmed, it is pushed up
by the colder denser air about
it, which takes its place, thus
creating a circulation of the air
in the room.   (See Fig. 137.)

The heated currents of air give
up their heat to the objects in the room as the circulation
continues.   These air currents may be observed readily
by using the smoke from burning "touch paper" (unglazed
paper that has been dipped into a solution of potassium
nitrate ["saltpeter"] and dried).

156. Draft of a Chimney.—When a fire is started in a
stove or a furnace the air above the fire becomes heated,
expands, and therefore is less dense than it was before.
This warm air and the heated gases which are the products

of the combustion of the fuel weigh less than an equal
volume of the colder air outside. Therefore they are
pushed upward by a force equal to the difference between
their weight and the weight of an equal volume of the
colder air.

The chimney soon becomes filled with these heated
gases. (See Fig. 138.) These are pushed upward by the

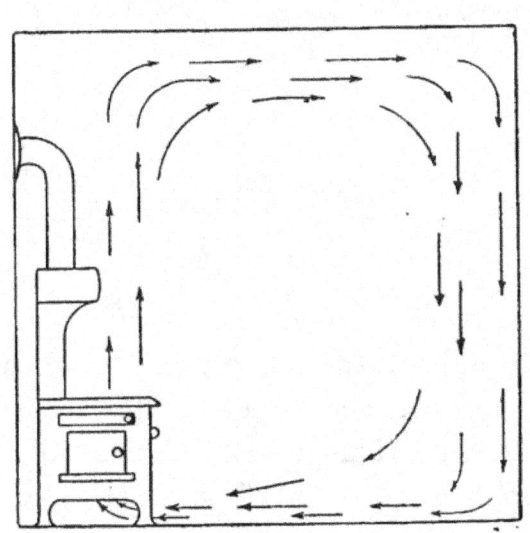

FIG. 137.—Convection currents in a room.

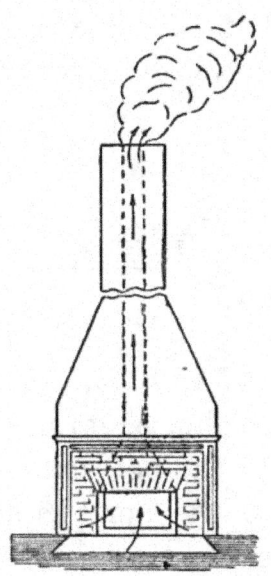

FIG. 138.—Fire place
showing draft of a
chimney.

pressure of the colder, denser air, because this colder air
is pulled downward more strongly by the force of gravity
than are the heated gases in the chimney.

Other things being equal, the taller the chimney, the
greater the draft, because there is a greater difference
between the weight of the gases inside and the weight of
an equal volume of outside air.

157. Convection Currents in Nature.—Winds are produced by differences in the *pressure* or *density* of the air,
the movement being from places of high toward places of
low pressure. One of the causes of a difference in density

of the air is a difference in temperature. This is illustrated by what are called the *land* and *sea breezes* along the sea shore or large lakes. During the day, the temperature of the land becomes higher than that of the sea. The air over the land expands and being lighter is moved back and upward by the colder, denser air from the sea or lake. This constitutes the *sea breezes* (Fig. 139). At night the land becomes cooler much sooner than the sea and the current is reversed causing the *land breeze*. (See Fig. 140.)

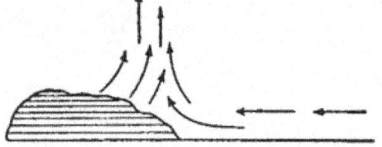　　

FIG. 139.—Sea breeze.　　　FIG. 140.—Land breeze.

The trade winds are convection currents moving toward the hot equatorial belt from both the north and the south. In the hot belt the air rises and the upper air flows back to the north and the south. This region of ascending currents of air is a region of heavy rainfall, since the saturated air rises to cool altitudes where its moisture is condensed. The *ocean currents* are also convection currents. Their motion is due to prevailing winds, differences in density due to evaporation and freezing, and to the rotation of the earth, as well as to changes in temperature.

158. The heating and ventilation of buildings and the problems connected therewith are matters of serious concern to all who live in winter in the temperate zone. Not only should the air in living rooms be comfortably heated, but it should be continually changed especially in the crowded rooms of public buildings, as those of schools, churches, and assembly halls, so that each person may be supplied with 30 or more cubic feet of fresh air per minute. In the colonial days, the *open fire place* afforded the ordinary means for heating rooms. This heated the room mainly by *radiation*. It was wasteful as most of the heat passed up the chimney. This mode of heating secured ample *ventilation*. Fire places are sometimes built in modern homes as an aid to ventilation.

Benjamin Franklin seeing the waste of heat in the open fire places devised an iron box to contain the fire. This was placed in the room and provided heat by conduction, convection, and radiation. It was called *Franklin's stove* and in many forms is still commonly used. It saves a large part of the heat produced by burning the fuel and some ventilation is provided by its draft.

159. Heating by Hot Air.—The presence of stoves in living rooms of homes is accompanied by the annoyance of

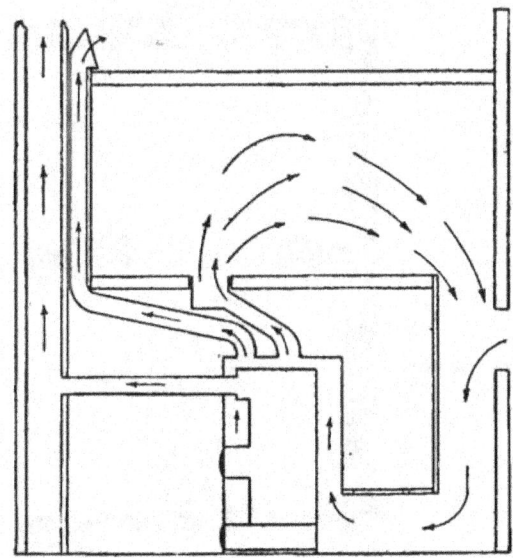

FIG. 141.—Heating and ventilating by means of a hot-air furnace.

scattered fuel, dust, ashes, smoke, etc. One attempt to remove this inconvenience led to placing a large stove or fire box in the basement or cellar, surrounding this with a jacket to provide a space for heating air which is then conducted by pipes to the rooms above. This device is called the hot-air furnace. (See Fig. 141.) The heated air rises because it is pushed up by colder, denser air which enters through the cold-air pipes. The *hot-air furnace* provides a good circulation of warm air and also ventilation, provided some cold air is admittd to the furnace from the outside. One objection to its use is that it may not

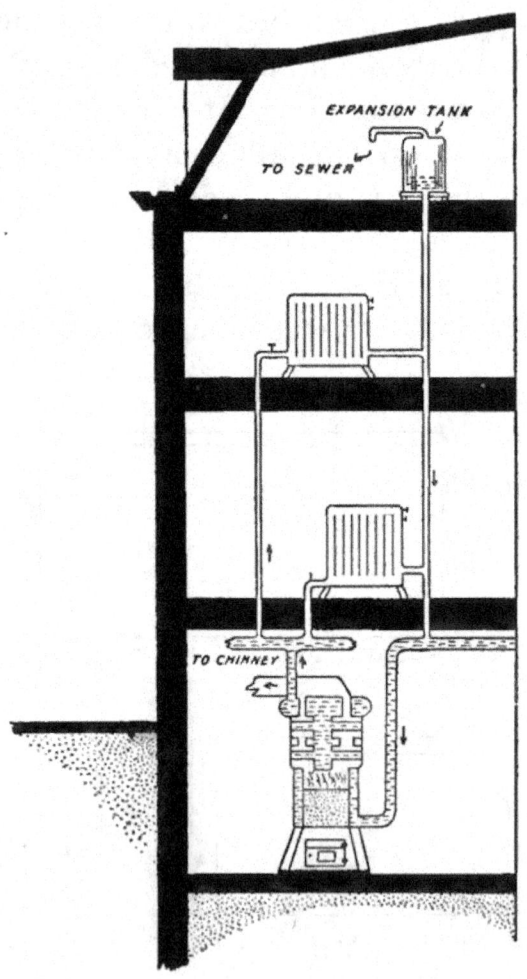

FIG. 142.—A hot-water system of heating.

heat a building evenly, one part being very hot while another may be cool. To provide even and sufficient heat throughout a large building, use is made of *hot water* or *steam heating*.

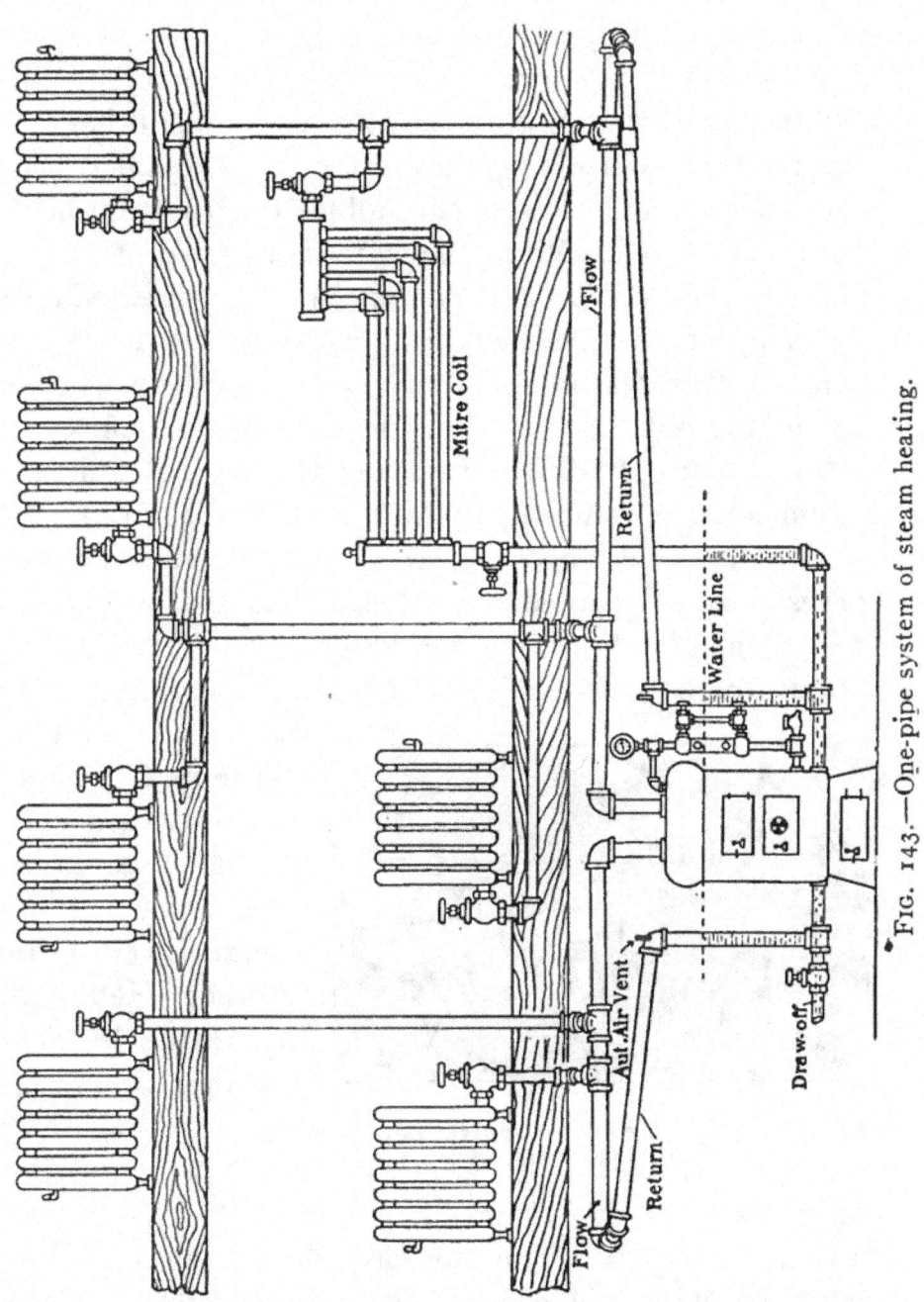

FIG. 143.—One-pipe system of steam heating.

160. Hot-water Heating.—In hot-water heating a furnace arranged for heating water is placed in the basement.
(See Fig. 142.)   Attached to the top of the heater are pipes
leading to the radiators in the various rooms; other pipes
connect the radiators to the bottom of the boiler.   The
heater, pipes, and radiators are all filled with water before
the fire is started.   When the water is warmed, it expands
and is pushed up through the pipes by the colder water in
the return pipe.   The circulation continuing brings hot
water to the radiator while the cooled water returns to the
heater, the hot radiators heating the several rooms.

161. Steam Heating.—In *steam heating* a steam boiler
is connected to radiators by pipes.   (See Fig. 143.)   The
steam drives the air out of the pipes and radiators and
serves as an efficient source of heat.   Heating by steam is
quicker than heating with hot water.   It is therefore pre-
ferred where quick, efficient heating is required.   Hot
water is less intense
and more economical
in mild weather and
is often used in pri-
vate homes.

**162. Direct and
Indirect Heating.**—
In heating by *direct
radiation* (Figs. 142,
143), the steam or
hot-water radiators

FIG. 144.—Heating by an indirect radiator
with side-wall register.

are placed in the rooms to be heated.   With direct
radiation, ventilation must be provided by special means,
such as opening windows, doors, and ventilators.   Sometimes radiators are placed in a box or room in the basement.   Air from out of doors is then driven by a fan
over and about the hot radiators.   The air thus heated

is conducted by pipes to the several rooms.   This arrangement is called *indirect heating*. (See Fig. 144.)   The latter method, it may be observed, provides both heat and ventilation, and hence is often used in schools, churches, court houses, and stores.   Since heated air, especially in cold weather, has a low *relative humidity* some means of moistening the air of living rooms should be provided.   Air when too dry is injurious to the health and also to furniture and wood

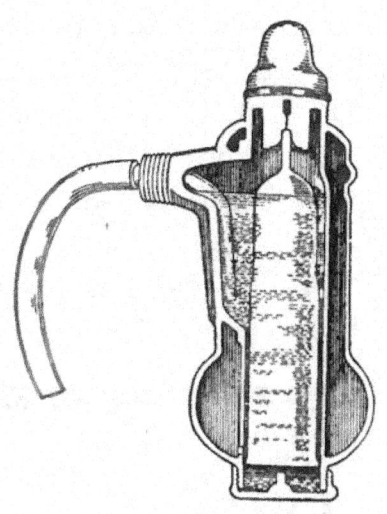

Fig. 145.—An automatic air valve.

work.   The excessive drying of wood and glue in a piece of furniture often causes it to fall apart.

163. Vacuum Steam Heating.—In steam heating, air valves (Fig. 145) are placed on the radiators to allow the air they contain to escape when the steam is turned on. When all the air is driven out the valve closes.   Automatic vacuum valves (Fig. 146) are sometimes used.   When the fire is low and there is no steam pressure in the radiators the pressure

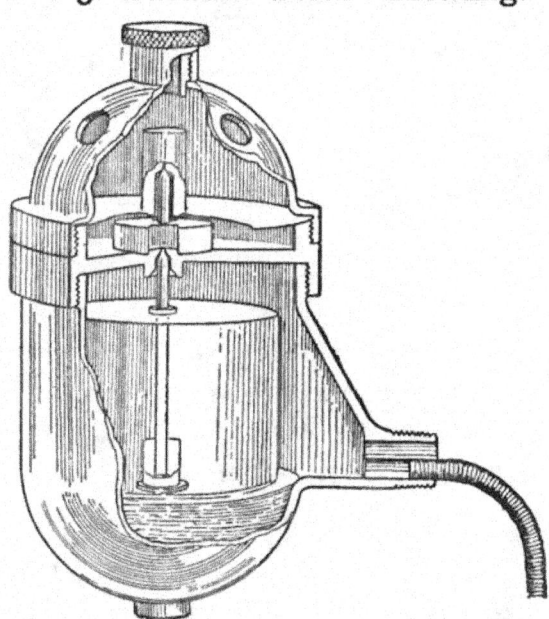

Fig. 146.—An automatic vacuum valve.

of the air closes the valve, making a partial vacuum

inside. The boiling point of water falls as the pressure upon it is reduced. As water will not boil under ordinary atmospheric pressure until its temperature is 100°C. (212°F.), it follows that by the use of vacuum systems, often called vapor systems, of steam heating, water will be giving off hot vapor even after the fire has been banked for hours. This results in a considerable saving of fuel.

164. The Plenum System of Heating.—In the plenum

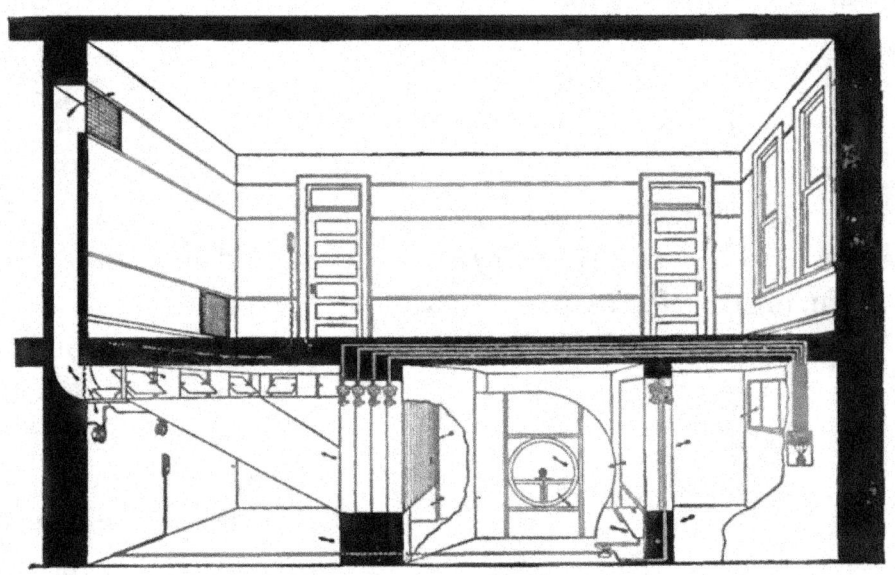

Fig. 147.—Plenum hot-blast system with temperature regulation.

system of heating (see Fig. 147) fresh air is drawn through a window from outdoors and goes first through tempering coils where the temperature is raised to about 70°. The fan then forces some of the air through heating coils, where it is reheated and raised to a much higher temperature, depending upon the weather conditions. Both the hot and tempered air are kept under pressure by the fan in the plenum room and are forced from this room through galvanized iron ducts to the various rooms to be heated. The foul air is forced out of the room through vent ducts which lead to the attic where it escapes through ventilators in the roof.

A thermostat is placed in the tempered-air part of the plenum room to maintain the proper temperature of the tempered air. This thermostat operates the by-pass damper under the tempering coils, and sometimes the

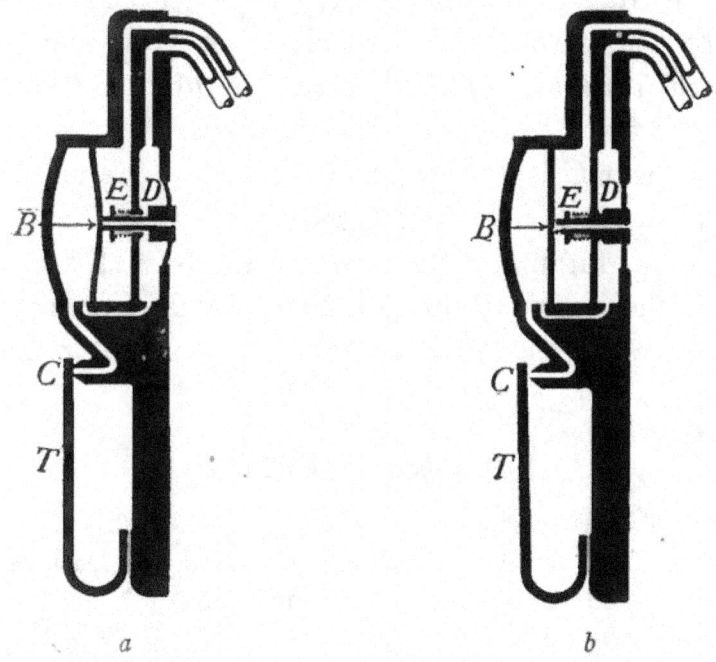

a   *b*

Fig. 148.—A thermostat. (Johnson System.)

valves on the coils. The mixing dampers at the base of the galvanized-iron ducts are controlled by their respective room thermostats. Attic-vent, fresh-air, and return-air dampers are under pneumatic switch control. A humidifier can be provided readily for this system. This system of heating is designed particularly for school houses where adequate ventilation is a necessity.

165. The Thermostat.—One of the many examples of the expansion of metals is shown in one form of the thermostat (Fig. 148) in which two pieces of different metals and of unequal rates of expansion, as brass and iron, are securely fastened together.

The thermostatic strip T moving inward and outward, as affected by the room temperature, varies the amount of air which can escape through the small port C. When the port C is completely closed (Fig. 148a) the full air pressure collects on the diaphragm B which forces down the main valve, letting the compressed air from the main pass through the chamber D into chamber E as the valve is forced off its seat. The air from chamber E then passes into the branch to operate the damper.

When port C is fully open (Fig. 148b) the air pressure on diaphragm B is relieved, the back pressure in E lifts up the diaphragm and the air from the branch escapes out through the hollow stem of the main valve, operating the damper in the opposite direction from that when C is closed.

Important Topics

1. Transmission of heat in fluids.
2. Convection. Drafts of a chimney. Land and sea breezes.
3. Heating and ventilation of buildings.
    (a) By hot air.
    (b) Hot-water heating.
    (c) Steam heating.
    (d) Direct and indirect heating.
    (e) Vacuum steam heating.
    (f) The plenum system.
    (g) The thermostat.

Exercises

1. Is a room heated mainly by conduction, convection, or radiation, from (a) a stove, (b) a hot-air furnace, (c) a steam radiator?
2. Name three natural convection currents.
3. Explain the *draft* of a chimney. *What* is it? *Why* does it occur?

4. Make a *cross-section* sketch of your living room and indicate the covection currents by which the room is heated. *Explain* the heating of the room.

5. Make a sketch showing how the water in the hot-water tank in the kitchen or laundry is heated. Explain your sketch, indicating convection currents.

6. Is it economical to keep stoves and radiators highly polished? Explain.

7. If you open the door between a warm and a cool room what will be the direction of the air currents at the top and at the bottom of the door? Explain.

8. If a hot-water heating system contains 100 cu. ft. of water how much heat will be required to raise its temperature 150°F.?

9. Why does a tall chimney give a better draft than a short one?

10. Explain how your school room is heated and ventilated.

11. Should a steam or hot-water radiator be placed near the floor or near the ceiling of a room? Why?

12. In a hot-water heating system an open tank connected with the pipes is placed in the attic or above the highest radiator. Explain its use

(6) The Moisture in the Atmosphere, Hygrometry

166. Water Vapor in the Air.—The amount of water vapor present in the air has a marked effect upon the weather and the climate of a locality. The study of the moisture conditions of the atmosphere, or hygrometry, is therefore a matter of general interest and importance. The water vapor in the atmosphere is entirely due to evaporation from bodies of water, or snow, or ice. In the discussion of evaporation, it is described as due to the gradual escape of molecules into the air from the surface of a liquid. This description fits exactly the conditions found by all careful observers. Since the air molecules are continually striking the surface of the liquid, many of them penetrate it and become absorbed. In the same manner many vapor molecules reenter the liquid, and if enough vapor molecules are present in the air so that as many vapor molecules reenter the liquid each second as leave it, the space above the liquid is said to be *saturated* as previously described. (See Art. 18.)

167. Conditions for Saturation.—If a liquid is evaporating into a vacuum, the molecules on leaving find no opposition until they reach the limits of the vessel containing the vacuum. Evaporation under these conditions goes on with great rapidity and the space becomes saturated almost instantly. If, however, air be present at ordinary pressure, many of the ordinary water vapor molecules on leaving are struck and returned to the water by the air molecules directly above. Those escaping gradually work their way upward through the air. This explains why it is that our atmosphere is not often saturated even near large bodies of water, the retarding effect of the air upon the evaporation preventing more than the layers of air near the water surface becoming saturated.

Just as the amount of salt that can be held in solution in a liquid is lessened by cooling the solution (Art. 26), so the amount of water vapor that can be held in the air is lessened by lowering its temperature. If air not moist enough to be saturated with water vapor is cooled, it will, as the cooling continues, finally reach a temperature at which it will be saturated or will contain all the water vapor it can hold at this temperature. If the air be still further cooled some of the water vapor will condense and may form fog, dew, rain, snow, etc., the form it takes depending upon where and how the cooling takes place.

168. The Formation of Dew.—If the cooling of the atmosphere is at the surface of some cold object which lowers the temperature of the air below its saturation point, some of its moisture condenses and collects upon the cold surface as *dew*.   This may be noticed upon the surface of a pitcher of ice-water in summer.   At night, the temperature of grass and other objects near or on the ground may fall much faster than that of the atmosphere owing to the radiation of heat from these objects.   If the temperature falls below the saturation point, dew will be formed. This natural radiation is hindered when it is cloudy, therefore little dew forms on cloudy nights.   Clear nights help radiation, therefore we have the most dew on nights when the sky is clear.   If the temperature is below freezing, *frost* forms instead of dew.

169. Formation of Fog.—If the cooling at night is great enough to cool the body of air near the earth below the saturation temperature, then not only may dew be formed, but some moisture is condensed in the air itself, usually upon fine dust particles suspended in it.   This constitutes a *fog*.   If the cooling of the body of air takes place above the earth's surface as when a warm moist current of air enters a colder region, *e.g.*, moves over the top of a cold mountain, or into the upper air, then as this air is cooled below its saturation point, condensation upon fine suspended dust particles takes place, and a *cloud* is formed. If much moisture is present in the cloud, the drops of water grow in size until they begin to fall and *rain* results; or if it is cold enough, instead of rain, snowflakes will be formed and fall.   Sometimes whirling winds in severe thunderstorms carry the raindrops into colder and then warmer regions, alternately freezing and moistening the drops or bits of ice.   It is in this way that *hail* is said to be formed.

170. The Dew Point.—The temperature to which air must be cooled to saturate it or the temperature at which condensation begins is called the *dew point*.   This is often determined in the laboratory by partly filling a polished

13

metal vessel with water and cooling the water by adding
ice until a thin film of moisture is formed upon the outer
surface.   The temperature of the surface when the moisture first forms is the dew point.

171. The Humidity of the Atmosphere.—After the dew
point has been obtained, one may compute the *re'ative
humidity* or *degree of saturation of the atmosphere*, from the
table given below.   This is defined as the *ratio of the
amount of water vapor present in the air to the amount that
would be present if the air were saturated at the same
temperature.*

For example, if the dew point is 5°C. and the temperature of
the air is 22°C., we find the densities of the water vapor at the two
temperatures, and find their ratio: 6.8/19.3 = 35 per cent. nearly.
Determinations of humidity may give indication of rain or frost and
are regularly made at weather bureau stations.   They are also
made in buildings such as greenhouses, hospitals, and schoolhouses
to see if the air is moist enough.   For the most healthful conditions
the relative humidity should be from 40 per cent. to 50 per cent.

WEIGHT OF WATER (w) IN GRAMS CONTAINED IN 1 CUBIC METER OF
SATURATED AIR AT VARIOUS TEMPERATURES ($t°$)C.

$t°$C.	w	$t°$C.	w	$t°$C.	w	$t°$C.	w
−10	2.1	0	4.9	10	9.4	20	17.2
− 9	2.4	1	5.2	11	10.0	21	18.2
− 8	2.7	2	5.6	12	10.6	22	19.3
− 7	3.0	3	6.0	13	11.3	23	20.4
− 6	3.2	4	6.4	14	12.0	24	21.5
− 5	3.5	5	6.8	15	12.8	25	22.9
− 4	3.8	6	7.3	16	13.6	26	24.2
− 3	4.1	7	7.7	17	14.5	27	25.6
− 2	4.4	8	8.1	18	15.1	28	27.0
− 1	4.6	9	8.8	19	16.2	29	28.6
0	4.9	10	9.4	20	17.2	30	30.1

172. Wet and Dry Bulb Hygrometer.—A device for
indicating the relative humidity of the air is called an

hygrometer.   There are various forms.   The *wet* and *dry bulb hygrometer* is shown in Fig. 149.   This device consists of two thermometers, one with its bulb dry and exposed to the air, the other bulb being kept continually moist by a wick dipping into a vessel of water.   An application of the principle of cooling by evaporation is made in this instrument.   Unless the air is saturated so that evaporation is prevented, the wet-bulb thermometer shows a lower temperature, the difference depending upon the amount of moisture in the air, or upon the relative humidity. Most determinations of relative

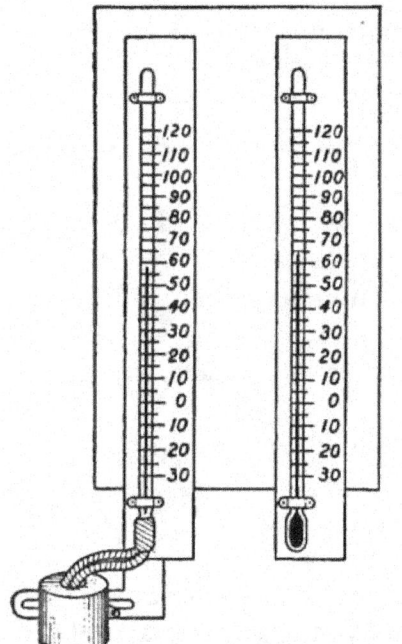

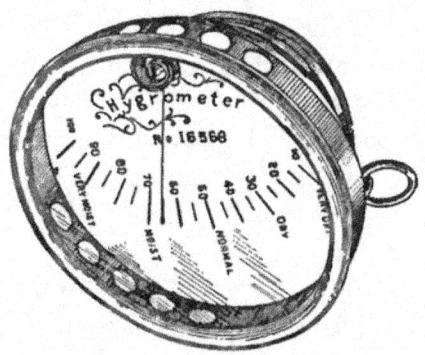

FIG. 149.—Wet and dry bulb   FIG. 150.—A dial hygrometer.
        hygrometer.

humidity are made with this kind of instrument.   It is necessary in order to make an accurate determination, to fan or set the air in motion about the thermometers for some time before reading them.   The relative humidity is then found by using tables giving the relative humidity that corresponds to any reading of the thermometers.

A form of hygrometer in common use is shown in Fig. 150.   In this device, a thin strip of hygroscopic material (as a piece of goose quill) is formed into a spiral coil.   One end of this is fastened to a post.   The other end carried a hand or pointer.   The latter moves over a printed scale and indicates directly the relative humidity. Its indications should be tested by comparing its readings with the results of dew-point determinations. The position of the pointer may be adjusted by turning the post.

Important Topics

1. Water vapor in the air. Cause and effect.
2. Formation of dew, fog, rain, and snow.
3. Dew point, relative humidity.
4. Use of the dry- and wet-bulb hygrometer. Goose-quill hygrometer.

Exercises

1. How is the relative humidity of the air affected by warming it? Explain.
2. How does the white cloud of steam seen about a locomotive in cold weather differ from fog? Explain.
3. In cold weather is the relative humidity of air out of doors and indoors the same? Explain.
4. Compare the relative humidity of air in a desert and near the ocean.
5. Look up the derivation of the term "hygrometer." Give the use of the instrument.
6. Find the relative humidity of air at 20°C. if its dew point is at 10°C.
7. How may the relative humidity of the air in a home be increased?
8. What is the effect of high humidity in the summer upon human beings? How do you explain this?
9. Does dew fall? Explain how dew is formed?
10. In what respects is a cloud similar to a fog? In what respects different?
11. Why are icebergs frequently enveloped in fog?
12. Does dew form in the day time? Explain.

(7) EVAPORATION

173. Effects of Evaporation.—In Art. 19 the cooling effect of evaporation is mentioned and some explanation is made of the cooling effect observed. Since evaporation is employed in so many ways, and since its action is simply explained by the study we have made of molecular motions and molecular forces, it may be well to consider this subject further.

Take three shallow dishes, and place in one a little water, in another some alcohol, and some ether in the third, the liquids being taken from bottles that have stood several hours in the room so that all are at the same temperature. After a short time take the temperature of the three liquids. Each will be at a lower temperature than at first, but of the three the ether will be found to be the coolest, alcohol next, and the water nearest its first temperature. It will be noticed also that the ether has evaporated most in the same time. Similar effects may be observed by placing a few drops of each of these three liquids upon the back of one's hand, or by placing a few drops in turn upon the bulb of a simple air thermometer

174. Cooling Effect of Evaporation.—The molecules that leave an evaporating liquid are naturally the swiftest moving ones, that is, the ones having the highest temperature, so their escape leaves the liquids cooler than before, and the one whose molecules leave fastest is naturally the one that becomes coldest, that is, the ether, in the experiment of Art. 173. If no air pressure were exerted upon the surface of the liquid, the escape of the molecules would be much increased and the temperature of the liquid would be lowered rapidly.

To test this, fill a thin watch glass with ether and place it over a thin slip of glass with a drop of cold water between the two. Now place this apparatus under the receiver of an air pump and exhaust the air. The rapid evaporation of the ether so lowers its temperature, that often the drop of water is frozen. The lowest temperatures are obtained by evaporating liquids at reduced pressure.

Onnes by evaporating liquid helium at a pressure of about 1.2 mm. reached the lowest temperature yet attained, $-456°F.$, or $-271.3°C.$

If four thermometers are taken, the bulbs of three being wetted respectively with ether, alcohol, and water the fourth being dry, on vigorously fanning these, the moistened thermometers show that they have been cooled while the dry one is unaffected.

This indicates that fanning a dry body at the temperature of the air does not change its temperature. Fanning does increase evaporation by removing the air containing the evaporated molecules near the surface of the liquid so that unsaturated air is continually over the liquid. If a pint of water is placed in a bottle and another pint in a wide pan the latter will become dry much sooner because of the greater surface over which evaporation can take place. Application of this is made at salt works where the brine is spread out in shallow pans.

175. Rate of Evaporation.—The rate of evaporation is affected by several factors. These have been illustrated in the preceding paragraphs. To briefly summarize:

The rate of evaporation of a liquid is affected by—

(a) The nature of the liquid.

(b) The temperature of the liquid.

(c) The pressure upon its evaporating surface.

(d) The degree of saturation of the space into which the liquid is evaporating.

(e) The rate of circulation of air over its surface.

(f) The extent of surface exposed to evaporation.

176. Molecular Motion in Solids.—Evidence of molecular motion in liquids is given by expansion on heating, evaporation, and diffusion. Do any of these lines of evidence apply to solids? It is a fact of common experience that solids do become larger on heating. Spaces are left between the ends of rails on railroads so that when they expand in summer they will not distort the track. Iron tires are placed on wheels by heating them until they slip on easily. Then on cooling, the iron shrinks and presses the wheel tightly. Many common demonstrations of expansion are found in lecture rooms. The fact of the evaporation of a solid is often detected by noticing the odor of a substance. The odor of moth balls is one example. Camphor also evaporates. Heated tin has a characteristic odor noted by many. Ice and snow disappear in winter even though the temperature is below freezing. Wet clothes "freeze dry," that is, dry after freezing, by evaporation. A few crystals of iodine placed in a test-tube and gently heated form a vapor easily seen, even though none of the iodine melts. Where the vapor strikes the side of the tube, it condenses back to dark gray crystals of iodine. This change from solid directly to gas and back again without becoming liquid is called *sublimation*. A number of solids are purified by this process.

Important Topics

1. Cooling effect of evaporation, rate of evaporation affected by six conditions.
2. Effects of molecular motion in solids: (a) Expansion, (b) Evaporation, (c) Sublimation.

Exercises

1. Does sprinkling the streets or sidewalks cool the air? Why?
2. Give an illustration for each of the factors affecting evaporation.
3. Give an illustration for each of the three evidences of molecular motions in solids.
4. Since three-quarters of the earth's surface is covered with water, why is not the air constantly saturated?
5. If the air has the temperature of the body, will fanning the perfectly dry face cool one? Explain. Will the effect be the same if the face is moist? Explain.
6. What is the cause of "Cloud Capped" mountains?
7. Why does the exhaust steam from an engine appear to have so much greater volume on a cold day in winter than on a warm one in summer?
8. What causes an unfrozen pond or lake to "steam" on a very cold day in winter, or on a very cool morning in summer?
9. As the air on a mountain top settles down the sides to places of greater pressure, how will its temperature be affected? its relative humidity? Explain.
10. On our Pacific coast, moist winds blow from the west over the mountains. Where will it rain? Where be dry? Explain.

CHAPTER VIII

HEAT AND WORK

(1) Heat Measurement and Specific Heat

177. Specific Heat.—In the study of density and specific gravity it is made clear that different substances differ widely in the amount of matter contained in equal volumes, *e.g.*, lead is much denser than water. The study of the relative densities of substance is usually considered under the subject of *specific gravity*.

Specific heat as distinguished from specific gravity is concerned with the *capacity* for heat possessed by different substances. The definition for specific heat is: *The ratio of the amount of heat required to change the temperature of a given mass of a substance 1 C. degree to the amount of heat required to change the temperature of the same mass of water 1 C. degree.* By definition, it requires 1 calorie to raise the temperature of the gram of water 1°C. The *specific heat* therefore of water is taken as one. The specific heat of most substances except hydrogen, is *less* than that of water, and as a rule, the denser the body the less its specific heat, as may be observed in the following table:

	Specific gravity	Specific heat		Specific gravity	Specific heat
Gold..........	19.3	0.032	Aluminum.....	2.67	0.218
Mercury......	13.6	0.033	Glass.........	2.5–3.6	0.19
Copper.......	8.9	0.093	Ice......... ..	0.918	0.504
Brass........	8.4–8.9	0.094	Air..........	0.00129	0.237
Nickel.......	8.57	0.11	Steam........	0.00061	0.480
Iron.........	7.5+	0.1125	Hydrogen.....	0.00009	3.409

178. Method of Determining Specific Heat.—The specific heat of a body is usually determined by what is called the *method of mixtures*.

For example, a definite weight of a substance, say a 200-g. iron ball, is placed in boiling water until it has the temperature of the hot water, 100°C. Suppose that 300 g. of water at 18°C. be placed in a calorimeter, and that the hot iron ball on being placed in the water raises its temperature to 23.5°C. The heat received by the water equals $5.5 \times 300 = 1650$ calories. This must have come from the heated iron ball. 200 g. of iron then in cooling 76.5°C. (100° − 23.5°) gave out 1650 calories. Then 1 g. of iron in cooling 76.5°C. would give out 8.25 calories or 1 g. of iron cooling 1°C. would yield about 0.11 calorie. The specific heat of the iron is then 0.11. For accurate determination the heat received by the calorimeter must be considered.

179. Heat Capacity of Water.—The large capacity for heat shown by water is useful in regulating the temperature of the air near lakes and the ocean. In hot weather the water rises slowly in temperature absorbing heat from the warm winds blowing over it. In winter the large amount of heat stored in the water is slowly given out to the air above. Thus the climate near the ocean is made more moderate both in winter and summer by the large capacity of water for heat. This large heat capacity of water may seem to be a disadvantage when one is warming it for domestic purposes since it requires so much heat to warm water to boiling. However, it is this capacity that makes hot-water bottles and hot-water heating effective.

If one takes a pound of ice at 0°C. in one dish and a pound of water at 0°C. in another, and warms the dish of ice by a Bunsen flame until the ice is just melted, and then warms the water in the other dish for the same time, the water will be found to be *hot* and at a temperature 80°C., or 176°F.

180. The Heat of Fusion of Ice.—This experiment indicates the large amount of heat required to change the ice to water without changing its temperature. As indicated by the experiment, it requires 80 calories to melt 1 g. of ice without changing its temperature or, in other words, if one placed 1 g. of ice at 0°C. in 1 g. of water at 80°C., the ice would be melted and the water would be cooled to 0°C.

181. Heat Given out by Freezing Water.—Just as 80 calories of heat are required to melt 1 g. of ice, so in freezing 1 g. of water, 80 calories of heat are given out.

The fact that heat is set free or given out when a liquid solidifies may be strikingly shown by making a strong solution of sodium acetate. On allowing this to cool quietly it will come to the room temperature and remain liquid. If now a small crystal of sodium acetate is dropped into the liquid the latter quickly becomes a solid mass of crystals, at the same time rising markedly in temperature. The amount of heat now liberated must enter the sodium acetate when the mass of crystals is melted again.

The large amount of heat that must be liberated before water freezes accounts for the slowness of the formation of ice. It is also the reason why the temperature never falls so low in the vicinity of large lakes as it does far inland, the heat given out by the freezing water warming the surrounding air.

The heat that disappears on melting and reappears on solidifying is called the *heat of fusion*. It is sometimes called *latent heat* since the heat seems to become hidden or latent. It is now believed that the heat energy that disappears when a body melts has been transformed into the *potential energy* of partially separated molecules. The heat of fusion therefore represents the work done in changing a solid to a liquid without a change of temperature.

182. Melting of Crystalline and Amorphous Substances. —If a piece of ice is placed in boiling hot water and then removed, the temperature of the unmelted ice is still 0°C. There is no known means of warming ice under atmospheric pressure above its melting point and maintaining its solid state. Ice being composed of ice crystals is called a crystalline body. All crystalline substances have fixed melting points. For example, ice always melts at 0°C. The melting points of some common crystalline substances are given below:

Melting Points of Some Crystalline Substances

1. Aluminun............................ 658 C.
2. Cast iron............................. 1200 C.
3. Copper.............................. 1083 C.
4. Ice................................. 0 C.
5. Lead................................ 327 C.
6. Mercury............................ —39 C.
7. Phenol (carbolic acid).............. 43 C.
8. Platinum............................ 1755 C.
9. Salt (sodium chloride)............... 795 C.
10. Saltpeter (potassium nitrate)....... 340 C.
11. Silver.............................. 961 C.
12. Sodium hyposulphite (hypo).......... 47 C.
13. Zinc............................... 419 C.

Non-crystalline or amorphous substances such as glass, tar, glue, etc., do not have well defined melting points as do crystalline bodies. When heated they gradually soften and become fluid. For this reason glass can be pressed and molded.

183. Change of Volume During Solidification.—The fact that ice floats and that it breaks bottles and pipes in which it freezes shows that water expands on freezing. How a substance may occupy more space when solid than when liquid may be understood when we learn that ice consists of masses of star-shaped crystals. (See Fig. 151.) The formation of these crystals must leave unoccupied spaces between them in the ice. When liquefied, however, no spaces are left and the substance occupies less volume.

Most substances contract upon solidifying. Antimony
and bismuth, however, expand on solidifying while iron
changes little in volume. Only those bodies that expand,
or else show little change of volume on solidifying, can
make sharp castings, for if they contract they will not

FIG. 151.—Ice crystals.

completely fill the mold. For this reason gold and silver
coins must be stamped and not cast. Type metal, an
alloy of antimony and lead, expands on solidifying to form
the sharp outlines of good type. Several important

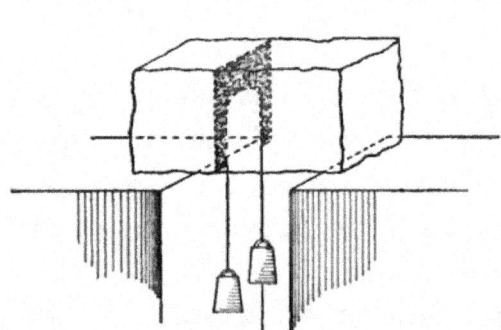

FIG. 152.—Melting ice by pressure.

effects of the expansion
of water when freezing
should be noted. (a)
Ice floats, (b) if it sank
as soon as formed, lakes
and rivers would freeze
solid, (c) freezing water
is one of the active
agents in the disinteg-
ration of rocks.

Since water expands on freezing, pressure would on compressing ice at o°C., tend to turn it into water. Pressure
does lower the melting point of ice, so that a little ice may
melt when it is subjected to pressure. On removing the
pressure the water freezes. This may be shown by placing
a loop of fine piano wire (see Fig. 152) over a piece of ice

supported so that a weight may be hung upon the wire. The wire will be found to gradually cut through the ice, the melted ice refreezing above the wire.

Important Topics

1. Specific heat.
2. Heat of fusion of ice.
3. Crystalline substances have fixed melting points.
4. Expansion on freezing, importance.

Exercises

1. What are two advantages in the high heat of fusion of ice?
2. What are two advantages in the expansion of water while freezing?
3. How much heat will be required to melt 1000 g. of ice and warm the water to 20°C.?
4. How many grams of ice at 0°C. can be melted by 400 g. of water at 55°C.?
5. What are two advantages of the high specific heat of water? Two disadvantages?
6. If the specific heat of iron is 0.1125, how much ice at 0°C. can be melted by a 200-g. ball of iron heated to 300°C?
7. What is the temperature of a hot ball of iron weighing 80 g., if when placed on a piece of ice at 0°C. it melts 90 g. of ice?
8. If 500 g. of copper at 400°C. are placed into 3000 g. of water at 10°C. what will be the resulting temperature?
9. What weight of water at 90°C. will just melt 10 kg. of ice at 0°C.?
10. If the smooth dry surface of two pieces of ice are pressed together for a short time the two pieces will be frozen into one piece. Explain.
11. Tubs of hot water are sometimes placed in vegetable cellars to prevent the vegetables from freezing. Explain.
12. How many B.t.u. are given out when 2 lbs. of water freeze?

(2) HEAT AND CHANGE OF STATE

184. Heat of Vaporization.—In our study of evaporation in Art. 174 we considered the more rapidly moving or vibrating molecules in the liquid escaping to the air above and the slower moving molecules being left behind in the liquid; this means that a loss of heat will result upon evaporation, the liquid remaining becoming cooler as the process continues. Now just as a ball thrown up in the air loses its kinetic energy as it rises, and acquires energy of position or potential energy, so molecules escaping from a liquid lose a certain amount of kinetic energy or heat and acquire a corresponding amount of *energy of position* or

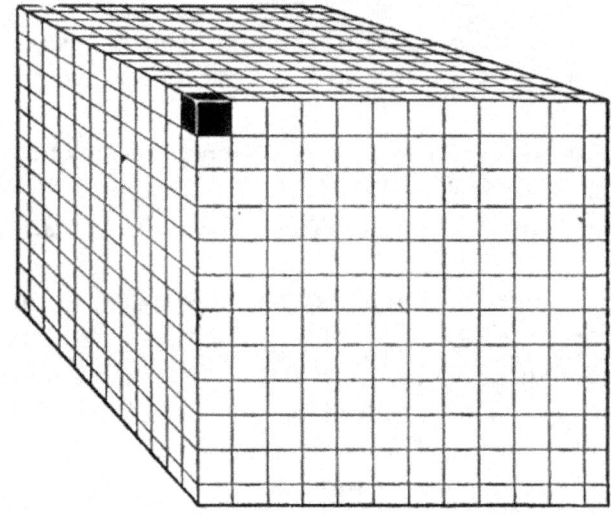

FIG. 153. –The black cube in the upper corner represents one cubic inch of water. The entire cube represents the space occupied by the cubic inch of water in the form of steam. The reduced spaces at the bottom and sides show how much short the cube is of being one cubic foot. (American Radiator Co.)

potential energy. *Conversely*, as the ball returns to the ground its potential energy is changed to kinetic energy. Similarly when vapor molecules return to the liquid condition they lose their energy of position and acquire kinetic energy. In other words, when a liquid evaporates a certain amount of heat disappears, or becomes *latent* and when the vapor condenses the heat reappears, or becomes *sensible* heat. *The amount of heat that disappears when* 1 g. *of a substance is vaporized is called the heat of vaporiza-*

tion. In the case of water at its boiling point, 536 calories of heat disappear when 1 g. of water turns to vapor, and this same amount of heat reappears when the vapor condenses.

The change of volume of water on turning to steam is shown in Fig. 153.

185. The Boiling Point.—The boiling temperature depends upon the pressure. The boiling point may be defined as *the temperature at which bubbles of vapor are formed within the liquid*. These bubble increase the

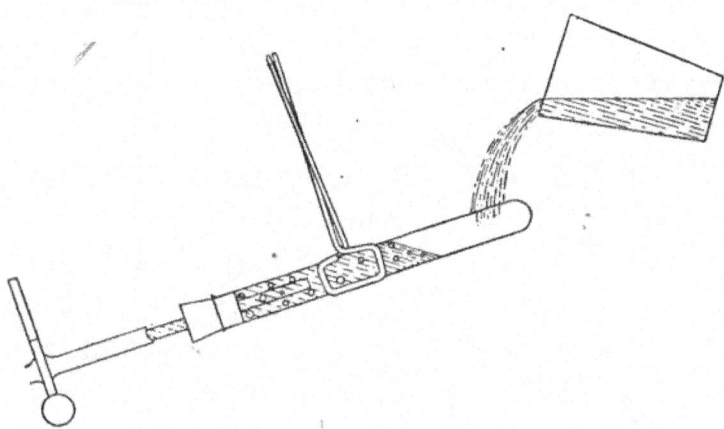

FIG. 154.—Effect of pressure on the boiling point.

surface at which evaporation can take place in the liquid, and the principal reason why rapid application of heat to a liquid does not raise its temperature above the boiling point is that as more heat is applied more bubbles form so that the increase of evaporating surface supplies a correspondingly greater surface for cooling. The variation of the boiling temperature with changing pressure may be shown by partly filling a strong ⅞-in. test-tube with water. Close the neck with a one-hole rubber stopper through which passes a glass tube to which is attached a soft rubber tube. (See Fig. 154.) Support the tube by a holder, heat the water and boil until all the air is driven from the tube, then close the soft rubber tube with a pinch cock and hold the tube in an inverted position. On cooling the end of the tube above the water with cold water or snow, the vapor within is condensed and the pressure upon the water is reduced. Vigorous boiling begins at once. By condensing the vapor repeatedly the water may be made to boil at the room temperature. At the top of Mt. Blanc water boils at 84°C. while in steam boilers at 225 lbs. pressure to the square inch the boiling point is nearly 200°C.

186. Laws of Boiling.—The following statements have been found by experiments to be true.

1. Every liquid has its own *boiling* point which under the same conditions of *pressure* is always the same.

2. The temperature of the boiling liquid remains at the boiling point until all the liquid is changed into vapor.

3. The boiling point rises with increased pressure and falls if the pressure is diminished.

4. A boiling liquid and the vapor formed from it have the same temperature. On cooling, a vapor will liquefy at the boiling point.

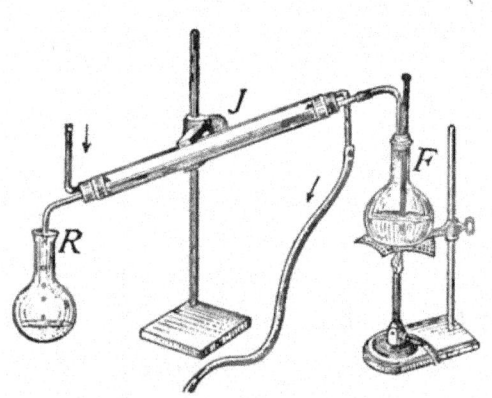

FIG. 155.—Distilling apparatus.

5. The solution of solid substances in a liquid raises its boiling point, additional energy being needed to overcome the adhesion involved in the solution. The boiling point is also affected by the character of the vessel containing the liquid. In glass the boiling point is 101°.

187. Distillation of Water.—Usually when solids are dissolved in liquids the vapor coming from the liquid contains none of the dissolved solid. Thus by evaporating salt sea water, and collecting and condensing the vapor, pure water is obtained. *Distillation* is the process of boiling a liquid and condensing the vapor formed back again into a liquid. (See Fig. 155.) The liquid to be distilled is placed in vessel F and boiled. The vapor is conducted into the tube J which is surrounded by a larger tube containing cold water. The vapor is condensed on the cold walls of the tube. The resulting liquid is collected in the vessel R. Distillation is employed for two purposes: (a) To remove impurities from a liquid (water is purified in this way). (b) Mixtures of different liquids having different boiling points may be separated by distillation. The one

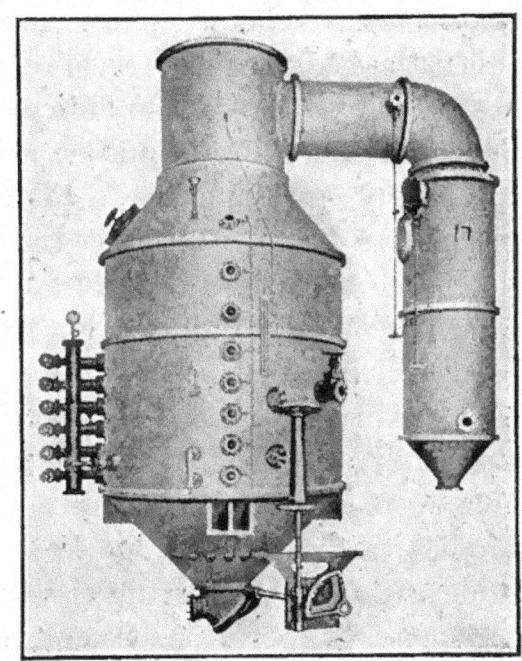

FIG. 156.—A vacuum pan.

having the lower boiling point will be vaporized first. Thus a mixture of alcohol and water, on distillation yields a distillate having a much larger percentage of alcohol than at first. Repeating this process which is called *fractional distillation* yields alcohol of increasing strength of purity. Distilled liquor such as alcohol, brandy, and whisky are made by distilling fermented liquor, alcohol being made from fermented grains. Gasoline and kerosene are distilled from crude petroleum. Sometimes as in the production of sugar or evaporated

14

milk the object is to remove the water by evaporation in order to obtain the solid material. Since the two substances named are injured by heating, the syrup, or milk is evaporated under reduced pressure in a *vacuum pan*, that is in a boiler from which air and vapor are removed by an air pump. (See Fig. 156.)

188. Artificial Cooling.—The fact has been brought out that when a solid is melted, a certain amount of heat, called the heat of fusion, is absorbed or disappears. This absorption of heat is also noticed when a solid is liquefied by dissolving it in a liquid as well as when it is liquefied by simply applying heat. Thus if some table salt is placed in a tumbler of water the temperature of the solution is lowered several degrees below that of the salt and water used. The liquefaction or solution of the salt has been accompanied by an absorption or disappearance of heat. This heat has been taken from the salt and from the water, resulting in a lowered temperature. Sal ammoniac or ammonium nitrate when dissolved in water produce a much more marked cooling effect than does table salt. The dissolving of a crystal in a liquid is something like evaporation, except that the molecules of the liquid attract the molecules of the solid and thus assist the change of state.

189. Freezing Mixtures.—If one attempts to freeze a solution of salt and water, ice will not form at 0°C. but several degrees lower. The ice formed however is pure. Evidently the attraction of the molecules of salt for the water molecules prevented the formation of ice until the motions of the water molecules had been reduced more than is necessary in pure water. As the temperature of freezing water is that of melting ice, ice in a salt solution melts at lower temperature than in pure water. In a saturated salt solution this temperature is − 22°C. It is for this reason that the mixture of ice and salt used in freezing cream is so effective, the salt water in melting the ice, being cooled to a temperature many degrees below the freezing point of the cream. The best proportion for a freezing mixture of salt and ice is one part salt to three parts of finely powdered or shaved ice.

190. Refrigeration by Evaporation.—Intense cold is also produced by permitting the rapid evaporation of liquids under pressure. Carbon dioxide under high pressure is a liquid, but when allowed to escape into the air

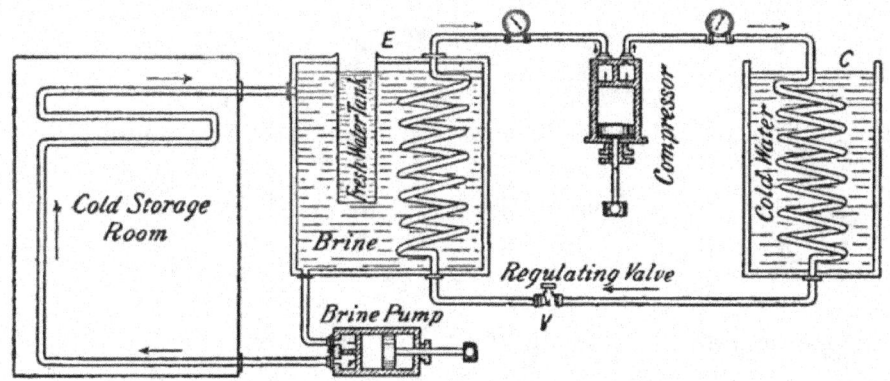

FIG. 157.—Diagram of a refrigerating system.

evaporates so rapidly that a portion of the liquid is frozen into solid carbon dioxide which has a temperature of $-80°$ C. The evaporation of liquid ammonia by permitting it to escape into a pipe, under reduced pressure, is used on a large scale as a means of producing cold in cold storage and refrigeration plants. (See Fig. 157.)

The essential parts of the refrigerating system employing ammonia is represented in Fig. 157. The *compressor* exhausts ammonia gas from the coiled pipe in "*E*" and compresses the gas in "*C*," where under 150 pounds pressure and the cooling effect of water it condenses to liquid ammonia. This is allowed to pass slowly through the regulating valve, whereupon it evaporates and expands in the long coiled pipe in "*E*" on its way back to the compressor. This evaporation and expansion causes a large amount of heat to be absorbed from the brine, cooling the latter below the freezing point of pure water and thus permitting the freezing of cans of water suspended in the brine. The chilled brine may also be sent through pipes in order to cool storage rooms containing meat or other food products. The ammonia absorbs heat when it vaporizes and gives up heat when it is compressed and liquified.

Important Topics

1. Heat of vaporization, of water 536 calories per gram.
2. Boiling point, effect of pressure upon boiling point, laws of boiling.
3. Distillation, artificial cooling, freezing mixtures, refrigeration by evaporation.

Exercises

1. How much heat is required (a) to melt 1 g. of ice at 0°C., (b) to raise the temperature of the water resulting to 100°C., (c) to change this water to steam?
2. If the water leaving a steam radiator is as hot as the steam how is the room warmed?
3. What is the effect of placing salt upon icy sidewalks in cold weather?
4. Is rain water distilled water? Is it perfectly pure?
5. What are two advantages of the high heat of vaporization of water?
6. If the heat from 1 g. of steam at 100°C. in changing to water and cooling to 0°C. could be used in melting ice at 0°C. how much ice would be melted?

(3) HEAT AND WORK

191. Necessity for Heat Energy.—From early times man has been able to transform motion into heat, and has used this ability in many directions as in starting fires and warming himself by friction. It took man many centuries, however, to devise an effective machine for transforming heat into mechanical energy or to use it in doing work.

The *power* of a man is small and as long as the work of the world had to be done by man power, progress was retarded. When man began the use of beasts of burden, he took a long step in advance since one man could then employ and direct the power of many men in the animals he controlled. Man also built water-wheels and windmills thus gaining power directly from the forces of nature and these added much to his working ability. But he took the greatest step in gaining control over his surroundings when he learned to use heat energy and to make it drive his machines.

192. Heat Engines.—At the present time there is a great variety of *heat engines* in use such as *steam, hot air, gas,* and *gasoline* engines, all using *heat energy* to produce motion. The expansive power of steam when confined has been observed for hundreds of years and many different machines have been invented to use it in doing work.

193. The Steam-engine.—The man who perfected the steam-engine, and devised its modern form was *James Watt* (1736–1819). The essential parts and the action of the steam engine may be readily understood by studying a diagram. In Fig. 158, *S* stands for *steam chest, C* for *cylinder, P* for *piston* and *v* for *slide valve.* The first two are hollow iron boxes, the latter are parts that slide back and forth within them. The action of the steam engine is as follows: Steam under pressure enters the steam chest, passes into the cylinder and pushes the piston to the other end. The slide valve is moved to its position in Fig. 159. Steam now enters the right end of the cylinder, driving the piston to the left, the "dead" steam in the left end of

the cylinder escaping at E to the air. The slide valve is
now shifted to its first position and the process is repeated.
It will assist the student to understand this action if he
makes a cardboard model of these parts, the piston and

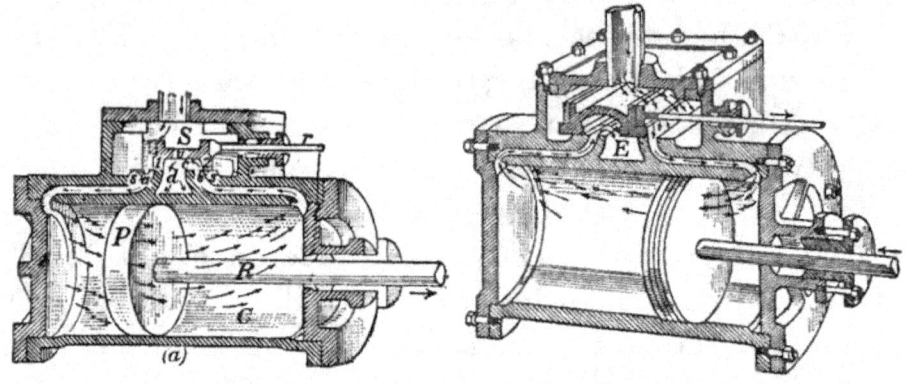

FIG. 158.—Cross-section view of
cylinder and steam chest of a steam
engine.

FIG. 159.—The steam drives
the piston to the left.

slide valve being movable. In practical steam-engines,
the piston rod is attached to a *crank rod* fastened to a crank
which turns a wheel. (See Fig. 160.) The back and
forth, or *reciprocating* motion of the piston is by this means

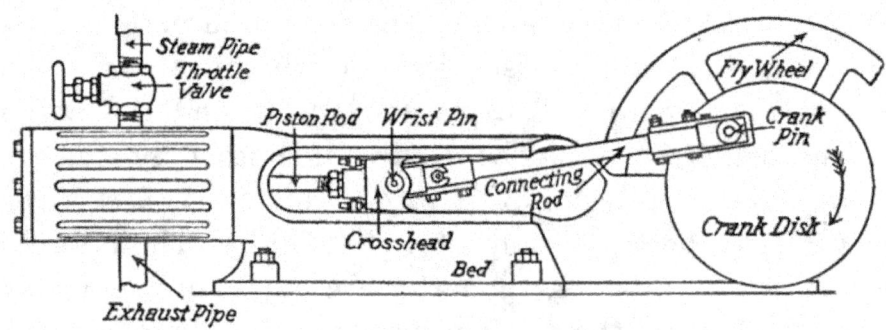

FIG. 160.—External view of steam engine.

transformed into *rotary* motion, just as in the sewing-
machine the back-and-forth motion of the treadle produces
rotary motion of the large wheel. Upon the shaft of the
steam engine is fastened an *eccentric* (see Fig. 163) which

moves the slide valve. The steam engine acts continuously as long as steam is supplied to it. Since it shifts the position of the slide valve automatically, it is called an automatic steam engine. And because the steam drives the piston both ways, it is called a *double-acting* steam engine. See Fig. 161 for a length-section of a modern locomotive.

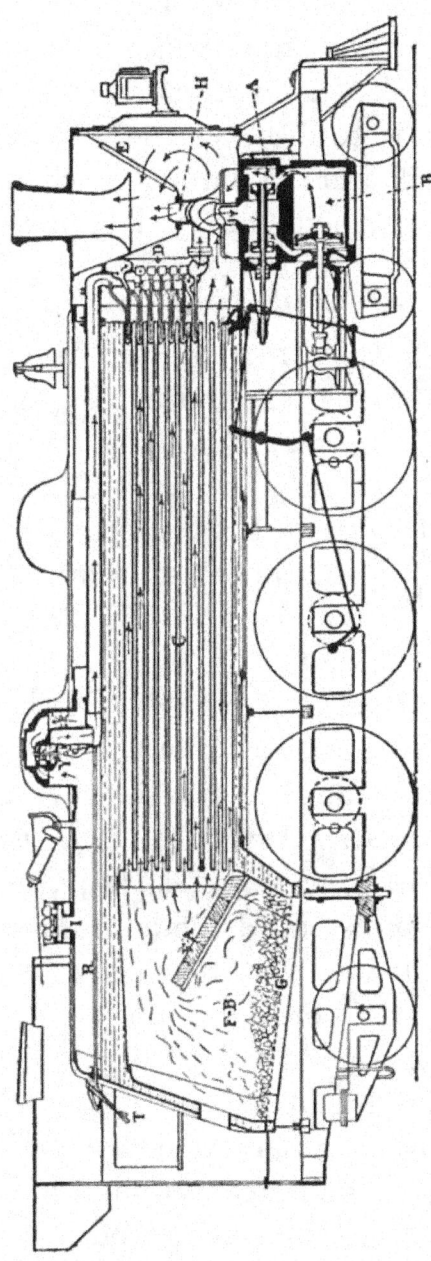

Fig. 161.—Length-section of modern, fast-passenger locomotive. *A*, cylinder valve—piston type valve; *B*, cylinder—piston at out end of stroke; *C*, boiler tubes—flues from fire-box; *D*, fire-tube type superheater; *E*, draught screen; *F–A*, fire-brick arch to protect tubes from direct heat; *F–B*, firebox; *G*, grate; *H*, exhaust nozzle; *I*, safety valve nest; *R*, throttle lever; *T*, throttle rod; *Y*, throttle valve.

194. The Mechanical Equivalent of Heat.—While watching workmen bore holes in cannon, Count Rumford, 1753–1814, noticed with much interest the large amount of heat produced in the process. He observed that the heat developed seemed to have some relation to the work done upon the drill in boring the holes. Later experiments performed by many men indicated that a definite relation exists between the heat produced by friction and the amount of work done in overcoming the friction. This discovery indicates that in some way heat is related to energy and that heat is probably a form of energy. Later experiments have confirmed this idea, and it is now considered well established that *heat is a form of energy*. Many attempts have been made to discover the relation between the units of heat energy and the units of mechanical energy. To illustrate one method employed, suppose one measures a given *length* in inches and in centimeters; on dividing one result by the other, it will be found that a certain relation exists between the two sets of measurements, and that in every case that 1 in. equals 2.54 cm. Similarly, when

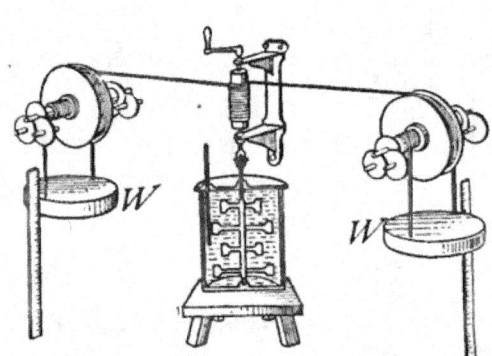

FIG. 162.—Apparatus for determining the mechanical equivalent of heat.

the same amount of *energy* is measured both in heat units and in work units a constant relation is always found between the units employed. *One B.T.U. is found equivalent to 778 ft.-lbs. 1, calorie being equivalent to 42,700 g. cm. (427 g. m.).* This relation is called the *mechanical equivalent of heat*, or in other words it represents the number of work units equivalent to one heat unit.

One of the first successful experiments in determining the relation between work units and heat units was devised by Joule in England. (See portrait p. 217.) The experiment consisted in taking a can of metal containing water (Fig. 162) in which was placed a thermometer, and a rod carrying paddles. The rod was turned by a cord connected through suitable apparatus to heavy weights, W and W. The energy represented by the downward

William Gilbert (1540–1603), "Father of magnetic philosophy." Especially noted for his experiments and discoveries in magnetism; first to use the word "electricity." First man to practically emphasize experimental science.

Dr. William Gilbert
(Popular Science Monthly)

James Prescott Joule (1818–1889), England, determined the mechanical equivalent of heat; discovered the relation between an electric current and the heat produced; first proved experimentally the identity of various forms of energy.

James Prescott Joule
(Popular Science Monthly)

motion of the weights through a given distance was compared with the heat energy developed in the water as shown by its rise in temperature. Careful experiments showed that when 778 ft.-lbs. of work had been done by the moving weights the heat produced at the same time would warm one pound of water 1 Fahrenheit degree. If the experiment was performed using metric units, it was found that the expenditure of 42,700 gram centimeters (427 gram meters) would result in producing enough heat to warm one gram of water one centigrade degree. The facts just given may be summarized as follows: *778 foot-pounds of energy are equivalent to 1 British thermal unit and 42,700 gram centimeters, or 427 gram meters, of energy are equivalent to 1 calorie.* This relation of work units to heat units is called the *mechanical equivalent of heat.*

195. The Heat Equivalent of Fuels and Efficiency Tests of Engines.—To determine the efficiency of a steam engine it is necessary to know not only the mechanical equivalent of heat but also the heat produced by burning coal or gas; 1 lb. of average soft coal should produce about 12,600 B.t.u. Now since 778 ft.-lbs. are equivalent to one B.t.u. the energy produced when 2 lbs. of average soft coal is burned is $778 \times 12,600 \times 2 = 19,605,600$ ft.-lbs. In actual practice 2 lbs. of average soft coal burned will develop about 1 horse-power for 1 hour. 1 horse-power-hour $= 33,000$ ft.-lbs. $\times 60 = 1,980,000$ ft.-lbs. Now efficiency equals $\dfrac{\text{work out}}{\text{work in}} \dfrac{1,980,000}{19,605,600} = \dfrac{1}{10}$ or 10 per cent. This is the efficiency of a good steam engine. Ordinary ones require 3 lbs. of coal burned to each horse-power-hour produced or they are but $\frac{2}{3}$ as efficient or have but about 7 per cent. efficiency.

Heat of Combustion of Various Fuels

Data in this table are taken from U. S. Geological Survey, Bulletin, No. 332, and U. S. Bureau of Mines, Bulletin No. 23.

	B.T.U. per lb.	Calories per gram
Alcohol, denatured..........................	11,600	6,450
Coal, anthracite, average.....................	13,500	7,500
Coal, bituminous, average....................	12,600	7,000
Gasoline..................................	19,000	10,550
Illuminating gas...........................	18,000	10,000
Kerosene.................................	19,900	11,050

Constants for Heat Transmission

Data from "Ideal Fitter," American Radiator Co.

B.t.u. transmitted per square foot per hour per degree (Fahrenheit) difference in temperature between inside and outside air.

Brick Work

4 in. thick = 0.68 ⎱
8 in. thick = 0.46 ⎰ concrete ⎱ 50 per cent. more than brick.
12 in. thick = 0.33 ⎰ cement ⎰

stone 33⅓ per cent. more than brick.

Window = 1.090 ⎱
Wood as wall = 0.220 ⎰ concrete ⎱ 20 per cent. more than
Double window = 0.560 ⎰ reinforced ⎰ brick.

Important Topics

1. Heat a manifestation of energy.
2. Steam-engine and its action.
3. Mechanical equivalent of heat and heat equivalent of fuels and efficiency of engines.

Exercises

1. Construct a working model of the cylinder and steam chest of a steam engine and be prepared to explain its action.
2. At $5.00 per ton how many B.T.U.'s should be produced from 1 cent's worth of bituminous coal?

3. Try the following experiment: Place a quart of water in a tea-kettle and place it over the fire for 5 minutes, and note the rise in temperature and compute the number of B.T.U.'s entering the water. Place another quart of water at the same temperature in an aluminum or tin dish and heat for 5 minutes, note the rise in temperature and compute the heat used before. Which of the dishes shows the greater efficiency? How do the efficiencies of the two dishes compare? How do you account for any differences in the efficiencies found?

4. How high would 8 cu. ft. of water be lifted if all of the energy produced by burning 1 lb. of coal were used in raising it?

5. What is the mechanical equivalent of a pound of coal expressed in horse-power hours?

6. If a furnace burns 100 lbs. of coal a day and its efficiency is 50 per cent. how many B.T.U.'s are used in warming the house?

7. How many B.T.U.'s can be obtained by burning ½ ton of bituminous coal?

8 When a pound of water is heated from 40°F. to 212°F., how many foot-pounds of energy are absorbed by the water?

9. How many loads of coal each weighing 2 tons, could be lifted 12 ft. by the energy put into the water in problem 8?

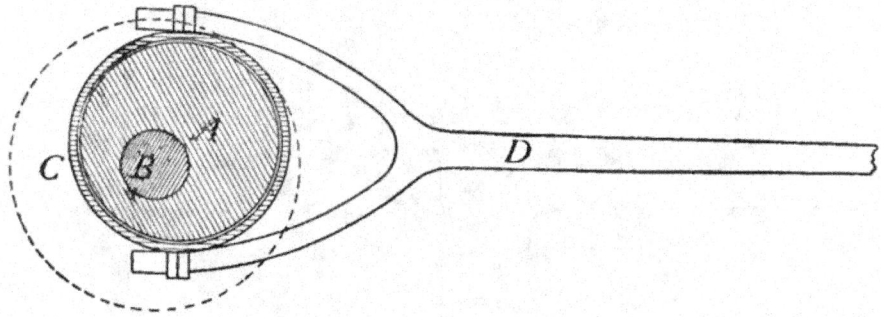

FIG. 163.—An eccentric.

10. When 3 cu. ft. of water are used for a hot bath and the water has been heated from 50°F. to 112°F., how many B.T.U.'s have been absorbed by the water?

11. If the average temperature of water at the surface of Lake Michigan is 50°F., how many B.T.U.'s would be given off by each cubic foot of water at the surface, if the temperature of the water should drop 5°F.?

12. In a cold storage plant carbon dioxide gas is used. The pipe

leading from the compression pump to the expansion valve passes through a condensing tank of cold water. Why?

13. When the gas is compressed in a cold storage plant, what becomes of the energy used by the compression pump?

14. An eccentric (Fig. 163), is a round disc mounted a little to one side of its center, *A*, on the engine shaft *B*. *A* band, *C*, on the circumference of the disc is connected by a rod, *D*, with the slide valve in the steam chest. How is the rotary motion of the shaft changed into a backward and forward motion of the slide valve?

(4) HEAT ENGINES

196. The Gas Engine.—One of the heat engines in common use to-day is the gasoline engine. It is used to propel automobiles and motor boats, to drive machinery,

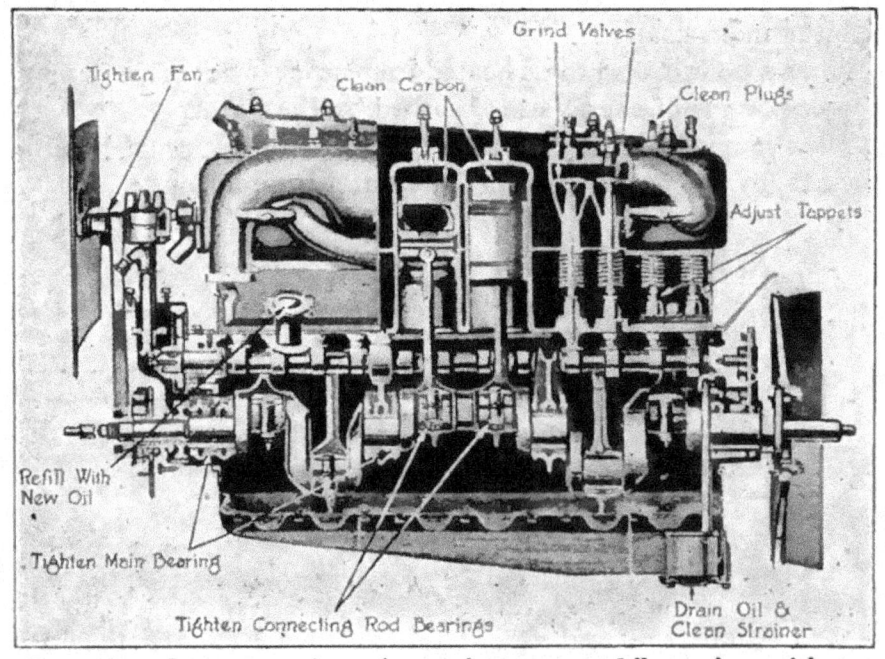

FIG. 164—Cut away view of a modern automobile engine, with parts requiring attention most frequently, indicated. (Courtesy of the "Automobile Journal.")

etc. The construction and action of a gasoline engine may be understood by studying a working model, or by proper diagrams.

The common gasoline or gas engine is called a four-cycle (better four-part cycle) engine (see Fig. 164) since it requires four movements of the piston to complete one cycle or series of changes. This is illustrated in Fig. 165 1, which represents a cross-section of the *cylinder* of the gasoline engine with the *piston* moving downward. At the upper end of the cylinder are two *ports* or openings. One, the *exhaust* port, is closed, the *inlet* port is open and a mixture of gas and air is entering. Fig. 165 2 shows the piston returning; both ports are closed and the "charge" of air and gas is being compressed. As the piston reaches the end of its stroke in compressing the charge, an electric spark explodes or "fires" the charge of gas and air. The hot burning gas expands suddenly driving the piston downward with great force (Fig. 165 3). The piston rod is attached to the crank of a heavy fly-wheel and this is given sufficient energy or momentum to keep

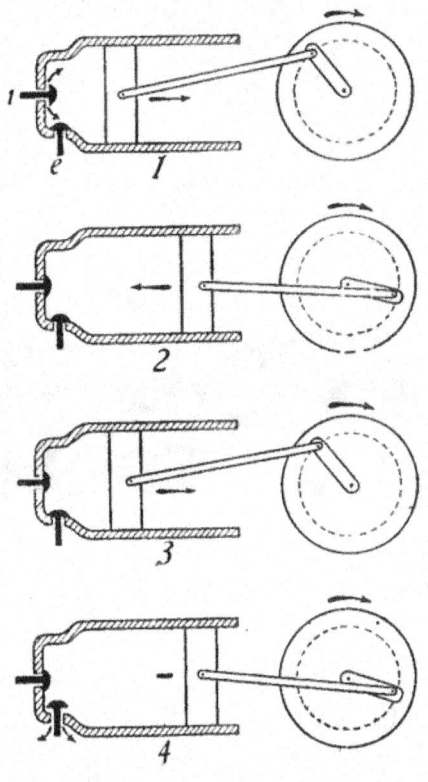

FIG. 165.—The four strokes of a gas engine cycle.

it going through the next three strokes. Fig. 165 4 represents the returning piston pushing out the burnt "charge" through the open exhaust valve *e*. On the next downward motion of the piston the valve *e* closes. It opens, and new charges of gas and air enter and the "cycle" is repeated.

In order to make the motion more even and continuous and also to secure more power, more than one cylinder is attached to the same shaft and fly-wheel. Two, three, four, six, eight and even more cylinders have been attached to one shaft. Four or six cylinders are commonly used in automobile gasoline motors. To lessen the sound of the "exhaust," the latter is sent through a "muffler" which often reduces the noise to a low throbbing. (See Fig. 166.) The gasoline engine is more efficient than the steam-engine, since the fuel, gas, is burned in the cylinder and not in a separate furnace. The combustion of the fuel in the cylinders makes some special cooling device necessary to prevent their overheating. This usually consists of a casing about the cylinders. Between the cylinder and this

Fig. 166.—An efficient automobile muffler. (*Courtesy Popular Science Monthly.*)

casing is water which on being heated passes to a tank or radiator. In the radiator the water cools and then returns to the space between the cylinders and casing thus keeping up the circulation.

197. Efficiency of Gas Engines.—One may test the *efficiency* of a gas engine by determining the amount of power developed and comparing it with the mechanical equivalent of the fuel burned. Illuminating gas is sometimes employed to drive gas engines. One cubic foot of illuminating gas should produce 600 B.T.U. when burned. The efficiency of the gas or gasoline engines is sometimes as high as 25 per cent. This engine is free from smoke and is also compact and quickly started. While the fuel, gas or gasoline, is somewhat expensive it is light and easily carried. Suppose a gas engine produces 1 horse-power and uses 20 cu. ft. of gas an hour, what is its efficiency? 1 horse-power-hour = 550 × 60 × 60 = 1,980,000 ft.-lbs. 20 cu. ft. of gas = 20 × 600 × 778 = 9,336,000 ft.-lbs.

$$\text{Efficiency} = \frac{\text{work in}}{\text{work out}} = \frac{1,980,000}{9,336,000} = 21.2 \quad \text{per cent.}$$

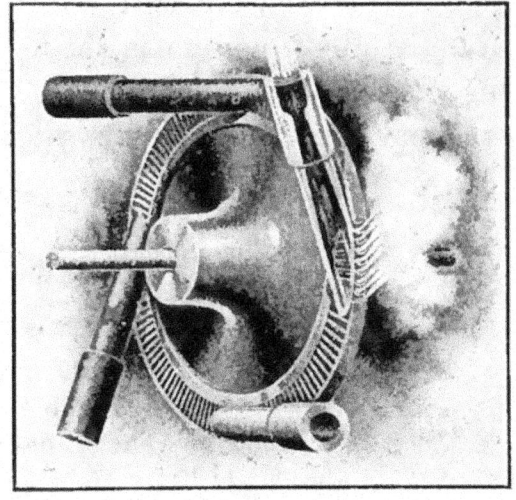

FIG. 167.—The principle of the steam turbine.

198. Steam Turbine. One form of the steam-engine that is coming into general use is the turbine. (See Fig. 167.) This consists of a shaft to which are attached blades, the shaft and blades being contained in a closed case. Steam is admitted by nozzles and strikes the blades so as to set them and the shaft in motion. There are also stationary blades (see Fig. 168), which assist in directing the steam effectively against the rotating parts. The

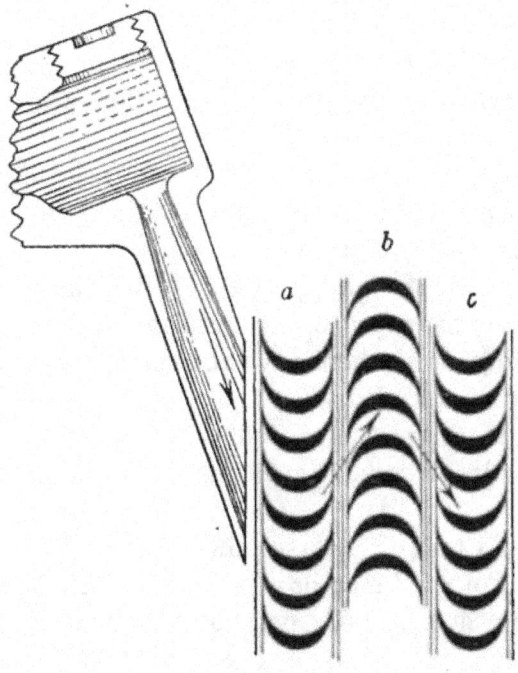

FIG. 168.—Path of steam in DeLaval steam turbine. (a) and (c) movable blades, (b) stationary.

steam turbine is used for large power plants. (See Fig. 293.) It is very efficient, makes very little vibration, and occupies about one-tenth the floor space that a reciprocating engine of equal power uses. Some large ocean steamers are now driven by steam turbines.

Important Topics

1. The gas engine, its construction, action and efficiency.
2. The steam turbine.

Exercises

1. If coal costs $4.00 a ton, and gas, $0.80 per 1000 cu. ft. what amounts of heat can be secured from 1 cent's worth of each?

2. What will it cost to heat 30 gallons of water (1 gal. of water weighs about 8⅓ lbs.) from 40°F. to 190°F. with coal costing $4.00 per ton and yielding 12,000 B.T.U. per lb. if the heater has an efficiency of 50 per cent.

3. What will it cost to heat 30 gallons of water from 40°F. to 190°F. with gas at $0.80 per 1000 cu. ft. if the heating device has an efficiency of 75 per cent.

4. Construct a cardboard working-model showing the action of the gas engine and be prepared to explain the action of the various parts.

5. If 500 lbs. of iron should fall 2000 ft. and all of the resulting mechanical kinetic energy should be transformed into heat, what would be the amount of heat produced?

6. What are the special advantages of (a) the gasoline engine? (b) the turbine? (c) of the reciprocating steam engine?

7. Do you burn coal or gas in your kitchen stove at home? Which is for you the more economical? Why?

8. What are the advantages of using a fireless cooker?

9. What is the efficiency of a locomotive that burns 3.2 lbs. of coal per horse-power-hour?

10. A gas engine developed in a test 0.34 horse-power for 1 minute. and 50 seconds, 0.5 cu. ft. of gas being used. The heat of combustion of the gas was 600 B.T.U. per cu. ft. Find the efficiency of the engine.

11. Find the horse-power of an engine, the diameter of the piston being 19 in., stroke 26 in.; it uses steam at an average pressure of 200 lbs. per square inch and makes 100 strokes a minute.

12. What is the efficiency of an engine and boiler that develops 200 horse-power, while burning 390 lbs. of soft coal per hour?

13. If a locomotive has an efficiency of 6 per cent. and develops 1700 horse-power how much coal is burned in an hour?

14. If an automobile engine burns 1 gallon of gasoline in an hour and develops 10 horse-power, what is its efficiency?

15. The A.L.A.M.[1] formula for horse-power is $\dfrac{N\,B^2}{2.5}$, when the piston speed is 1000 ft. per minute, N being the number of cylinders and B, their diameter. Find the horse-power of a 4-cylinder engine, the cylinders having a diameter of 4 in.

16. Find the horse-power of a 6-cylinder automobile engine, if the cylinder diameter is 4.5 in.

17. A 4-cylinder automobile having 4-in. cylinders, uses 1 gallon of gasoline in 1 hour. Find its efficiency, if its average horse-power developed is 6.

18. The motor boat Disturber III, has 24 cylinders each with diameter 3.5 in. If the piston speed is 1000 ft. per minute, what is the horse-power? (See problem 15.)

Review Outline: Heat

Heat; sources (4), effects (5), units (2).

Temperature; thermometer scales (3), absolute temperature, $9C°/5 + 32° = F°$.

Expansion; gases, Law of Charles ($V_1/V_2 = T_1/T_2$), liquids, peculiarity of water, solids, coefficient of expansion, uses, results.

Heat Transference; conduction, uses of good and poor conductors, convection, in nature, heating and ventilating systems, radiation, 3 peculiarities, value of sun's radiation.

Heat and Moisture; relative humidity, dew point, formation of dew, fog, rain, snow, etc., evaporation, effects, conditions.

Heat Measurement; specific heat, heat of fusion, of vaporization, combustion.

Vaporization; Boiling point, laws of boiling, distillation, artificial cooling.

Heat Engines; steam, gas,—construction, action, efficiency, mechanical equivalent of heat. Heat equivalent of fuels.

[1] American League of Automobile Manufacturers.

CHAPTER IX

MAGNETISM

(1) General Properties of Magnets

199. Magnets.—Since the times of the early Greek philosophers men have known of certain stones that have the property of attracting to themselves objects of iron and steel. Such stones are called *natural magnets*. It is thought by many that the name magnet is derived from Magnesia in Asia Minor, where these stones are abundant, though this is but tradition.

It was also learned long ago that iron and steel objects when rubbed with natural magnets become magnetized, that is, acquire the properties of magnets. These are said to be *artificial magnets*.

Some 800 years ago it was discovered that magnets, natural or artificial, when

Fig. 169.—A bar magnet.

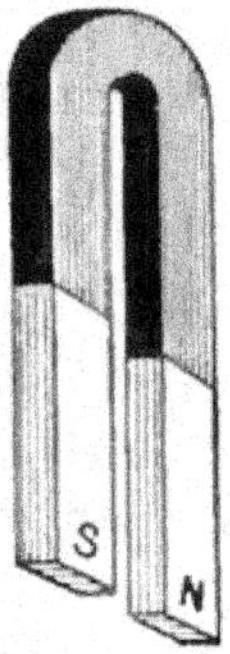

Fig. 170.—A horse-shoe magnet.

suspended so as to turn freely, always come to rest in a definite position pointing approximately north or south. This is especially noticeable when the magnet is long and narrow. Because of this property of indicating direction, natural magnets were given the name of *lode-stone* (lode-leading).

Artificial magnets are made by rubbing steel bars with a

magnet or by placing the steel bar in a coil of wire through which a current of electricity is flowing. The magnetized steel bars may have any form, usually they are either straight or bent into a "U" shape. These forms are known as *bar* and *horseshoe* magnets. (See Figs. 169 and 170.) Magnets retain their strength best when provided with soft-iron "*keepers*," as in Fig. 171.

200. Magnetic Poles.—If a magnet is placed in iron filings and removed, the filings will be found to cling strongly at places near the ends of the magnet, but for a portion of its length near the middle no attraction is found. (See Fig. 172.) These places of greatest attraction on a magnet are called *poles*. If a bar magnet is suspended so as to swing freely about a vertical axis the

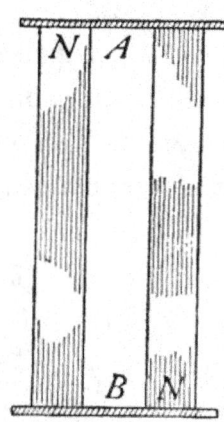

Fig. 171.—Bar magnets with keepers.

magnetic pole at the end pointing north is called the *north-seeking* pole; at the other end, is the *south-seeking* pole. In most places the needle does not point to the true north, but somewhat to the east or west of north. The direction taken by a magnetic needle is parallel to the *magnetic meridian*.

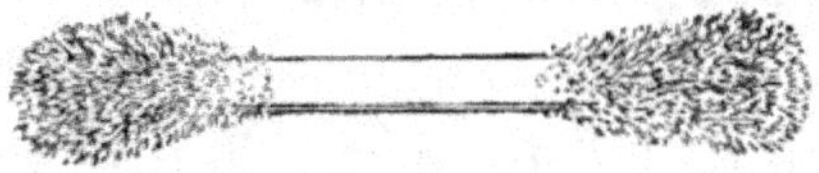

Fig. 172.—Iron filings attracted to the poles of a magnet.

201. Law of Magnetic Action.—The north pole of a magnet is usually marked. If a marked bar magnet be held in the hand and its north-seeking pole be brought near the north-seeking pole of a freely suspended bar magnet, the two poles will be found to repel each other, as will also two south-seeking poles, while a north-seeking and a south-seeking pole attract each other. (See Fig. 173.) This action leads to the statement of the *Law of Magnetic*

Action: Like poles repel, while unlike poles attract each other. The force of attraction or repulsion lessens as the distance increases. *The force of the action between magnetic poles is inversely proportional to the square of the distance between them.* Compare this with the law of gravitation (Art. 88).

202. Magnetic Substances and Properties.—It is found that if an iron or steel magnet is heated *red hot* that its magnetic properties disappear. Accordingly one method of *demagnetizing* a magnet is to raise it to a red heat. If

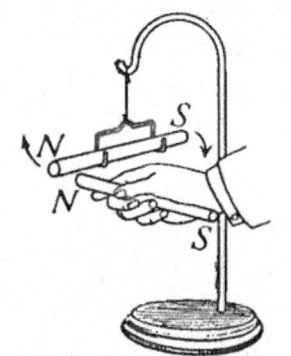

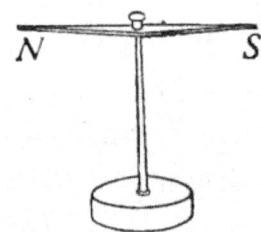

Fig. 173.—Like poles of two magnets repel.   Fig. 174.—A magnetoscope.

a magnet that has been heated red hot and then cooled is brought near a suspended bar magnet, it is found to *attract either* end, showing that it has regained *magnetic properties* even though it has lost its *magnetic polarity.* A suspended bar magnet used to test the magnetic properties of a body is called a *magnetoscope.* (See Fig. 174.) The needle of a *magnetic compass* serves very well as a magnetoscope. Magnetic properties are most strongly exhibited by iron and steel, though nickel and cobalt show some magnetic effects. There is a peculiar alloy of copper, aluminum, and manganese, known as *Heusler's Alloy,* that is also magnetic. However, of all substances, iron and steel show the strongest magnetic effects.

203. Magnetic Induction.—Let the north-seeking pole of a bar magnet support an iron nail by its head. (See Fig. 175.) Test the point of the nail for polarity. See whether a second nail can be attached by its head to the point of the first. Test the polarity of the point of this nail. Find by trial how many nails can be suspended in succession from the magnet. Test in each case for polarity. Withdraw carefully the magnet from the first nail—the string of nails will fall apart. Repeat the test with a thickness of paper between the magnet and the first nail. Results similar to those secured at first will be found, though probably fewer nails will be supported. The presence of paper between the magnet and nails simply weakens the action. Test the action of the magnet upon the nail when there is between them a piece of glass, one's thumb, thin pieces of wood, copper, zinc, etc. *The magnetizing of a piece of iron or steel by a magnet near or touching it is called magnetic induction.* This action takes place through all substances except large bodies of iron or steel hence these substances are often used as *magnetic screens*. The pole of the new *induced magnet*

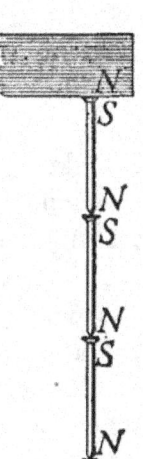

Fig. 175.— Nails magnetized by induction.

adjacent to the bar magnet is just opposite to the pole used. Thus the N.-pole of the magnet used will produce a S.-pole at the near end of the nail and a N.-pole at the end farther away. (See Fig. 175.) On removing the magnet, the nails are found to retain a part of their induced magnetism.

204. Retentivity.—In several of the foregoing paragraphs it has been seen that a piece of iron or steel when once magnetized does not entirely lose its magnetism when the magnetizing force is removed. Different pieces of iron and steel vary greatly in this respect, some remaining strongly magnetized, others losing much of their magnetism. *This property of retaining magnetism is called retentivity.* Hardened steel has a high degree of retentivity, while soft iron retains but little magnetism.

Important Topics

1. Magnet; natural, artificial, bar, horseshoe.
2. Magnetic poles; north seeking, south seeking.
3. Law of action, magnetoscope, retentivity, induced magnet.

Exercises

1. Make a summary of the facts of magnetism presented in this lesson.
2. Is magnetism matter, force, or energy? How do you decide? To what other phenomenon that we have studied is it similar? How?
3. Make a simple magnetoscope for yourself by suspending a thin steel needle or rod 5 to 10 cm. long, with a light thread or silk fiber at its center, so that it will hang level. Then magnetize the needle, and keep the magnetoscope in your book.
4. Name three uses for magnets or magnetism.
5. Mention three uses for a magnetoscope.
6. Are all magnets produced by induction? Explain.
7. In what magnetic devices is a high retentivity desirable?

(2) THE THEORY OF MAGNETISM AND MAGNETIC FIELDS

205. The Theory of Magnetism.—If a magnetized watch spring is broken in two, *each part* is found to be a magnet. If one of these parts be broken and this process of breaking be continued as far as possible, the smallest part obtained has two poles and is in fact a complete magnet. (See Fig. 176.) It is supposed that if the division could be continued far enough that each of the *molecules of the steel spring* would be found to have *two poles*

and to be a magnet. In other words, magnetism is believed to be *molecular*. Other evidence supporting this
idea is found in the fact that when a magnet is heated red
hot, to a temperature of violent molecular motion, its
magnetism disap-
pears. Also if a
long, fine soft iron
wire be strongly
magnetized, a
light jar causes its
magnetism to dis-

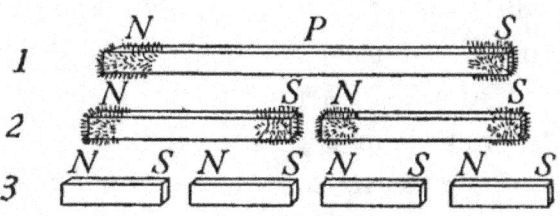

FIG. 176.—Effect of breaking a magnet.

appear. This would lead us to believe that magnetism
is not a property of the surface of the body, but that it
depends upon molecular structure or the arrangement
of the molecules.

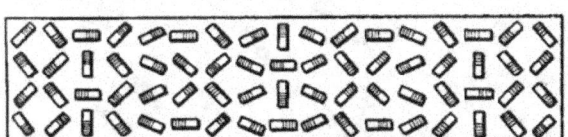

FIG. 177.—Possible arrangement of molecules
in an unmagnetized iron bar.

It is believed
also that the *mol-
ecules* of a mag-
netic substance are
magnets at al
times; that before
the body is magnetized the molecules are arranged
haphazard (see Fig. 177) but that when a magnet is
brought near, the molecules tend to arrange themselves
in line, with their
north-seeking
poles pointing in
the same direction.
(See Fig. 178.) If
the magnet is

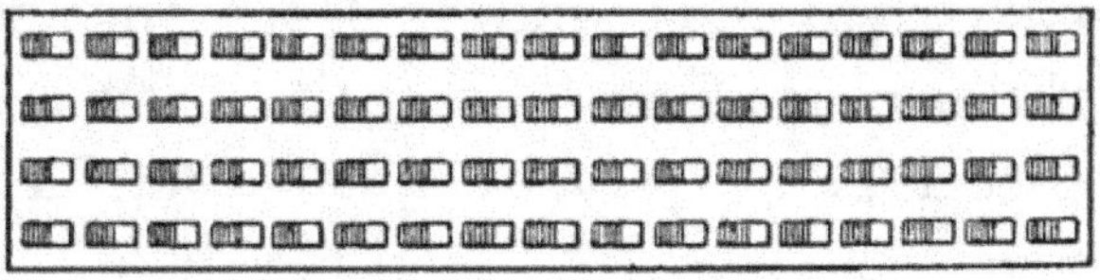

FIG. 178.—Arrangement of molecules in a
saturated magnet.

jarred some of the molecules tend to get out of line, perhaps
to form little closed chains of molecules. (See Fig. 177.)

206. Magnetic Fields and Lines of Force.—The behavior of magnets is better understood after observing and

studying the *lines of force* of a magnet. The earliest
descriptions of these are by William Gilbert, the first
Englishman to appreciate fully the value of making
experimental observations. He wrote a book in 1600
called *De Magnete* in which he published his experiments
and discoveries in magnetism. (See p. 217.)

Magnetic lines of force may be observed by placing a
magnet upon the table, then laying upon it a sheet of paper
and sprinkling over the latter fine iron filings. On gently
tapping the paper, the filings arrange themselves along

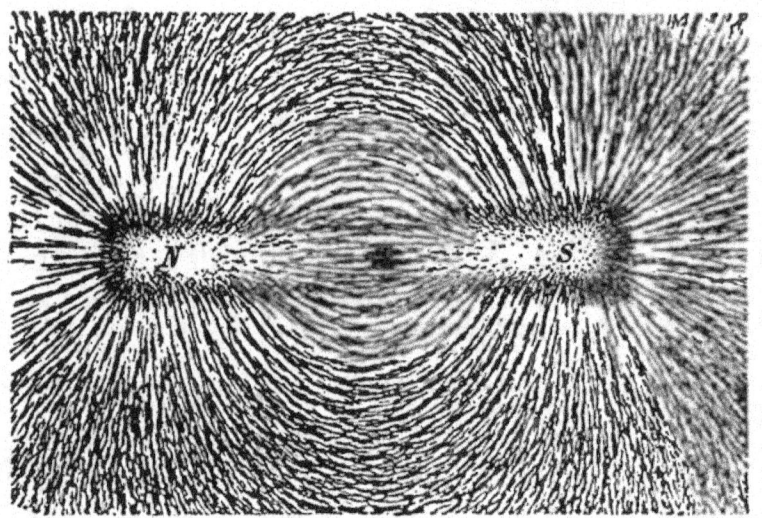

FIG. 179.—Iron filings on paper over a bar magnet

curved lines extending from one end of the magnet to
the other. These are called the *magnetic lines of force.*
(See Fig. 179.) The space about a magnet in which the
magnetic lines are found is called the *magnetic field.* (See
Fig. 180.)

Many interesting things have been discovered concerning the lines of force. Some of the facts of magnetic action
are given a simple explanation if we think of them as due
to the magnetic lines of force. A summary of several
discoveries concerning magnetic fields follows:

(A) Magnetic lines of force run side by side and do not cross one another. (See magnetic fields.)

(B) Magnetic lines of force are believed to form "*closed curves*" or to be continuous. The part outside of the magnet is a continuation of the part within the magnet. (See Fig. 180.)

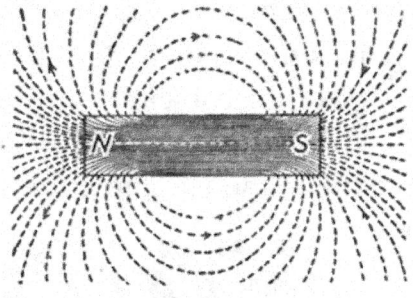

(C) The attraction of a magnet is strongest where the magnetic lines are thickest, hence they are believed to be the means by which a magnet attracts.

FIG. 180.—Diagram of the field of a bar magnet.

ce like poles repel and unlike poles attract. it hat the action along a line of force i th directions It has therefore been

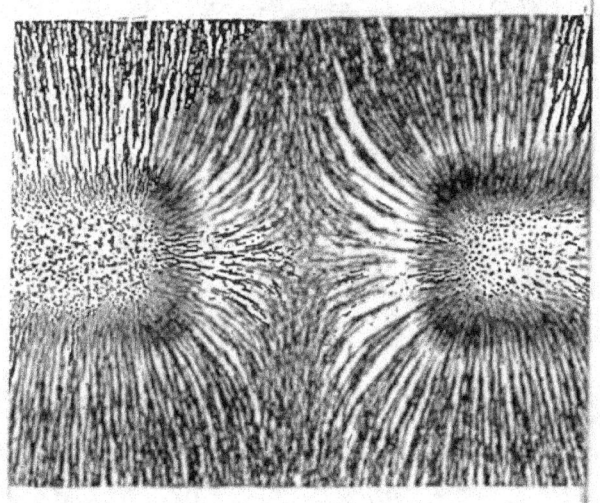

.—Magnetic field between like poles showing

to indicate by an arrow head (Fig hat a north-seeking pole tends to m ce. The lines of force are considered seeking pole of a magnet and entering le. (See Figs. 181 and 182.)

(E) A freely suspended small magnet in a magnetic field places itself parallel to the lines of force. (Test this by holding a magnetic compass in different portions of a magnetic field). Note the position of the needle and the lines of force. This fact indicates that the compass needle points north on account of its tendency to turn so as to be parallel to the earth's magnetic field.

(F) Each magnet is accompanied by its own magnetic field. When a piece of iron is brought within the field

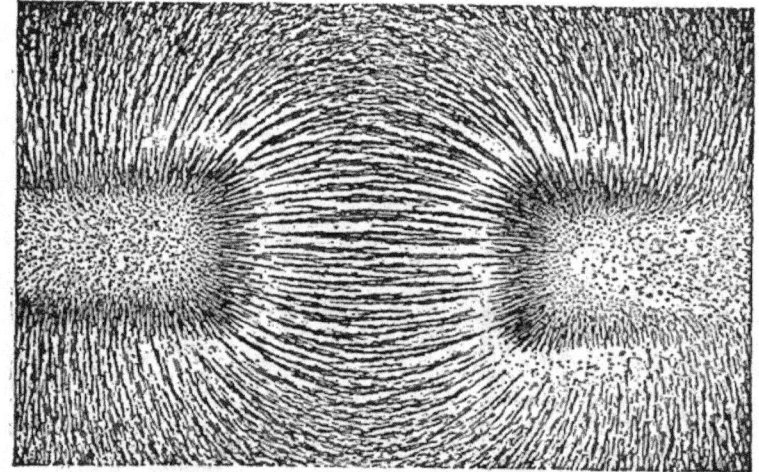

Fig. 182.—Magnetic field between unlike poles showing attraction.

of a magnet the lines of force passing through the iron tend to arrange the iron molecules in line or to magnetize the iron.

207. Magnetic Induction.—The action of magnetic lines of force in magnetizing iron when they pass through it, is called *Magnetic Induction*. This may now be defined as *the production of magnetism in a body by placing it within a magnetic field*. Freely suspended magnets place themselves parallel to the lines of force in a magnetic field, therefore when an iron rod is placed in a weak field, or one with few lines of force, the iron is but slightly magnetized; that is, but few molecules are brought into line. Increasing the strength of the magnetizing field, gives stronger magnetization to the iron up to a certain point. After this, stronger fields give no increase in magnetizing effect. When iron exhibits its greatest magnetization it is said to be *saturated*.

208. Permeability.—If a piece of iron is placed between the poles of a horseshoe magnet, the "field" obtained by sprinkling iron filings upon a sheet of paper over the magnet resembles that shown in Fig. 183. The lines in the space between the poles of the magnet seem to crowd in to the piece of iron. The *property* of the iron by which it tends to concentrate and increase the number of lines of force of a magnetic field is called *permeability*. Soft iron shows high permeability. Marked differences in behavior are shown by different kinds of iron and steel

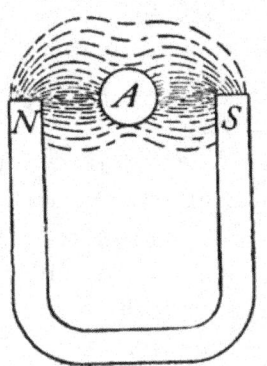

FIG. 183.—Effect of a piece of iron in a magnetic field.

when placed in a magnetic field. Very pure iron, or *soft* iron, is strongly magnetized by a magnetic field of medium strength. Its magnetism, however, is quickly lost when the magnetizing field is removed. This indicates that soft-iron molecules are easily swung into line, but also disarrange themselves as easily when removed from a magnetizing force. Soft-iron magnets having high permeability quickly lose their magnetism. They are therefore called temporary magnets. On the other hand a hardened steel bar is difficult to magnetize, but when once magnetized retains its magnetism permanently, unless some action weakens the magnet. Such magnets are called *permanent* magnets.

NOTE.—The term "line of force" as used in this text means the same as "line of induction" as used in more advanced texts.

Important Topics

1. Molecular theory of magnetism, saturation, permeability.
2. Magnetic fields and lines of force.
3. Six facts concerning magnetic fields.

Exercises

1. Name an object whose usefulness depends upon its retentivity. Explain.
2. How do you explain the rententivity of hard steel?
3. Are the molecules of a piece of iron magnetized at all times? Explain.
4. When a piece of iron is magnetized by induction does any magnetism enter the iron from the magnet? Does the magnet lose as the iron gains magnetism? Explain.
5. Have all magnets been produced by induction? Explain.
6. Why will tapping a piece of iron when in a magnetic field increase the amount it will be magnetized?
7. Express in your own words the theory of magnetism.
8. Place two bar magnets in a line 5 cm. apart, *unlike* poles adjacent; obtain the magnetic field with iron filings. Sketch it.
9. Repeat Exercise No. 8 using *like* poles. Describe the appearance of a field that gives attraction; of a field that gives repulsion.

(3) The Earth's Magnetism

209. The Earth's Magnetic Field.—Dr. William Gilbert's famous book, *De Magnete*, contains many helpful and suggestive ideas, none perhaps more important than his explanation of the behavior of the compass needle. He assumed that the earth is a magnet, with a *south-seeking* pole near the geographical north pole, and with a *north-seeking* pole near the geographical south pole. This idea has since been shown to be correct. The north magnetic (or south-seeking) pole was found in 1831, by Sir James Ross in Boothia Felix, Canada. Its approximate present location as determined by Captain Amundsen in 1905 is

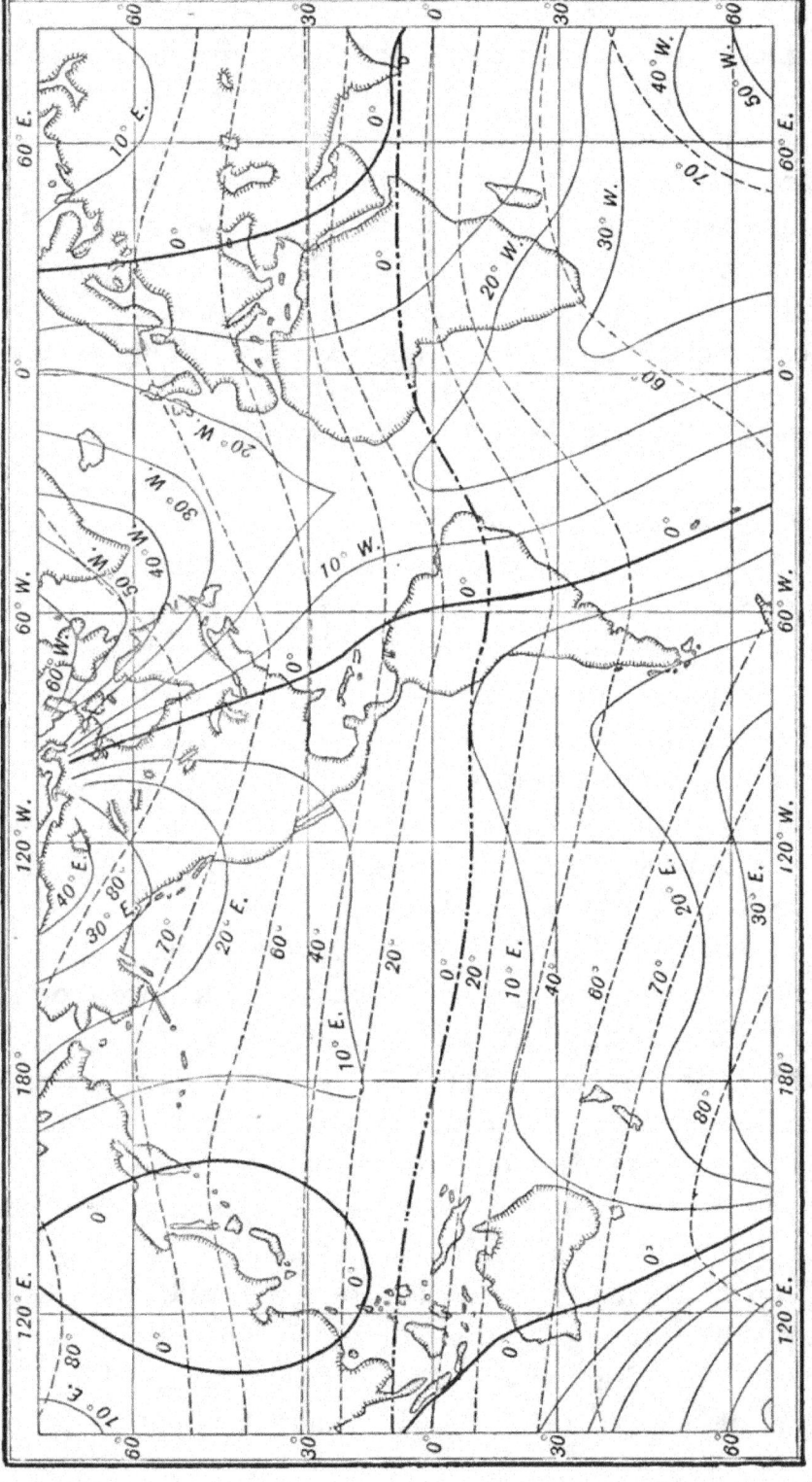

FIG. 184.—Magnetic map of the earth for 1910. Isogonic lines —— Isoclinic lines........

latitude 70° 5' N. and longitude 96° 46' W. The south magnetic pole is in latitude 72° S., longitude 155° 16' E. The north magnetic pole is continually changing its position. At present it is moving slowly westward.

210. Direction of the Earth's Magnetic Field.—Reference has been made to the fact that the compass does not always point exactly north. This indicates that the earth's magnetic field varies in its direction. Columbus discovered this fact upon his first voyage. The discovery alarmed the sailors since they feared they might come to a place where the compass would be unreliable. This variation is called *declination*. It is defined as the *angle between the direction of the needle and the geographical meridian*. Declination is due to the fact that the geographical and magnetic poles do not coincide. What is meant by a declination of 90°? Lines drawn upon a map so as to pass through places of the same declination are called *isogonic* lines. The line passing through points where the needle points north, without declination, is the *agonic* line. The agonic line is slowly moving westward. It now passes near Lansing, Michigan; Cincinnati, Ohio; and Charleston, S. Carolina. (See Fig. 184.) At all points in the United States and Canada east of the agonic line the declination is *west*, at points west of the agonic line the declination is *east*.

211. The Dipping Needle.—Mount an unmagnetized steel needle on a *horizontal* axis so as to be in neutral equilibrium, that is, so as to remain balanced in any position in which it is left. Upon being magnetized and placed so that it can swing in a north and south plane, the north-seeking pole will now be found to be depressed, the needle forming an angle of nearly 70° with the horizontal. (See Fig. 185.) The position assumed by the needle indicates that the earth's magnetic field instead of being horizontal in the United States *dips* down at an angle of about 70°. Over the magnetic pole, the *dipping needle* as it is called, is vertical. At the earth's equator it is nearly horizontal. *The angle between a horizontal plane and the earth's magnetic lines of force is called the inclination or dip.*

212. Inductive Effect of the Earth's Magnetic Field.—The earth's magnetic lines of force are to be considered as filling the space above the earth, passing through all objects on the surface and into and through the earth's interior. The *direction* of the earth's field is shown by the compass and the dipping needle. Magnetic lines of

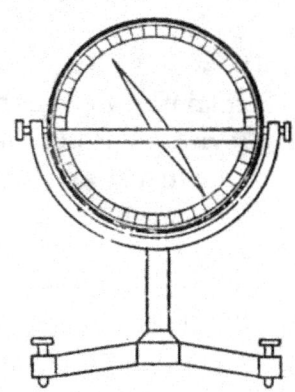

FIG. 185.—A dipping needle.

force tend to crowd into and follow iron and steel objects on account of their permeability. Therefore, iron or steel objects, such as posts, columns, etc., are permeated by the earth's lines of force, which in the United States enter at the top of these objects and leave at the bottom. The lines of force passing through these bodies arrange their molecules in line or magnetize the bodies. The *inductive effect* of the earth's magnetism indicates how lodestones or natural magnets acquire their magnetized condition. So far as is known, magnetism produces no effect upon the human body. It can therefore be studied only by observing its effects upon magnets or bodies affected by it.

Important Topics

The earth's magnetic field, dip, declination, agonic line, induction by the earth's field.

Exercises

1. How would a dipping needle be of assistance in locating the magnetic poles of the earth?
2. Will a dipping needle weigh more before or after it is magnetized? Explain.
3. It is said that *induction precedes attraction*. Using this idea, explain how a magnet attracts a piece of soft iron.
4. Devise an experiment to show that a piece of iron attracts a magnet just as a magnet attracts a piece of iron.
5. Give two methods for determining the poles of a magnet.
6. State three of the most important points in the theory of magnetism. What evidence supports each?
7. Why is a permanent magnet injured when it is dropped?
8. Name two important uses of the earth's magnetic field.
9. What magnetic pole would you find at the top of an iron post that has stood for some time in the ground? What pole at the bottom? How would you test this?

CHAPTER X

STATIC ELECTRICITY

(1) ELECTRIFICATION AND ELECTRICAL CHARGES

213. Electrical Charges.—The ideas gained in the study of magnetism are of assistance in the study of electricity in giving some fundamental ideas and principles that will often be referred to as a basis for comparing the actions of magnetized and electrified bodies. The process of electrifying a body is very different from that of magnetizing it. Thus if a rubber comb or rod be rubbed with a woolen cloth the object rubbed is able to attract to itself light bits of paper, thread, etc. This peculiar attraction was noticed and recorded by the ancient Greeks, 600 B. C., when it was found that amber when rubbed wou'd attract light objects to itself. For a long time it was supposed that amber was the only substance showing this property. Dr. William Gilbert, however, discovered that the electrified condition could be produced by rubbing a great variety of substances. He named the *result* produced, *electrification*, after the Greek name for amber (*elektron*). A body like hard rubber or amber which will attract light objects when rubbed is said to be *electrified*, or to have been given a *charge* of electricity.

214. Law of Electric Action.—Let a vulcanite rod be electrified by rubbing with a woolen cloth until it will attract light objects; then place it in a wire stirrup suspended by a silk thread. If a second vulcanite rod is similarly electrified and brought near the first, the two

will be found to repel.   (See Fig. 186.)   If now a glass rod
be rubbed with silk and brought near the suspended rod,
the two will *attract*.   This difference in behavior indicates
a difference in the electrification or charge upon the rods.
The two charged vulcanite rods repelling and the charged
glass and vulcanite attracting indicate
*the law of electric action.   Like charges
repel each other and unlike charges attract
each other.*   Extensive experiments with
all kinds of substances indicate that
there are but two kinds of electrical
charges.   The electrical charge upon
glass when rubbed with silk or wool is

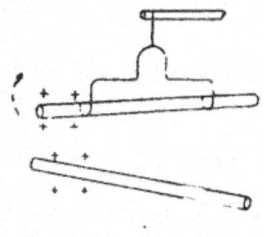

FIG. 186.—Repulsion
of like charges.

called *positive*, and that upon hard rubber or vulcanite
when rubbed with wool is called *negative*.

215. The Electroscope and its Uses.—An electroscope
is a device employed to test the pres-
ence of an electrical charge.   The
aluminum foil electroscope consists of a
flask closed by a rubber stopper
through which passes a rod which ends
at the top in a ball or plate and below
is attached two narrow leaves of thin
aluminum-foil.   Ordinarily the two

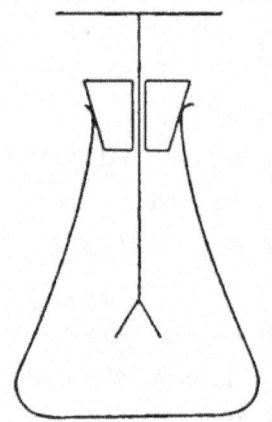

FIG. 187.—An aluminum
foil electroscope.

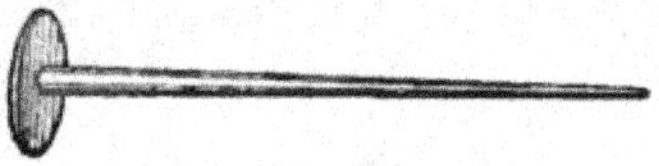

FIG. 188.—A proof plane.

leaves hang close together and parallel but if a charged
body is brought near the electroscope the leaves
spread apart at the bottom. (See Fig. 187.)   The
kind of charge upon a body may be determined with an
electroscope as follows: Make a *proof-plane* by sealing a

small metal disc on the end of a hard rubber rod. (See Fig. 188.) Touch the disc of the proof-plane first to a charged rubber rod and then to the top of the electroscope. The leaves of the latter will separate showing that the electroscope is charged. This charge remains after the proof-plane is removed. If the charged vulcanite rod is brought near the electroscope, the leaves separate further That is, a charge *like* that on the electroscope makes the leaves separate further. But if an *unlike* charge, as that on a positively charged glass rod, is cautiously brought near, the leaves will be seen to move together.

216. Two Charges are Produced at the Same Time.— A closely fitting woolen cover or cap some 3 in. long is made for the end of a vulcanite rod. A silk thread attached to the cap enables one to hold the latter while the rod is turned within it. (See Fig. 189.) If the rod bear-ing the cap is held near a

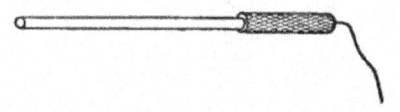

FIG. 189.—Rod with woolen cap.

charged electroscope little or no effect is noticed. If now the cap is removed by the silk thread and held near the electroscope, it will be found to be positively charged while the rod is negatively charged. The fact that no result is seen when the cap and rod are together, indicates that one charge neutralizes the other. In other words, *the charges* must *be equal*. This illustrates the truth that *when electrification is produced by friction, the two objects rubbed together acquire equal and opposite charges.*

217. Charging by Contact and Conduction.—If a small pith ball is suspended by a silk thread, a charged rod brought near is at first attracted, but after contact is repelled (see Fig. 190) showing that the ball has become charged with the same kind of electrification that is upon the rod. That is, a charge given to an object by *contact* with a charged body is of the *same kind* as that upon the charged one. The proof-plane in Art. 215 carries the same kind of charge that is upon the rod it is charged from. Some substances have the ability to transfer charges of electrification. These are called *conductors*, those that do not conduct electrification are *insulators*. The conducting power of a body is readily tested by placing one end of a rod of the material upon the top of an electroscope and the other end upon an insulated support, as in Fig. 191. If now a charge be put in contact

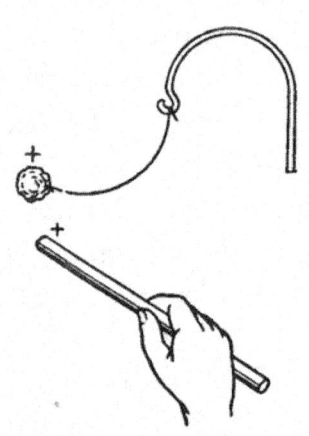

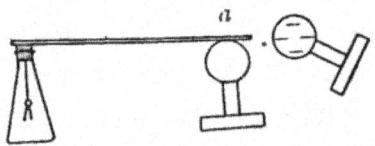

FIG. 190.—The pith ball charged by contact is repelled.

FIG. 191.—Testing for conductivity.

with the body of *a*, the electroscope will show by its leaves whether the rod tested conducts or not. The leaves separate instantly when conducting substances are tested, while no action results with insulators. In testing some materials for conductivity the leaves are found to diverge gradually. Such bodies are said to be *poor* conductors. All degrees of conductivity are found. The metals are the best conductors. The best insulators are rubber, mica, shellac, glass, silk, porcelain, parrafin, and oils.

Important Topics

1. Positive and negative changes. Law of electric action.
2. Electroscope and its uses.
3. Conductors and insulators.

Exercises

1. Is air a conductor? Give reasons for your answer.
2. Mention two points of likeness and two points of difference between magnetism and electrification.
3. If you were testing the electrification of a body with a charged pith ball suspended by a silk thread, would attraction or repulsion be the better test? Give reasons.
4. Have you ever produced electrification by friction outside of a laboratory? Explain.
5. Are the rods upon which we produce electrification by friction, conductors or insulators? How do you explain this?
6. Are conductors or insulators of the greater importance in practical electricity? Explain.

(2) Electric Fields and Electrostatic Induction

218. Electrical Fields.—In our study of magnetism we learned that a magnet affects objects about it by its mag-

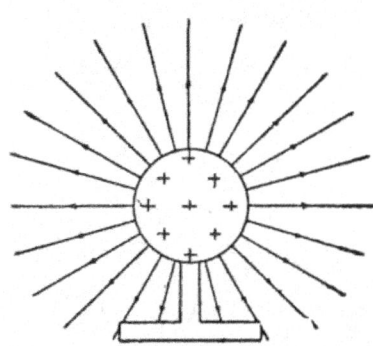

netic lines of force. In a similar way it is assumed that a charged body produces electrical effects upon its surroundings by *electric lines of force.* For example,

Fig. 192.—An electric field about a positively charged shell.

Fig. 193.—A "detector."

the attraction that a charged body exerts upon light objects through short distances or the influence of a charge upon an electroscope several feet away, is said to be due to the *electric field* about the charged body. (See Fig. 192.) The presence of the electric lines of force may be shown by placing a perforated, slender, diamond-shaped piece of tissue paper upon a light glass pointer (Fig. 193). When placed in an electric field the tissue paper "detector" places itself parallel to the lines of force. Electric lines of force are said to extend from a positive to a negative charge. (See Fig. 194.) The direction shown by the arrow upon the

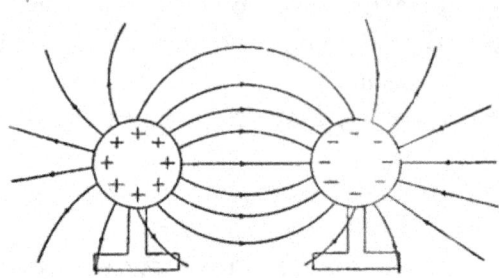

lines is that along which a small positive charge tends to move. Electric lines of force unlike those from magnets are not continuous. They extend from a positive charge to a negative charge. Therefore each

FIG. 194.—Electric field between unlike charges.

positive charge is connected by lines of force to a negative charge somewhere. These ideas of electric fields are of much assistance in explaining many electrical effects. Electrical fields between *oppositely* charged shells will be found similar to Fig. 194, while between shells with like charges, fields are found as in Fig. 195.

219. Electrostatic Induction.—If a charged body is brought near an aluminum-foil electroscope, the leaves separate. (See Fig. 198.) The nearer the charge is brought the wider the leaves spread,

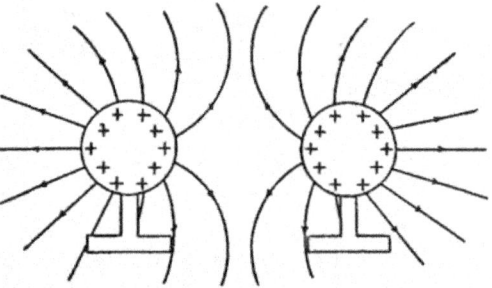

FIG. 195.—Electric field between like charges.

but when the charge is removed, the leaves collapse showing that nothing was given to the electroscope. It was simply affected by the charge in its vicinity. *This production of an electrified condition in a body by the influence of a charge near it is called electrostatic induction.* Placing insulators, such as a sheet of glass, between the charge and the electroscope does not affect the result, which is apparently brought about by the action of the electric lines of force. These lines of force extend without difficulty *through uncharged insulators* and terminate often at the surface of a conductor, where their influence causes a charge to accumulate. *Charged insulators*, however, do affect inductive action. This may be noticed by using a sensitive electroscope.

220. Electrical Separation by Induction.—The action just described may be illustrated further by taking two insulated, uncharged brass shells, *A* and *B*. (See Fig. 196.) Bring a charged vulcanite rod near shell "*A*"

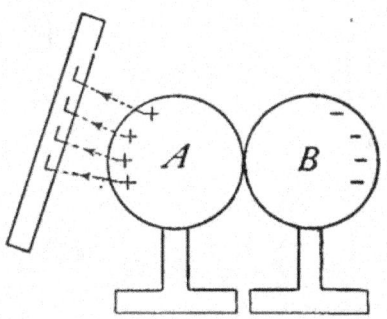
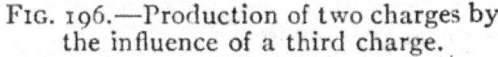
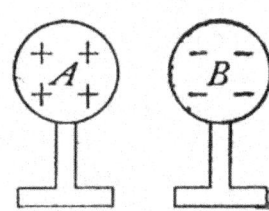

FIG. 196.—Production of two charges by the influence of a third charge.

FIG. 197.—The two charges separated.

while the shells are touching each other. Then remove shell *B* (Fig. 197) while the rod remains near *A*. On testing the shells for electrification, *A* is found to possess a positive charge. This action is in some respects similar to magnetic induction, for if one places a north-seeking pole near a piece of iron, the iron develops by induction a south-seeking pole at the end nearest the magnet and a north-seeking at the other end. There is, however, one striking difference. If the magnetized iron be separated into two parts, each part is a complete magnet possessing two unlike poles; while if the object affected by electrostatic induction is separated into two parts *one part* has a *positive* charge and the other a *negative* charge.

221. Charging a body by induction is easily accomplished. To charge an aluminum-foil electroscope by induction bring *near* (say 10 cm.) from the top of the electroscope a charged rubber rod. (See Fig. 198.) The separated leaves show the presence of the repelled or *negative* charge, the *positive* charge being on the disc at

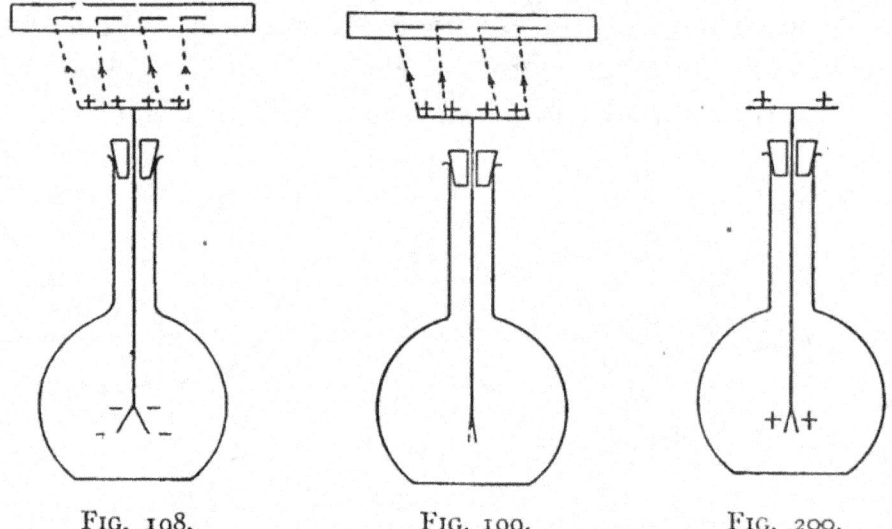

FIG. 198. FIG. 199. FIG. 200.

FIG. 198.—Effect of a charged rod near an electroscope.
FIG. 199.—When a finger is touched to the top of the electroscope, the repelled negative charge escapes.
FIG. 200.—The electroscope is now positively charged.

the top. If while the charged rod is held near, the metal top of the electroscope is touched by the finger the leaves at once fall together showing that the repelled negative charge has escaped from the electroscope (Fig. 199). On removing *first* the *finger* and next the charged rod, the positive charge spreads over the metal parts of the electroscope, as is shown by the separation of the leaves (Fig. 200). The electroscope is now *charged positively* by induction. If the charged rubber rod is brought to about 30 cm. from the electroscope, its leaves tend to move together. If a body charged similarly to the electroscope or *positively*, is moved toward the electroscope the leaves separate further. This behavior of the electroscope enables one to determine the *kind* of charge upon a body.

Two principles of *electrostatic* induction may now be stated: (1) Two *equal, unlike* charges are always produced by *electrostatic* induction.

(2) If the body affected by induction is connected to the earth by a conductor, the repelled or *"free"* charge is conducted away from the body while the *"bound"* charge is held by the inducing charge.

These principles apply in every case of induction.

Important Topics

1. Electric lines of force. Characteristics (3).
2. Electrostatic induction. Principles (2).
3. Charging by induction. Explanation.

Exercises

1. What are elctric lines of force? Where are they found? What does the arrow mean upon the lines?
2. Name three effects produced by electric fields.
3. Does electrostatic induction occur outside of laboratories? Where? When?
4. Given a charged rubber rod, how may one charge from it by induction, insulated brass shells, giving some a positive and some a negative charge?
5. How may the charges upon the shells be tested?
6. In charging an electroscope by induction, why must the finger be removed before the glass rod?
7. Why is it best to have the rubber and glass rods, used in electrification, warmer than the air of the room in which the experiments are being performed?
8. When a sharp metallic point is held near the knob of a charged electroscope the leaves quickly come together. Explain.

9. Might one of the members of your class in physics be charged with electricity, if he should stand on a board supported by dry glass insulators? Explain.

10. If a metal can is charged strongly while standing on an insulator, tests made by means of the proof-plane and electroscope show no charge on the inside. Explain.

(3) ELECTRICAL THEORIES AND DISTRIBUTION OF CHARGES

222. Franklin's Theory of Electricity.—We have studied the production of electrification by friction and induction. It will be helpful now to consider some of the theories of electricity. From the ease with which electrification moves along a conductor, many have imagined that electricity is a fluid. Benjamin Franklin's *One Fluid Theory* held that a *positive* charge consisted in an accumulation or an excess of electricity while a *negative* charge implies a deficiency or less than the usual amount. This theory led to representing positive electrification by a plus (+) sign and *negative*, by a minus (−) sign. These signs are in general use to-day. The use and significance of these signs should be clearly fixed in mind.

223. The Electron Theory.—Various discoveries and experiments made in recent years indicate, however, that *negative* electricity consists of little *corpuscles* or *electrons* which may pass readily from one molecule of a conductor to another while their movement through an insulator is much retarded if not entirely prevented. This theory, sometimes called the *Electron Theory*, holds that each atom of a substance has as a nucleus a corpuscle of *positive* electricity, and surrounding it, minute negative corpuscles or electrons. It is thought that the electrons in the atom are very much smaller than the positive charges and are revolving about the latter with great rapidity. Ordinarily, the positive and negative charges are equal so that the atom is in a neutral or uncharged condition. By the action of various forces some of the *negative* corpuscles within a conductor may be moved from molecule to molecule. Thus if a negatively charged rod is brought near a conductor, many electrons stream away to the far end charging it *negatively*, while the nearer end of the conductor is left with fewer electrons than usual along with the fixed positive corpuscles. Hence the near end is positively charged. (See Fig. 198.) On the other hand, if a positive charge is used, it attracts the electrons from the far end, leaving the immovable positive corpuscles there, and that end becomes positively electrified, while the nearer end with its surplus of electrons is, of course, negatively electrified.

The Electron Theory is considered well founded since the electrons have (a) had their *mass* determined, (b) their *speed* measured, (c) their *electric charge* determined, (d) and their *behavior* while *passing through magnetic* and *electric fields* observed. These facts and other experimental evidence have demonstrated the existence of electrons. The positive corpuscle has not been directly observed but is assumed to exist to account for the effects observed in induction, charging by friction, etc.

224. Distribution of an Electric Charge upon a Conductor.—We have applied the electron theory in explaining the phenomenon of electrostatic induction. Let us now use it in studying the distribution of an electric charge upon a conductor. Let a cylindrical metal vessel open at the top and insulated by being placed upon pieces of sealing wax have a charge of negative electricity given it. (See Fig. 201.) On now taking a proof plane and attempting to obtain a charge from the *interior* of the vessel no result is found, while a charge is readily obtained from the *outside* of the dish. This result is explained by considering that the electrons are mutually self-repellant and in their attempt to separate as widely as possible pass to the outer surface of the vessel. This same condition is also true of a dish made of woven wire. If the charged conductor is not spherical in outline, an uneven distribution of the charge is observed. Thus if an *egg-shaped* conductor

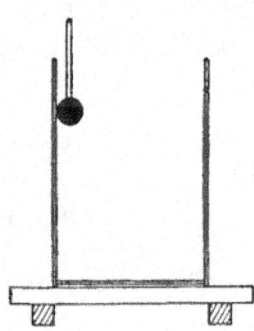

is insulated and charged (see Fig. 202), a proof plane touched to the broad end of the body and then to an electroscope causes a certain divergence of the leaves of the latter. If now a charge be taken from the *pointed* end by the proof plane to the uncharged electroscope, a greater spreading of the leaves than before will be noticed. This indicates that the electricity may be unevenly distributed over the surface of a body. It is found

FIG. 201.—No charge is found inside a hollow vessel.

that the *electric density*, as it is called, is greatest where the surface curves most sharply. At a very sharp curve, as at a point, the electric density may be so great that a part of the charge escapes into the air. (See Fig. 203.) For this reason electric conductors on which it is desired to *keep* an electric charge have round surfaces and all sharp points and corners are avoided. While conductors, such as lightning rods, which are designed to facilitate the escape of electric charges, are provided with a number of sharp points at the end or elsewhere. At such points, air particles are drawn forcibly against the point and after being charged are driven away strongly, creating

FIG. 202.—More charge at the pointed end.

the so-called *electrical wind* which carries away the charge at a rapid rate. (See Fig. 203.)

225. Lightning and Electricity.—The fact that lightning is an electrical discharge was first shown in 1752 by

Benjamin Franklin, who drew electric charges from a cloud by flying a kite in a thunderstorm. With the electricity which passed down the kite string he performed a number of electrical experiments. This discovery made Franklin famous among scientific men everywhere. Franklin then

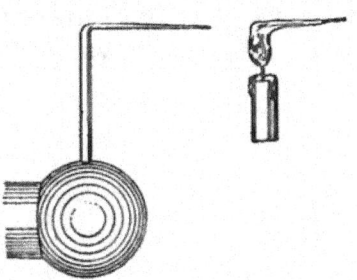 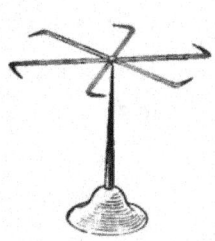

FIG. 203.—Electrical wind produced by a pointed conductor.

FIG. 204.—Electrical whirl. The reaction from the electrical wind causes it to revolve.

suggested the use of lightning rods to protect buildings from lightning. These rods act as conductors for the electric discharge and thus prevent it from passing through the building, with the risk of overheating some part and setting the latter on fire. The points provided at the top of lightning rods are believed to aid in preventing strokes of lightning by the *silent discharge* of the so-called electric wind which tends to quietly unite the charges in the clouds and on the earth beneath.

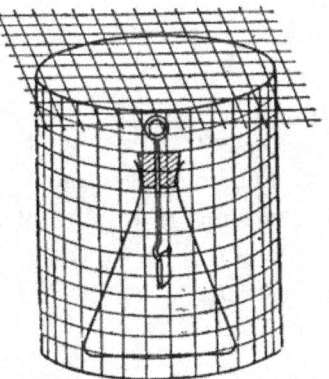

FIG. 205.—The wire screen protects the electroscope.

The charge in an electrified cloud acts inductively upon the earth beneath, attracting an opposite charge to the objects below. The discharge from the cloud often passes to the objects beneath, such as trees or buildings. *Thunder* is believed to be due to the sudden expansion of the air when intensely heated by the electric discharge and its sudden contraction, like a *slap*, as the track instantly cools. Thunder at a distance is usually followed by rumblings due to changes in the intensity of the sound mainly due to reflections of sound waves from clouds and other reflecting surfaces.

226. An electric screen is a device for cutting off the influence of an electric charge. Faraday found that if a sensitive electroscope is surrounded by a wire mesh screen (see Fig. 205), no evidence of electrification could be found inside. In other words, a network of conductors on a building makes the best protection against lightning, provided it is connected to the earth by good conductors at several places.

Important Topics

1. Electrical theories. Evidences for electron theory.
2. How is the theory used in explaining induction?
3. Charges, and distribution on conductors (effect of shape).
4. Lightning: cause, effects, lightning rods.

Exercises

1. In what respects is Franklin's one-fluid theory like the electron theory? In what respects different?
2. Consider two shells charged by induction from an electrified rubber rod, one positively and one negatively. Explain the process, using the ideas of the electron theory.
3. Should the metal top of an electroscope have sharp corners? Explain.
4. Would a tall steel tower have the same need of a lightning rod as a brick chimney of the same height? Explain.
5. Will a solid sphere hold a greater charge of electricity than a hollow one of the same diameter? Explain.
6. If a positively charged cloud floats over a tree which is a good conductor of electricity will the tree be charged? Show diagram. Explain.

(4) POTENTIAL, CAPACITY AND THE ELECTRIC CONDENSER

227. Conditions Causing a Movement of Electricity.—
In the study of conductors and insulators it was observed
that an electric charge moved along the conducting rod
to the electroscope. This *movement* of *electricity* along a
conductor is a result of great practical importance. We
will now consider the conditions that produce the "flow"
or "current" of electricity.
Let two electroscopes stand
near each other. Charge
one, C' (Fig. 206), strongly
and charge the other
slightly. If now a light
stiff wire attached to a
stick of sealing wax be
placed so as to connect the
tops of the electroscopes,
the leaves of C will partly
close while those of D will

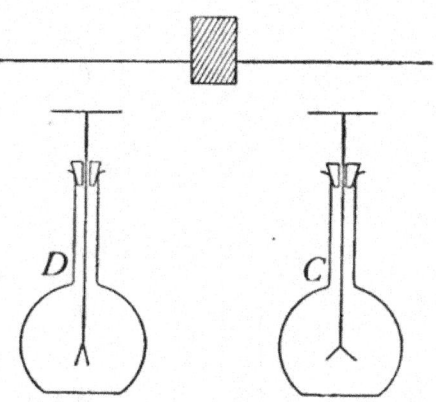

FIG. 206.—Electricity flows from
high to low potential.

open slightly, thus indicating a movement of electricity
from C to D along the wire. The movement was from
a place of greater degree of electrification to one of less.

228. Potential.—The *potential* of an electrified body is
its *degree* of *electrification*. Therefore, it is said that
electroscope C mentioned above has a greater potential
than electroscope D. The movement of electricity is from
a place of greater or *high* potential to one of lesser or *low*
potential. If two bodies are at the *same* potential there
will be found no movement of electricity between them.
A *difference* of *potential* between two points connected by
a conductor is therefore the *necessary condition* for an
electric current. Just as heat is transmitted along a con-

17

ductor from a place of high to one of lower temperature, so electricity is transmitted along a conductor from a place of high to one of low potential. Thus potential in electricity corresponds to temperature in heat. One is the "degree of electrification," the other, "the degree of hotness."

229. Electrical pressure is a term sometimes used for difference of potential. To better understand electrical

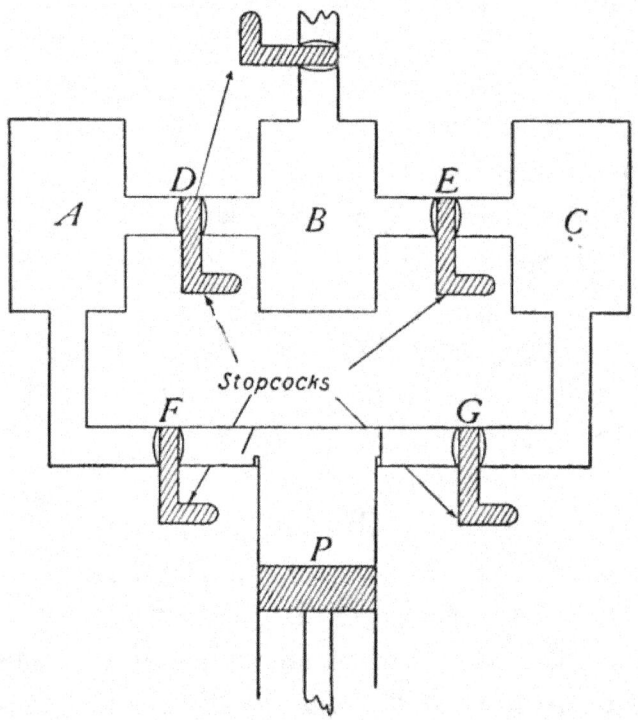

FIG. 207.—Air pressure apparatus to illustrate electrical pressure.

pressure, consider three round tanks (Fig. 207) containing air. *A* is a tank holding air at 10 lbs. pressure per square inch, above atmospheric pressure, *B* is open to the air and hence is at atmospheric pressure while *C* has a partial vacuum, with 10 lbs. less pressure than that of the atmosphere. If the valve at *D* or *E* is opened a flow of air sets up until the pressures are equalized. While if the pump at P is working a difference in pressure is easily maintained. Tank A corresponds to an insulated body charged to a high *positive* potential; tank B, open to the air, a body connected to the earth; while tank C represents a body having a *negative* potential. The earth is said to have *zero potential*.

Now just as compressed air will be pushed into the atmosphere (as from A to B) while air at atmospheric pressure will if possible be forced itself into a partial vacuum (as from B to C), so electricity at a positive potential will tend to move to a place at zero potential, while that at zero potential tends to move to a place of negative potential. Bodies at the *same* potential as the earth, or at zero potential, are also said to be *neutral*. Those positively electrified have a positive potential, those negatively electrified have a negative potential. As in gases, movement always tends from higher pressure (potential) to lower pressure (potential.)

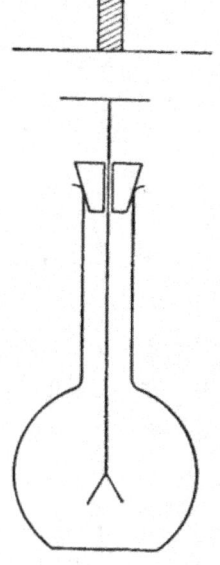

FIG. 208.—The metal plate gives the electroscope a greater surface and hence greater capacity.

230. Capacity.—If we have a 100-gallon tank and a 10-gallon tank connected by a pipe both filled with compressed air, the larger tank will contain ten times as much air as the smaller at the *same pressure* since it has ten times the capacity, or, if the two tanks are separated and the same amount of air is contained in each, the pressure of the air contained in the small tank will be ten times that in the large one.

The *electrical capacity* of a conductor is in some respects similar to the capacity of a tank for air. Since, however, electrical charges are upon the surface of a body, its capacity depends in part upon the extent of surface. For example, if a charge is taken from a charged rubber rod by a proof plane to an electroscope a certain divergence of the leaves will be noticed. If a circular metal plate several times the diameter of the top of the electroscope is laid upon the latter (see Fig. 208), and a charge equal to that used before is brought to the electroscope, the leaves show less divergence than before, showing that the *same charge gives a lower potential when placed upon a body of greater capacity.*

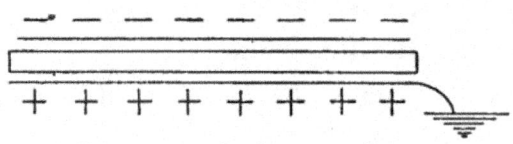

FIG. 209.—A plate condenser.

231. The electric condenser is a device having a large electrical capcatiy consisting of parallel conductors separated by good insulators. It has been devised to enable one to obtain a large electrical charge upon a body of convenient size. Such an apparatus is of great practical value in many experiments and operations. Its construction involves the principle of electrostatic induction in which a charge of one kind attracts and "holds" strongly a charge of opposite kind near it. In its simplest form it consists of two parallel conductors separated from each other (Fig. 209). The upper

FIG. 210.—A condenser of several plates.

plate has been charged negatively. This has given the lower plate a positive charge by induction, since the latter is connected to the earth. These positive and negative charges hold or "bind" each other so that a large quantity may be accumulated. To increase the capacity of a condenser, several plates are used connected as in Fig. 210.

It is a curious fact that the kind of insulator between the

charged conductors of a condenser affects its capacity. Thus if glass, paraffine, or beeswax is between the plates instead of air, the plates will "hold" more electricity at the same potential. For this reason condenser plates are often separated by sheets of glass, paraffined paper, or mica.

232. The Leyden Jar.—A convenient form of condenser, used as long ago as 1745, is the Leyden jar. It consists of a glass jar (Fig. 211) coated part way up, inside and out, with tinfoil. The inner coating is connected by a chain to a knob at the top. The Leyden jar is charged by connecting the outer coating to the earth while to the inner coating is given a charge of either kind of electricity. The other kind of charge is developed by induction upon the outer coating, and each charge binds the other. To discharge a jar, a

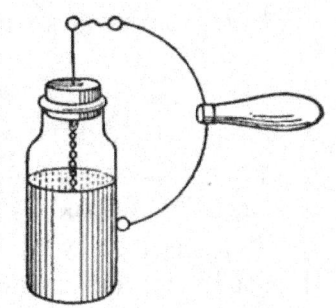

FIG. 211.—A Leyden jar and a discharger.

conductor, as a wire, is connected first to the outer coating and held there while the other end is brought to the knob at the top. A bright spark is produced when the two charges combine. It is best not to let the discharge from the jar pass through the body unless one is certain that only a very small charge is present.

233. Oscillatory Discharge.—The discharge from a Leyden jar is an interesting phenomenon. The rush of electricity from one coat to the other does not stop when the two coats are exactly neutralized but continues until the two plates are charged just oppositely to their condition at first, then a rush of electricity in the opposite direction occurs. This alternation continues several times and constitutes what is called the *oscillatory discharge*. (See Fig. 414.) This oscillatory discharge sets up waves in the ether. These are called *Hertzian* waves in honor of their discoverer, Heinrich Hertz. They are the ether waves used in wireless telegraphy. A *lightning flash* has been shown by photographs and by other means to be oscillatory. This fact supports the idea that the electrical conditions just preceding the stroke of lightning reproduce a condenser on a large scale. The charged cloud is the upper charged plate, the earth beneath, charged by induction from the cloud, is the lower charged plate, while the air between is the insulator or *dielectric* as it is sometimes called.

Important Topics

1. Potential: high, low, zero, positive, negative, similar to temperature and air pressure.
2. *Capacity* affected by (1) area, (2) induction.
3. Condensers, Leyden jar, parallel plate.
4. Oscillatory discharge, conditions, results.

Exercises

1. Is the air a conductor? Explain.
2. Can the Leyden jar be strongly charged if the outer coat is insulated? Explain.
3. Upon what two conditions does the capacity of a body depend? How in each case?
4. Would a lightning discharge produce wireless waves? Explain.
5. If a sharp tack be dropped point up on the plate of an electroscope the latter is quickly discharged. Explain.

(5) Electrostatic Generators

234. Static Electric Machines.—Many machines have been invented to produce larger quantities of static electricity than we have used in the experiments previously described. One of the earlier of these was the *plate friction machine* in which a large circular glass plate was rotated while a pad of some material was held against it. This machine was capable of producing powerful effects, but it took much work to turn it, and it has been abandoned for a more efficient device, the *static induction machine.*

235. The electrophorus is the simplest static induction generator, consisting simply of a flat circular plate of some insulating material, as paraffine, shellac, or

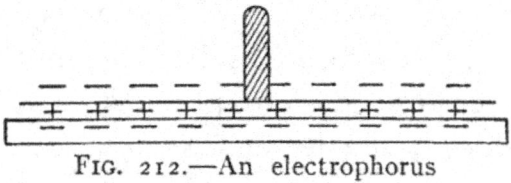

FIG. 212.—An electrophorus

rosin contained in a metal pan, and a flat circular metal *disc* having an insulating handle.

The electrophorus is used as follows: The plate is first electrified by rubbing or beating with fur or a woolen cloth. The plate will be found to be charged negatively. The metal disc is placed upon the plate by holding the insulating handle. The upper surface of the charged body is slightly uneven so that the disc touches but a few high points. The greater part of the charged surface is sepa-

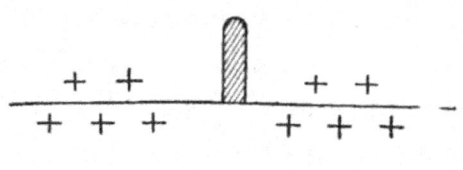

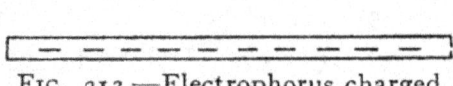

FIG. 213.—Electrophorus charged.

rated from the metal disc by air, a good insulator. The charge therefore acts inductively upon the disc *repelling negative* electricity to the upper surface of the disc, leaving the lower surface charged positively (Fig. 212). If now the finger is touched to the disc the repelled negative charge escapes and the whole disc is left positively charged. The disc is now removed (Fig. 213) and the charge upon it may be tested or used in any desired manner. The disc may be recharged many times without rubbing the plate again.

These electrical charges possess *energy.* What is the

source of this energy? The answer may be determined by the following experiment. Place the disc upon the charged plate. Touch the disc with the finger to remove the repelled charge. Connect an electroscope to the disc by a fine wire. Nothing appears on the electroscope, since the disc has been connected to the earth, and is therefore at zero potential. If now the disc is lifted slowly, the leaves of the electroscope gradually separate, showing that a charge of electricity appears when the disc is being lifted against the force of attraction between the two charges. Just as potential energy is developed in a weight when it is lifted against the earth's attraction so electrical energy appears in the disc while it is being separated from the plate. The electrical energy of the charge is therefore due to the work done in separating the two charges. This electrical energy appears as heat and light, when the disc is discharged. It may be employed to ignite gas, gunpowder, etc.

236. The Toepler-Holtz Induction Machine.—This is a type of induction or influence machine that is often used for producing a continuous supply of electricity as in the operation of "X" ray machines, in lecture demonstrations, etc. This machine (Fig. 214) consists of two discs: one fixed, the other mounted so as to revolve. Upon the back of the fixed plate are two sectors of tinfoil which become charged oppositely. Upon the revolving plate are six metallic discs. These discs act like the discs of the electrophorus. They become charged by induction from the charges upon the sectors fastened to the fixed plate. The brushes held by a rod touch the discs at just the right time to take off the repelled charge. The charges induced upon the discs are taken off by two metal combs whose points are held close to the revolving disc. The Leyden jars assist in accumulating a good strong charge before a spark passes between the terminal knobs. Some machines are built up of several pairs of plates and give correspondingly large amounts of electricity.

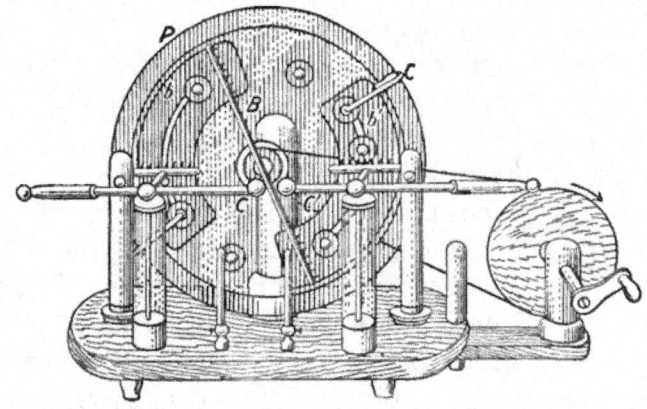

FIG. 214.—The Toepler-Holtz induction machine.

Important Topics

Static Electric Generators.—(a) plate friction machine, (b) electrophorus, (c) induction or influence machine.

Exercises

1. Potential is similar to what other terms that we have studied?
2. What three electrical phenomena are better understood from a study of the lines of force?
3. How many charges may be produced by an electrophorus before the plate needs to be electrified again? Explain.
4. The static induction machine is often called a "continuous electrophorous." Why?
5. The Leyden jars used with the induction machine cause much brighter sparks to be produced than without them. Explain.
6. With the Leyden jars removed, would the frequency with which the sparks pass between the knobs be increased or decreased? Explain.
7. Mention three likenesses and three differences between magnetism and static electricity.
8. Will you receive a greater shock by touching a knob of a charged Leyden jar when it is held in the hand or when it is standing on a sheet of glass? Explain.
9. In what way may an electric charge be divided into three equal parts?

Review Outline: Magnetism and Static Electricity

Comparison between Magnetism and Static Electricity.

Substances are: { magnetic,   { conductors,
             { non-magnetic. { insulators.

Produced by: induction.   friction, or induction.

Theory: molecular.   electron.   (fluid)

Fields of Force
Explain:
{ attraction,   { attraction,
  repulsion,   { repulsion,
  induction,   { induction.
  action of compass

Terms:
{ magnetoscope, dip   { electroscope, electron,
  declination, pole,   positive, negative,
  retentivity,   { potential, capacity,
  permeability,   condenser, electrophorus,
  lodestone,   oscillatory discharge,
  magnetic meridian. { lightning. ·

Likeness:
  both are:
{ *a*-produced by induction, *b*-attract and repel,
  c-have fields of force.

Differences:
{ *a*-electricity can be *conducted*, magnetism
     cannot.
  b-electricity in *all substances*, magnetism in few.
  c-magnetism with the compass indicates
     direction.

CHAPTER XI

CURRENT ELECTRICITY

(1) ELECTRICAL CURRENTS AND CIRCUITS

237. Sources of Electric Currents.—In studying the production and distribution of static electricity it was seen that if two bodies at *different potentials* are connected by a copper wire a *movement of electricity to the body* having the *lower potential* occurred along the conducting wire. This movement of electricity is called an *electric current* (Art. 227). *A difference of potential* is therefore often called an *electromotive force* (E.M.F.), since it produces the movement of electricity in a conductor. The current between two *oppositely charged* bodies lasts for so short a time as to be of little or no practical value unless some means are found for continually recharging the bodies. That is, some device must be used to restore the difference in potential as fast as the conducting wire equalizes it. The continual charging of the bodies takes work. In other words, it requires a continual expenditure of some form of energy (which is converted into electrical energy) to produce the electric current. Two forms of energy are commonly used for this purpose.

(A) *Chemical energy* is employed in *voltaic cells* for producing electric currents. (B) *Mechanical energy* is used for the same purpose in the *dynamo* and similar devices.

238. The voltaic cell is named after Volta, an Italian physicist, who in 1800 invented it. In its simplest form it consists of a strip of copper and a strip of zinc placed in dilute sulphuric acid (one part acid to fifteen or twenty of water) (Fig. 215). By the use of sensitive apparatus, it can be shown that the copper plate of the voltaic cell has a positive charge and the zinc plate a negative charge. For example, let a flat plate 10 cm. in diameter be placed upon the knob of an electroscope and a similar plate, coated with shellac and provided with an insulating handle, be set upon it to form a condenser. (See Fig. 216.) If now wires from the two plates of a simple voltaic cell be respectively connected to the plates of the condenser,

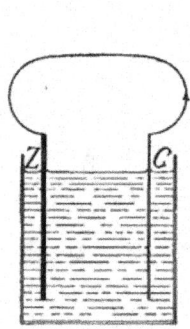

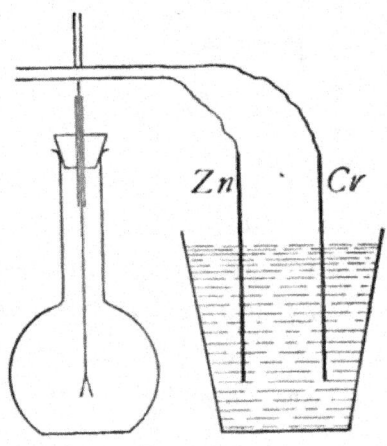

Fig. 215.—Cross-section of a simple voltaic cell.

Fig. 216.—Testing the charges upon the plates of a simple voltaic cell.

charges from the copper and zinc plates will accumulate upon the two condenser plates. Now remove the wires and lift the upper plate. The "bound" charge upon the lower plate will spread over the leaves and cause them to separate. Upon testing, the charge from the zinc plate will be found to be *negative* and that from the copper plate, *positive*. Since a positive charge is found upon the copper plate it is called the *positive electrode;* the zinc plate is called the *negative electrode.*

239. Test for an Electric Current.—If the copper and zinc plates of a voltaic cell are connected by a wire, a current of electricity is set up in the conductor. Evidence of the current may be obtained by holding the conducting wire over and parallel to the needle of a magnetoscope. The needle is deflected by the action of the current parallel to it (Fig. 217). This *magnetic effect* of a current is the means usually employed for the *detection* and *measurement* of an electric current. Such a device which detects an electric current by its *magnetic effect is called a galvanoscope*, in honor of Galvani, who in 1786 was the first to discover how to produce an electric current.

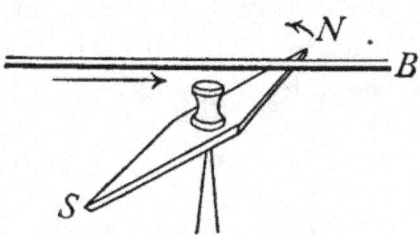

FIG. 217.—The magnetic needle is deflected by the current.

240. The Electric Circuit.—*The entire conducting path along which a current of electricity flows is called an electric circuit.* In the case of a voltaic cell, the circuit includes not only the wires connecting the plates but also the plates themselves and the liquid between them. When some device or apparatus is to receive current from the cell, it is attached to the plates and wires so that the device is a part of the electric circuit. Separating the circuit at any point is called *breaking* or *opening* the circuit, while connecting the ends of an open circuit is called *making* or *closing* the circuit. A device for opening and closing a circuit is called a *key* or *switch*. The electric circuit used in ringing a door bell is familiar to most boys and girls. This circuit is *open* most of the time. It is closed by pressing the *push-button* at the door, and the flow of current through the *electric bell* causes the latter to ring. Such a circuit is represented in Fig. 218. Here C is the

FIG. 218.—Diagram of an electric bell circuit.

voltaic cell, the two lines representing the plates of the cell. A cross-section view of the push-button (*P*), shows how the circuit is closed, (*B*) is the bell. Wherever current electricity is used the device in which it is employed forms a part of an electric circuit extending back to some electric generator. This generator must be able to continually produce an E.M.F., or a difference of potential between its terminals, in order that the movement of electricity may be continuous.

Important Topics

(a) Electric generators: (1) voltaic cell uses chemical energy; (2) dynamo uses mechanical energy.
(b) Electric circuits: (1) open, (2) closed, (3) key and switch.
(c) Voltaic and galvanic electricity (names).
(d) Galvanoscope, uses.

Exercises

1. In what *two* ways are static and current electricity alike? In what two different?
2. Draw a diagram of an electric bell circuit at your home. Give the location of the electric bell, the electric generator and the push-button. Show the connecting wires, and explain briefly how the circuit is operated.
3. Represent some other electric circuit, naming the generator and other devices in the circuit.
4. Look up the work of Volta and Galvani and write a statement of the electrical discoveries and inventions made by them.

(2) The Voltaic Cell and its Action

241. The simple voltaic cell consists of a strip of copper and a strip of zinc placed in dilute sulphuric acid. (See Fig. 219.) A short time after placing the plates in the acid, bubbles of a gas (hydrogen) appear on the surface of the zinc. These bubbles increase in size and some rise to the surface of the liquid. Nothing appears upon the copper plate. If the tops of the plates are connected by a wire, an electric current is set up through the wire and the cell, and bubbles of gas also appear upon the *copper* as well as on the zinc. In a short time the surface of the copper becomes coated with bubbles and the current becomes much *weaker*. If the plates are left in the acid for some time the zinc is found to be eaten away, having been dissolved in the acid through chemical action. The copper, however, remains practically unaffected.

242. How the Current is Produced.—To maintain the electric current a continual supply of energy is required. This is furnished by the *chemical action* of the acid upon the zinc. The chemical action is in several respects like *combustion* or *burning*, by means of which chemical energy is

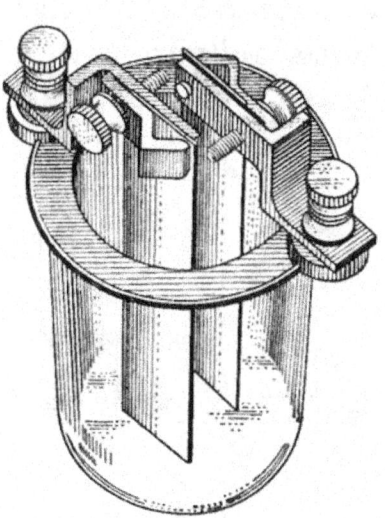

FIG. 219.—A simple voltaic cell.

transformed into heat energy. In the voltaic cell the chemical action of the acid upon the zinc *transforms* chemical energy into electrical energy. The E.M.F. or *difference of potential* may be considered as originating at the surface of the zinc where the chemical action takes place. At this point the zinc has the lower and the liquid in contact with it the higher potential. The molecules of the acid are believed to be separated or broken up into two parts called *ions;* one ion, the SO_4 or *sulphion*, combines with the zinc forming zinc sulphate, the other, or hydrogen (H) ion, passes over to the copper plate, and accumulates on the surface of this plate giving it a positive charge. It is therefore called the *positive ion*.

The sulphion ion, or SO_4 ion, carries a negative charge to the zinc. It is therefore called the *negative ion*.

243. The Direction of the Current. *—Beginning at the surface of the zinc the *direction* of the movement of *positive* electricity may be traced through the liquid to the copper plate, to the wire, to the zinc plate, to the starting point, thus completing the electric circuit. When the circuit is closed it is found that the movement of electricity starts in *all* parts of the circuit at practically *the same instant.*

244. The production of the current may be illustrated by describing a device for producing a continuous circulation of water. Thus let Cu and Zn represent two pipes connected by two horizontal tubes, one at V provided with a valve and one at P with a rotary *Pump*. (See Fig. 220.) Suppose the pipes filled to the level of V and the pump started. The pump will force water from Zn to Cu, through P, the level falling in Zn and rising in Cu.

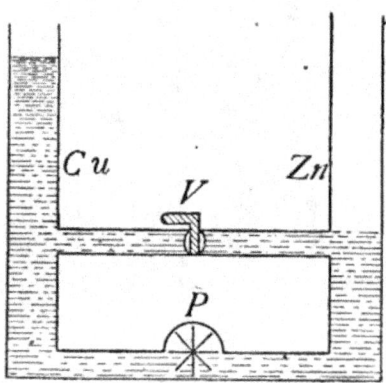

FIG. 220.—A comparison of a voltaic cell and circuit to a water pump and connecting pipes.

If the valve V is open the water will flow back through V as long as the pump is working. If V is closed, the level in Cu will rise as high as the driving force of the pump can send it. If now V is opened, the pump will maintain the water in circulation from Cu to Zn through V. In the illustration, the tubes Cu and Zn correspond to the conducting plates of *copper* and *zinc* of a voltaic cell. The pump P represents the chemical action which produces

* Many scientists consider that current in a conductor consists of *negative electrons* flowing in a direction *opposite* to that described in Art. 243. This is called the *electron current*, as distinguished from the *electric current* described above.

the electrical pressure. The upper pipe represents the part of the circuit outside of the cell, the valve *V* corresponds to an electric key or switch which is used to open and close the electric circuit.

245. Polarization.—In the simple voltaic cell, after the circuit is closed, bubbles of hydrogen collect upon the copper plate. This accumulation of hydrogen gas is called *polarization.* It acts as a non-conducting layer upon the surface of the plate and seriously interferes with the movement of electricity from the liquid to the copper plate not only in the simple voltaic cell but in many others as well. Some voltaic cells are made entirely free from this defect, either (a) *by the removal of the hydrogen as fast as* it is formed, or (b) *by the use of such chemicals that no hydrogen is produced.*

246. Local Action.—It is noticed that when a strip of zinc is placed in dilute acid that bubbles appear upon the surface of the zinc. The appearance of these bubbles indicates that some of the hydrogen ions carrying positive electricity have moved to the zinc plate. Careful examination of the plate after it has been in acid shows numerous black spots upon it. These are bits of carbon. They are always found in ordinary zinc. Small electric currents are set up which run from molecules of pure zinc into the liquid and back to the carbon particles, thus forming small closed circuits. (See Fig. 221.) The formation of these circuits from and to the zinc is called *local* action. This action is a defect in voltaic cells since a part of the current is thus kept from passing through the main outside circuit, and the zinc may be consumed even when no outside current is flowing.

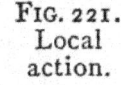

FIG. 221. Local action.

247. Amalgamation.—Local action is prevented by coating the zinc with mercury. This process is called *amalgamation.* The mercury covers the entire surface of the plate in the acid.   Its action is to dissolve pure zinc and bring it to the outer surface where it is acted upon by the acid.   The carbon particles are kept covered so that no local currents can be formed as long as the bits of carbon are below the surface.   Amalgamation therefore prevents local action.

Important Topics

The Simple Voltaic Cell

1. Two plates: zinc, copper; electrolyte, dilute sulphuric acid.
2. Ions: hydrogen, positive: sulphion, negative.
3. Current, where and how produced, direction, illustration.
4. Polarization: cure, local action, cure.

Exercises

1. Write in your own words an account of the production of an electric current by the simple voltaic cell. Use sketches.
2. Which plate has the higher potential?   How is it produced?
3. Would you expect to get an E.M.F. by forming a cell of two copper plates?   Why?

(3) PRACTICAL VOLTAIC CELLS

248. Advantages of Voltaic Cells.—Many forms of voltaic cells have been devised.   Several of the more common of these will be described and their electro-chemical action explained.

At the present time voltaic cells are employed only where small currents are needed, such as for electric bells and induction coils.   Where more than a small amount of current is required, the dynamo and the storage battery have generally taken their place as sources of electric current.

The advantages of voltaic cells as electric generators are: (a) they are inexpensive, (b) they are easily taken from place to place, (c) they may be ready for instant use.

The most desirable voltaic cell would be one having the

following qualities: (a) High electromotive force, (b) no polarization or local action, (c) very low internal resistance, d) small expense, both as to first cost and upkeep.

249. The Leclanché cell is the one commonly used for ringing door bells. It has two plates: one of zinc and the other of *carbon*. These are placed in a solution of sal ammoniac (Fig. 222). Take up the desirable qualities mentioned at the end of the preceding paragraph. (a) It may be shown that this cell has a good E.M.F. about 1.5 volts. (b) It *polarizes* easily yet it recovers well when left upon open circuit. Usually a substance called manganese dioxide is mixed with the carbon. This acts as a *depolarizer* that is, it combines with the hydrogen to form water. (c) Its resistance varies and is often considerable. (d) The expense for upkeep is small, since a

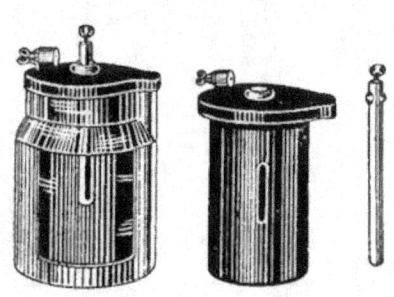

FIG. 222.—The Leclanché cell, "wet" type.

5-cent rod of zinc, and a 5-cent charge of sal ammoniac will keep the cell in action on a bell circuit from six months to a year or more. It is well suited for use on *open circuits* that is, where the circuit is open the greater part of the time and is closed only occasionally; as in ringing door bells, operating telephones, and other devices whose circuits are usually open.

250. The Dry Cell.—Many forms of Leclanché cells are made. One of these is called the *dry cell* (See Fig. 223.) In this cell the zinc plate is made into a jar or can and contains the other materials. At the center of the cell is a rod of carbon and manganese dioxide. The space between the carbon and zinc is filled with a porous material such as sawdust or plaster of Paris. A strong solution of sal ammoniac fills the porous material. The top of the cell is sealed with pitch or wax to prevent evaporation. The great advantage of this cell is that it may be used or carried in any position without danger of spilling its contents. Dry cells are often used to operate the spark coils of gas and gasoline engines. The Leclanché cell described in Art. 249 is commonly known as the "wet cell."

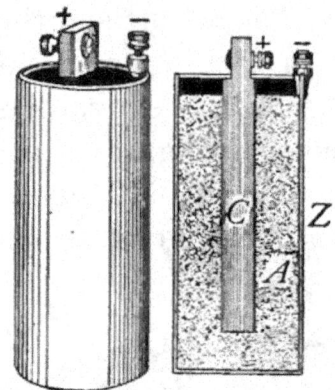

FIG. 223.—The Leclanché cell, "dry" type.

251. The Daniell Cell.—This cell is often used in laboratories, and on closed circuits such as those connected with fire and burglar alarms and telegraph lines. It has two plates of zinc and copper placed in two different liquids which are kept separated by a porous clay cup (Fig. 224). The zinc rod is kept in a solution of zinc sulphate contained in the porous cup. The copper plate is in a solution of copper sulphate filling the rest of the glass jar. Unlike the Leclanché cell, this one must be kept upon a *closed circuit* to do its best work, as the two liquids mix when the circuit is open. Taking its qualities in order, (a) its E.M.F. is about one volt, (b) it has no polarization since copper instead of hydrogen is deposited upon the copper plate. Therefore a uniform E.M.F. may be obtained

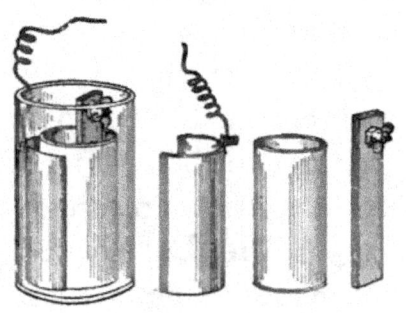

FIG. 224.—The Daniell cell.

from it, making it especially useful in laboratory experiments and tests. (c) Its resistance is considerable and (d) it is more expensive to operate than the Leclanché. It is sometimes used upon closed circuits outside of laboratories as in burglar and fire alarms, although in recent years, the storage battery is taking its place for these purposes.

252. The Gravity Cell.—Fig. 225 is like the Daniell cell in most respects, except that in this cell, the zinc plate is held at the top of the jar in a solution of zinc sulphate while the copper plate is at the bottom, surrounded by a solution of copper sulphate. The solutions mix but slowly as the copper sulphate solution is denser and remains at the bottom. This cell like the Daniell must also be kept upon closed circuit. On account of its simplicity and economy it is often used to operate telegraph instruments. Its qualities are similar to those of the Daniell cell.

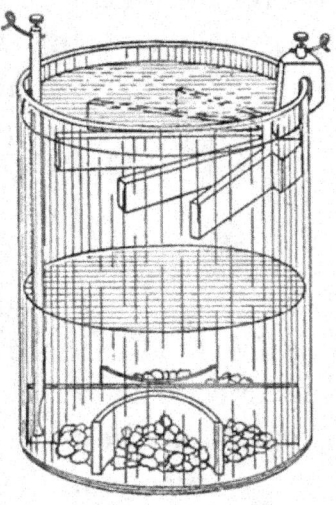

Fig. 225.—The gravity cell.

253. Symbol for Voltaic Cells.—In electrical diagrams, the symbol employed to represent a voltaic cell is a short thick line near to and parallel to a longer thin one. As in Fig. 226 If several cells are to be represented the conventional symbol of the combination is represented as

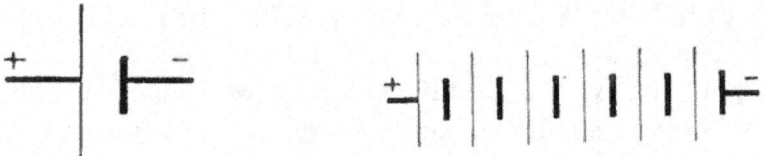

Fig. 226.—Diagram of a single cell.

Fig. 227.—Diagram of a group of cells.

in Fig. 227. A single cell and a group of cells are each frequently called a battery.

254. Effects of Electric Currents.—Having studied some of the devices for producing an electric current, let us now consider some of the *effects* caused by it. These effects will be studied under three heads: (a) *Magnetic*, (b) *Chemical*, and (c) *Heat* effects. Devices or articles showing these effects known to most high school students are respectively: (a) the *electromagnet* (b) *electro-plated silver ware* and (c) *electric heaters*, such as electric flat irons, electric toasters, etc. The *mag-

FIG. 228.—A gal-vanoscope.

netic* effect of an electric current was first detected by Oersted at the University of Copenhagen in 1819. It may be observed by holding a wire carrying a current from a voltaic cell above and *parallel* to the needle of a *magnetoscope*. The needle is at once deflected (Fig. 228). If the current is reversed in direction the magnetoscope needle is deflected in the reverse direction. This simple device is the most common means for detecting an electric current. It therefore constitutes a *galvanoscope*. (See Art. 239.)

Important Topics

1. Leclanché cells, (a), wet, (b), dry, construction, advantages, uses.
2. Daniell and gravity cells, construction, advantages, uses.
3. Three effects of electric currents, illustrations.
4. The galvanoscope, uses.

Exercises

1. Explain how the direction of current in a wire can be determined by a compass.
2. Would you expect to obtain a current from a zinc and copper cell containing a solution of common salt? Perform the experiment.
3. What conditions in a voltaic cell will give a steady electromotive force.
4. What conditions in a voltaic cell will give a strong electromotive force.
5. Name three different electric circuits that you know exist. Which are *open* and which are *closed* circuits?
6. Are voltaic cells used in your home? If so, for what purpose are they used? On open or closed circuits? Have you seen them? What kind are they?

CHAPTER XII

THE MAGNETIC EFFECT OF ELECTRIC CURRENTS. ELECTRICAL MEASUREMENTS

(1) THE MAGNETIC EFFECT OF ELECTRIC CURRENTS

255. The Magnetic Effect.—Of all the effects of electric currents, it is generally conceded that the *magnetic effect* is the one of *greatest practical importance*, and it is also the one most extensively used. An experiment illustrating this effect has been described in Art. 239. This experiment shows that an electric current, if *parallel* to a magnetic needle, and *near it* will deflect the north-seeking pole of the needle to the right or left depending upon the *direction* of the current flow. This *deflection* of the magnetic needle is due to the fact that surrounding every electric current are magnetic lines of force. It is this magnetic field of the current that causes the needle to turn. The position taken by the needle is the resultant of the forces of two magnetic fields; one, the earth's field, the other, that of the current.

256. Right-hand Rule for a Conductor.—To show the presence of the magnetic field about a current, pass a thick copper wire vertically through a sheet of paper, and connect the ends of the wire to a source of current. While the current (this should be as much as 10 amperes if possible) is flowing, sprinkle iron filings upon the paper and tap gently. The filings will arrange themselves in circles about the wire showing the magnetic field. (See Fig. 229.) The needle of a magnetoscope tends to place itself parallel to the lines of force of this field and from this action or tendency the *direction* of the magnetic lines about a current may be determined. The following rule is helpful and should be memorized: *Grasp the conductor with the right hand with the outstretched thumb in the direction that the current is flowing. The fingers will then encircle the wire in the direction of the lines of force.* This rule may be reversed, for, if the fingers of the right hand grasp the wire so as to point with the magnetic field, then the current flows in the direction in which the thumb points. (See Fig. 230.)

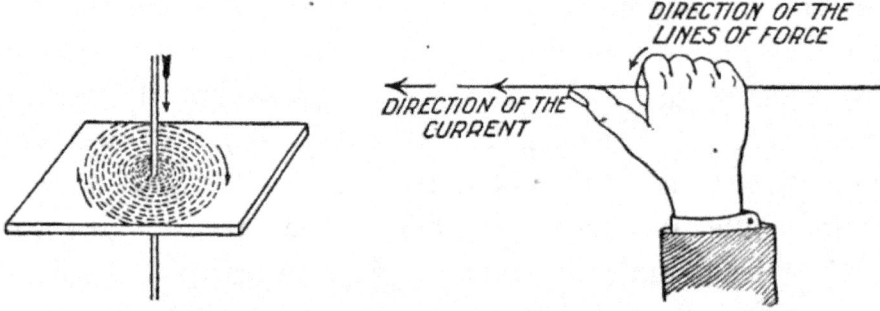

FIG. 229.—Magnetic field about a wire carrying an electric current.

FIG. 230.—Right-hand rule for the magnetic field of a current.

257. Magnetic Field of a Helix.—If a wire be wound about a cylinder to form a cylindrical coil with parallel turns, it forms a *helix* or *solenoid*. The shape of the magnetic field about a current depends upon the *form* of the conductor. If the latter is in the form of a *helix* its magnetic field resembles that of a straight bar magnet. (See Fig. 231). In fact the helix *has the properties of a magnet* with north- and south-seeking poles while a *current* is *flowing* through it. If such a coil is suspended so as to turn freely, it tends to turn until the field within it is parallel to the earth's magnetic field. Such a suspended helix may therefore be used as a compass. In order to strengthen the magnetic field of a helix or solenoid, the space within its turns is filled with iron, often in the form of small soft-iron wires.   This bundle of iron wire is called the *core* of the helix.   The core becomes strongly magnetized by the field of the helix while the current is flowing and quickly loses its magnetic force when the current is stopped.   *The direction of the current in a helix* (Fig. 232) or the *polarity* of its core may be determined by another *right-hand rule*.   *If the helix is grasped with the right hand so that the fingers point in the direction in which the current is flowing, the extended thumb will point in the direction of the north pole of the helix.*   On the other hand, if the

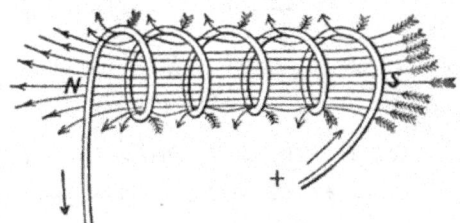

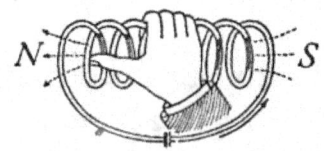

FIG. 231.—The magnetic field of a helix.   FIG.   232.—Right-hand   rule for a helix.

poles of the helix are known, then, when the helix is grasped with the right hand so that the thumb points to the north-seeking pole, the current is flowing in the wires in the direction that the fingers point.

258. The Electromagnet.—These "right-hand" rules are applied in many different devices. Among these, perhaps the most important is the electromagnet, which is used in the electric bell, the telegraph, the telephone, the dynamo, the motor, and many other electric contrivances.

The electromagnet is defined as a *mass of iron around which is placed a helix for conducting an electric current.* On account of its large permeability, the iron core of the helix adds greatly to the effectiveness of the electromagnet, since the magnetism of the iron is added to that of the current in the helix. The magnetism remaining in the iron after the current stops is called the *residual* magnetism. The residual magnetism is small when the core is made of small wires or thin plates, but is larger when the iron core is solid. Like artificial steel magnets,

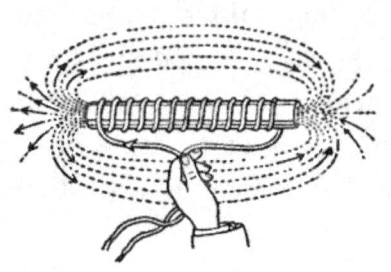
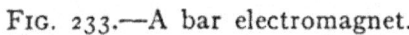

FIG. 233.—A bar electromagnet.

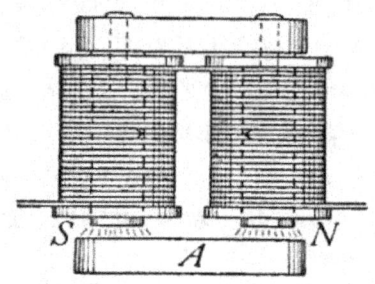

FIG. 234.—A horseshoe electromagnet.

electromagnets are usually of two forms, *bar and horseshoe.* (See Figs. 233 and 234.) For most purposes the horseshoe form is the more effective since it permits a complete iron circuit for the magnetic lines of force. (See Fig. 235.)

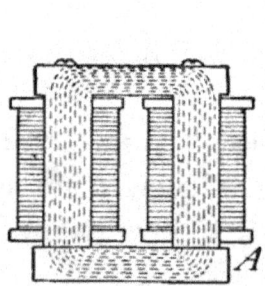

FIG. 235.—A horseshoe electromagnet may have a complete iron circuit for its lines of force.

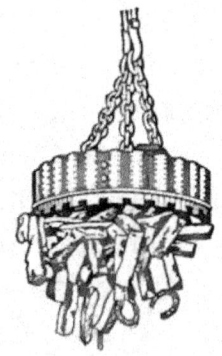

FIG. 236.—A lifting magnet.

This is the form used in the electric bell, in the telegraph sounder, and in lifting magnets. (See Fig. 236.)

259. Effective Electromagnets.—The *magnetic* effect of a current in a helix is small, hence the force usually is increased by inserting a core of iron. When at first man tried to signal with electromagnets at a distance it was found that the current would not work the electromagnet. An American by the name of Joseph Henry discovered the remedy for this condition.   He found that if the copper wire was insulated by wrapping silk thread about it, and then many layers of the silk insulated wire were wound upon a spool with an iron core, that the magnet would work at a great distance from the source of current.   If the current is increased, the magnet is stronger than at first. Thus an *electromagnet may be made stronger by* (a) *increasing the number of turns of wire in its coils* and by (b) *sending a stronger current through it.*

260. The Telegraph.—The invention of an effective electromagnet by Henry made possible the *electric telegraph.*   In its simplest form it consists of a battery, C, a key, K,   and a sounder, S, with connecting wires.   (See Fig. 237.) The *sounder* (Fig. 238) contains a *horseshoe electromagnet* and a bar of soft iron across its

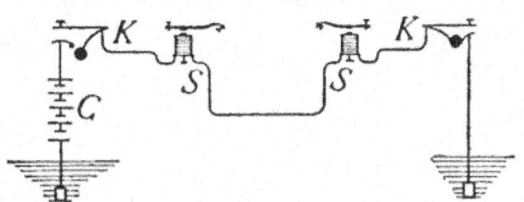

FIG. 237.—A simple telegraph circuit.

poles called an armature, A, attached to a lever L.   When the key is closed, the electromagnet draws down the armature and lever until the latter hits a stop O, making a click.   When the key is raised, the magnet releases the armature which is raised by the action of a spring at S until the lever hits a stop at T making another click. Closing and opening the circuit at K will start and stop the current which operates S which may be 100 miles or more from K.   One voltaic cell will work a sounder in the same room.   But if many miles of wire are in the circuit, the E.M.F. of a single cell will not force sufficient current through the long wire to operate the sounder.

A battery of several cells is then required. Even a large battery is insufficient to operate a long line containing many sounders in circuit. Recourse is therefore usually made to a more sensitive device called a *relay*. (See Fig. 239.) In the relay a very small current will magnetize its electromagnet enough to draw toward it the delicately hung armature thereby closing a second circuit which contains a sounder and a battery. (See Fig. 240.) When the current in the main circuit is stopped, the armature of the relay is drawn back by a light spring. This opens the *local* circuit. Thus the local cir-

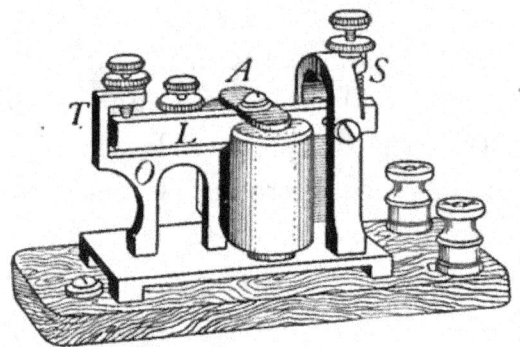

FIG. 238.—A telegraph sounder.

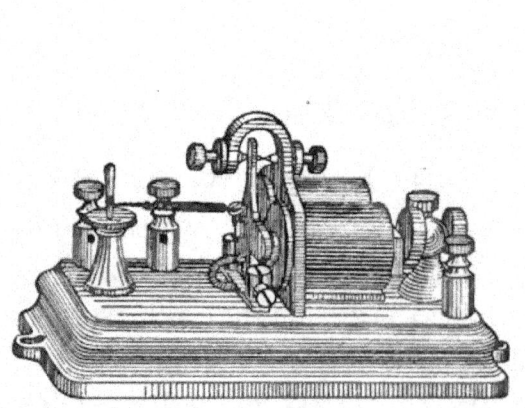

FIG. 239.—A telegraph relay.

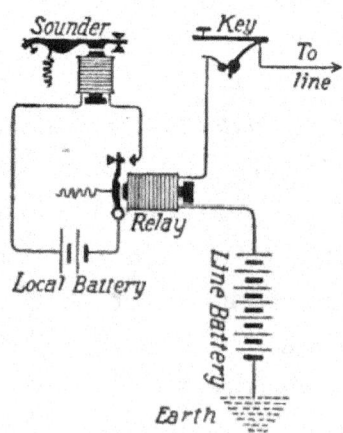

FIG. 240.—How the relay is used.

cuit is closed and opened by the relay just in time with the starting and stopping of the current in the main line. It is thus possible for a small current in the main line by the use of a relay, to close and open a second local circuit

Samuel F. B. Morse (1791–1872). Inventor of the electro magnetic recording telegraph and of the dot and dash alphabet.

SAMUEL F. B. MORSE
"From Appleton's Cyclopedia of American Biography, Copyright 1888 by D. Appleton & Co."

Thomas A. Edison, Orange, New Jersey. Invented the incandescent lamp; phonograph; moving picture; most noted inventor of electrical appliances of the present day.

THOMAS A. EDISON
"Copyright, Photographische Gessellschaft," and "By Permission of the Berlin Photographic Co., New York."

containing a local battery and sounder. Modern telegraph lines are operated in this manner.

261. The electric bell (see Fig. 241), consists of an electromagnet, *M*, a soft iron *armature*, *A*, attached to the *tapper*, *T*, and a post, *R*. When no current is flowing a spring at *S* holds the armature against the post *R*. When current flows through the helix, its core becomes magnetized and attracts the armature, drawing it away from the post, *R*, and causing the tapper to hit the bell. Drawing *A* away from the post, however, breaks the circuit at *R* and the current stops. The magnetism in the core disap-

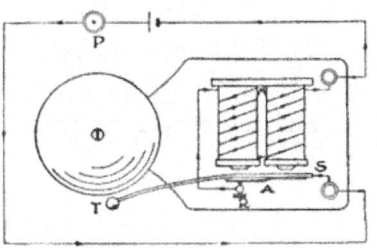

FIG. 241.—An electric bell and its circuit.

pears releasing the armature, which is then pulled back by the spring *S* against the post *R*. This completes the circuit and the process repeats itself several times a second as long as the current flows.

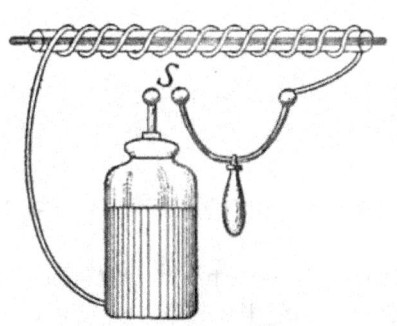

FIG. 242.—Magnetizing by the discharge of a Leyden jar.

262. Static and Current Electricity Compared.—The likeness between a discharge of static electricity and an electric current may be shown by winding a coil of insulated wire about a glass tube which contains a steel needle. If a Leyden jar (see Fig. 242) is discharged through the coil the steel needle is usually found to be magnetized, showing that the discharge of the static electricity has a magnetic effect similar to that of an electric current. Sometimes a given end of the needle has a north pole and at other times a south pole. This is believed to indicate that the charge of the Leyden jar is *oscillatory*, and that in different discharges sometimes a surge in one direction and at other times a surge in the reverse direction has been most effective in magnetizing the needle. Compare this action with that described in Art. 233.

Important Topics

1. Right-hand rules, for conductor, for helix.
2. The electromagnet, two forms, where used?
3. Likeness between static and current electricity.
4. The electric bell, parts, action.
5. The telegraph, key, sounder, relay.

Exercises

1. What is the difference between an electric charge and a current?
2. How can a magnetic effect be produced from an electric charge?
3. What is a magnetic field? Give two evidences of a magnetic field about a current in a wire?
4. A current is flowing north in trolley wire, what is the direction of the magnetic field under the wire? Explain.
5. What would be the result if a hard steel core were placed in the electromagnet? Explain.
6. If the north-seeking pole of a helix is facing you, does the current in the coils before you move in a clockwise or in a counter-clock wise direction? Explain.
7. A helix is placed horizontally with its north-seeking pole toward the north. Does the current in the wire at the top of the helix move east or west? Explain.
8. State at least six conditions any one of which will put an electric bell circuit out of commission.
9. If one desires to insert a battery into a telegraph circuit already in operation, how will he determine the direction of the current in the wire.
10. If a boy who had magnetized his knife blade in a physics laboratory, pointed end south-seeking, should lose his way in the woods on a cloudy day, how could he determine his way out?
11. At a certain point the earth's field acts north, that of an electric current, east. The magnetoscope needle points exactly northeast when placed at that point. How do the two magnetic fields compare?

(2) Electrical Measurements

263. Galvanometers.—In using electric currents it is often necessary or desirable to be able to know not only that a given current is weak or strong, but precisely what its strength is. We can determine the relative strengths of two currents by the use of a *galvanometer*.

The older or *moving-magnet* type of galvanometer is similar to the galvanoscope mentioned in Art. 239. It

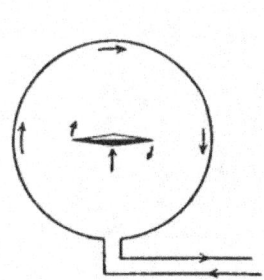

Fig 243.—The magnet is at the center of the coil.

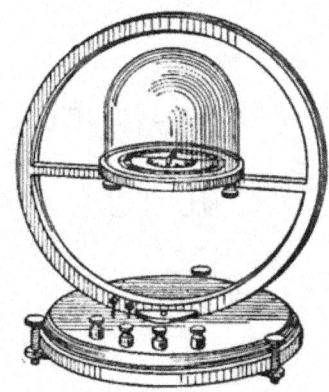

Fig. 244.—A moving-magnet (tangent) galvanometer.

consists of a magnetic needle mounted at the center of a coil of wire. The coil is placed facing east and west, so that the needle will be held by the earth's magnetic field parallel to the plane of the coil. When a current is sent through the coil a magnetic field is produced within it. This deflects the needle, its north end turning east or west depending upon the direction of the current. (See Fig. 243.) The *coils* of a moving-magnet or *tangent* galvanometer (see Fig. 244) are *large* and firmly fastened to the base, while the *magnet* is *small*.

The *moving-coil* type of galvanometer (see Fig. 245) consists of a *large magnet* fastened to the frame of the device. The magnet usually has a horseshoe form to

19

produce as strong a field as possible. The *coil* is wound on a *light* rectangular frame and is suspended between the two poles of the magnet. To concentrate the magnetic field, a cylinder of soft iron is usually placed within the coil. Fig. 246 represents a common form of moving-coil galvanometer.

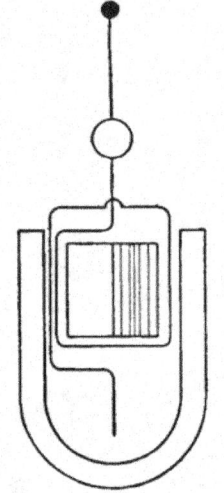

FIG. 245.—To illustrate the principle of the moving-coil galvanometer.

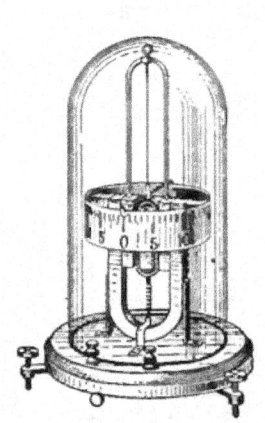

FIG. 246.—A moving-coil (D'Arsonval) galvanometer.

264. Measurement of Electric Currents.—A galvanometer enables one to *compare* electric currents. To *measure* electric currents it is necessary to employ a *unit* of electrical quantity, just as in measuring the quantity of water delivered by a pipe, a unit of liquid measure is employed; thus, *e.g.*, the current delivered by a given pipe may be 2 gallons of water per second, so in measuring the flow of an electric current one may speak of two *coulombs* per second. The *coulomb* is the unit quantity of electricity just as the unit of quantity of water is the gallon.

For most practical purposes, however, we are more interested in the *rate* or *intensity* of flow of current than in the actual *quantity* delivered. The unit of rate of flow or current is called the *ampere*.

In determining the exact *quantity* of an electric current, physicists make use of a device called a *coulomb meter*. (See Fig. 247.) This contains a solution of silver nitrate in which are placed two silver plates. The current to be measured is sent through the solution, in at one plate and out at the other. The plate where the current goes *in*, the *anode*, A (Fig. 247), loses in weight since some of the silver is dissolved. The plate where the current goes *out*, the *cathode*, C, increases in weight since some of the silver is deposited. By an international agreement, *the intensity of the current which deposits silver at the rate of 0.001118 g. per second is 1 ampere.* This is equal to 4.025 g. per hour.

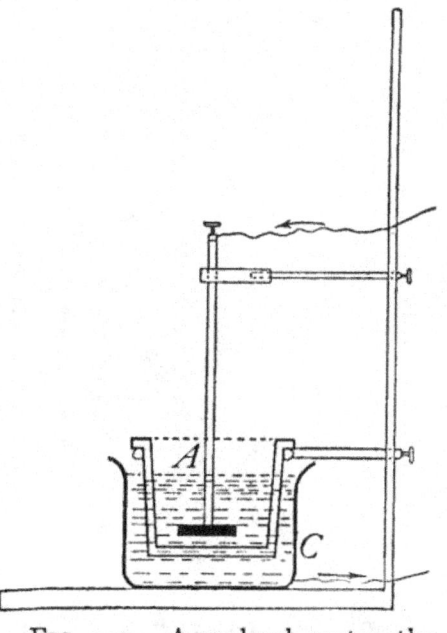

FIG. 247.—A coulomb meter, the anode A is separated from the cathode C by a porous cup.

The *coulomb* is defined as the quantity of electricity delivered by a current of one ampere during one second.

A 40-watt-incandescent lamp takes about 0.4 ampere of current. An arc lamp takes from 6 to 15 amperes. A new dry cell may send 20 amperes through a testing meter. A street car may take from 50 to 100 amperes.

265. The Ammeter.—The method described above is not used ordinarily for measuring current strengths on account of its inconvenience. The usual device employed is an *ammeter*. This instrument is a *moving-coil galva-*

nometer.   It contains, wound on a light form, a coil of fine
copper wire.   The form is mounted on jewel bearings
between the poles of a strong permanent horseshoe magnet.
(See Fig. 248.)   As in other moving-coil galvanometers, a
soft iron cylinder within the form concentrates the field
of the magnet.   The form and its coil is held in balance
by two spiral springs which also conduct current into and
out of the coil.

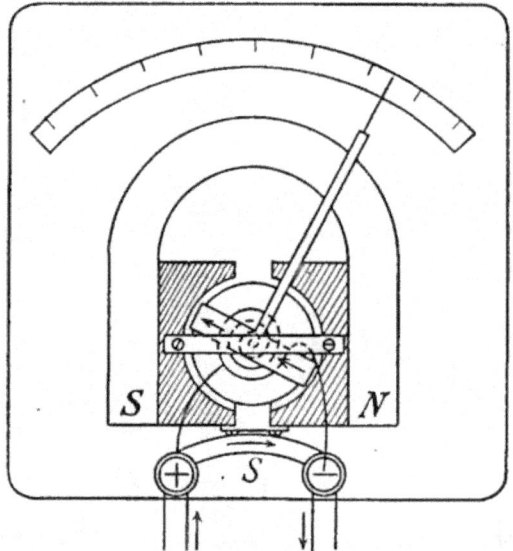

FIG. 248.—Diagram of a commercial ammeter.   *S* is the shunt.

Only a small part
of the whole current
measured, in s o m e
cases only 0.0001
passes through the
coil, the larger part
of the current pass-
ing through a metal
wire or strip called a
shunt[1] (see Fig. 248)
connecting the bind-
ing posts of the in-
strument.   A fixed
fraction of the whole
current flows through
the coil.   Its field crossing the field of the horseshoe
magnet, tends to turn until its turning force is balanced
by the spiral springs.   As the coil turns it moves a
pointer attached to it across a scale graduated to indicate
the number of amperes in the whole current.

It should be noted that while *all* of the current measured
passed through the ammeter, but a small *part* goes through
the coil.

[1] A shunt is a conductor or coil connected in parallel with another
conductor or circuit.   It carries a part of the current.

266. Resistance of Conductors.—With an ammeter one may study the change produced in the amount of current flowing in a wire when a change is made in the wire conducting the current. For example, if one measures with an ammeter the current flowing from a dry cell through a long and then through a short piece of fine copper wire, it will be seen that less current flows when the long piece is used. That is, the long wire seems to hinder or to *resist* the passing of the current more than the short piece. In other words, the long wire is said to have more *resistance*.

The resistance of a conducting body is affected by several conditions.

(a) It is *directly* proportional to the *length* of the conductor, one hundred feet of wire having twice the resistance of fifty feet.

(b) It is *inversely* proportional to the *square of the diameter;* a wire 0.1 inch in diameter has four times the resistance of a wire 0.2 inch in diameter.

(c) It differs with different substances, iron having about six times as much as copper.

(d) It varies with the temperature, metals having greater resistance at a higher temperature.

Since silver is the best conductor known, the resistances of other substances are compared with it as a standard.

The ratio of the resistance of a wire of any substance as compared to the resistance of a silver wire of exactly the same diameter and length is called its *relative* resistance.

Purified substances arranged in order of increasing resistance for the same length and sectional area (Ayrton-Mather) are given on p. 294.

Silver annealed............................		1.00
Copper annealed { from....................		1.04
{ to......................		1.09
Aluminum annealed.......................		1.64
Nickel annealed......		4.69
Platinum annealed.......................		6.09
Iron annealed............................		6.56
German silver { from....................		12.80
{ to......................		20.20
Mercury.................................		63.30
Nichrome...............................		67.50
Carbon { from...........................		2700.00
{ to...........................		6700.00

267. The ohm, the unit of resistance, is defined by international agreement as follows: *An ohm is the resistance of a column of pure mercury, 106.3 cm. long with a cross-section of a square millimeter and at a temperature of 0° C.*

It should be noted that each of the four conditions affecting resistance is mentioned in the definition, viz., length, cross-section, material, and temperature. Since it is inconvenient to handle mercury, *standard resistance coils*, made of an alloy of high resistance are used in comparing and measuring resistances.

A piece of copper wire No. 22 (diameter 0.644 mm.) 60.5 ft. long has a resistance of 1 ohm. See table p. 296.

The resistance of some telephone receivers is 75 ohms, of a telegraph sounder, 4 ohms, of a relay 200 ohms.

268. Resistance of Circuits.—Every part of an electrical circuit possesses resistance. In an electric-bell circuit, for instance, the wires, the bell, the push-button, and the cell itself, each offers a definite resistance to the passage of the current. The resistance *within the cell* is termed *internal resistance*, while the resistance of the parts outside of the electric generator is called *external resistance*.

269. Electromotive Force.—In order to set in motion anything, some *force* must be applied. This is as true of electricity as of solids, liquids, or gases. By analogy that which is exerted by a battery or by a dynamo in causing current to flow is called an *electromotive force.* The unit of electromotive force, the *volt*, may be defined as *the electromotive force that will drive a current of 1 ampere through the resistance of 1 ohm*. The electro-

motive force of a dry cell is about 1.5 volts, of a Daniell cell 1.08 volts. Most electric light circuits in buildings carry current at 110 or 220 volts pressure. Currents for street cars have an electromotive force of from 550 to 660 volts.

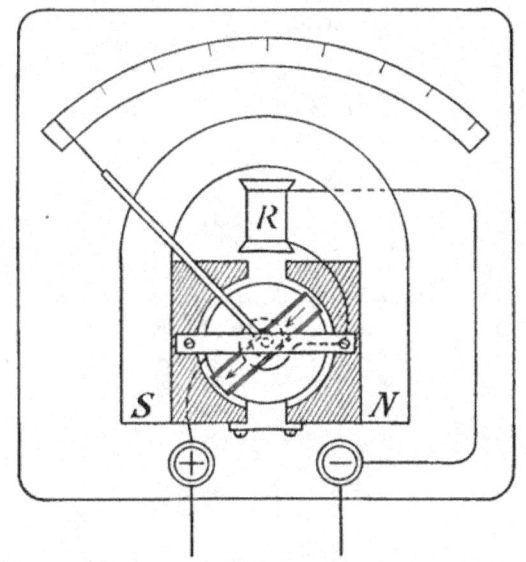

FIG. 249.—Diagram of a commercial voltmeter.

270. The Voltmeter.—An instrument for measuring the electromotive force of electric currents is called a *voltmeter* (Fig. 249). It is usually a moving-coil galvanometer, and is always of *high resistance*. It is like an ammeter in construction and appearance. In fact, a voltmeter is an ammeter which has had its shunt removed or disconnected. In place of a shunt, the voltmeter uses a coil of wire of high resistance (see *R*, Fig. 249) *in series* with the galvanometer coil. The high resistance of the voltmeter permits but a very small current to flow through it. Hence a voltmeter must be placed *across* a circuit and not in it. In other words *a voltmeter is connected in shunt*, while *an ammeter is in series with the circuit* as is shown in Fig. 250.

DIMENSIONS AND FUNCTIONS OF COPPER WIRES

B. & S. gauge number	Diameter		Circular mils	Sectional area in square milli-meters	Weight and length, Density = 8.9, feet per pound	Resistance at 24°C., feet per ohm	Capacity in amperes
	Mils	Milli-meters					
0000	460.000	11.684	211,600.00	107.219	1.56	19,929.700	312.0
000	409.640	10.405	167,805.00	85.028	1.97	15,804.900	262.0
00	364.800	9.266	133,079.40	67.431	2.49	12,534.200	220.0
0	324.950	8.254	105,592.50	53.470	3.13	9,945.300	185.0
2	257.630	6.544	66,373.00	33.631	4.99	6,251.400	131.0
4	204.310	5.189	41,742.00	21.151	7.93	3,931.600	92.3
6	162.020	4.115	26,250.50	13.301	12.61	2,472.400	65.2
8	128.490	3.264	16,509.00	8.366	20.05	1,555.000	46.1
10	101.890	2.588	10,381.00	5.260	31.38	977.800	32.5
12	80.808	2.053	6,529.90	3.309	50.69	615.020	23.0
14	64.084	1.628	4,106.80	2.081	80.59	386.800	16.2
16	50.820	1.291	2,582.90	1.309	128.14	243.250	11.5
18	40.303	1.024	1,624.30	0.823	203.76	152.990	8.1
20	31.961	0.812	1,021.50	0.5176	324.00	96.210	5.7
22	25.347	0.644	642.70	0.3255	515.15	60.510	4.0
24	20.100	0.511	504.01	0.2047	819.21	38.050	2.8
26	15.940	0.405	254.01	0.1288	1,302.61	23.930	2.0
28	12.641	0.321	159.79	0.08097	2,071.22	15.050	1.4
30	10.025	0.255	100.50	0.05092	3,293.97	9.466	1.0
32	7.950	0.202	63.20	0.03203	5,236.66	5.952	0.70
34	6.304	0.160	39.74	0.02014	8,328.30	3.743	0.50
36	5.000	0.127	25.00	0.01267	13,238.83	2.355	0.35
38	3.965	0.101	15.72	0.00797	20,854.65	1.481	0.25
40	3.144	0.080	9.89	0.00501	33,175.94	0.931	0.17

Important Topics

(1) *Galvanometers:* (1) moving magnet, fixed coil; (2) moving coil, fixed magnet, ammeter, voltmeter.

(2) *Unit of quantity, coulomb.*

(3) *Unit of current, ampere.*

(4) *Unit of resistance, ohm.*

(5) *Unit of electromotive force, volt.*

Exercises

1.. How will the resistance of 20 ft. of No. 22 German silver wire compare with that of 10 ft. of No. 22 copper wire?   Explain.

2. Where in a circuit is copper wire desirable?   Where should German silver wire be used?

3. Explain the action of the ammeter.   Why does not the needle or coil swing the full distance wth,a small current?

4. Why is a telegraph sounder more apt to work on a short line than upon a long one?

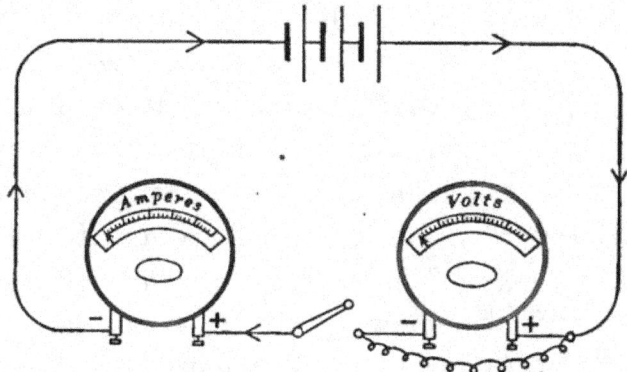

FIG. 250.—The ammeter is connected in series and the voltmeter in shunt.

5. Find the resistance of 15 miles of copper telephone wire No. 12. (See table p. 296.)

6. What will be the weight and resistance of 1,000 feet of No. 20 copper wire?

7. A storage battery sends 4 amperes of current through a plating solution.   How much silver will it deposit in 2 hours?

8. (a) Compare the diameters of No. 22 and No. 16 copper wire.
   (b) Compare the lengths of the same wires giving 1 ohm resistance.
   (c) What relation exists between (a) and (b)?

9. Why is an electric bell circuit usually open while a telegraph line circuit is usually closed?

10. A copper wire and an iron wire of the same length are found to have the same resistance.   Which is thicker?   Why?

11. Why are electric bells usually arranged in parallel instead of in series?

12. What would happen if a voltmeter were put in series in a line?

(3) Ohm's Law and Electrical Circuits

271. Conditions Affecting Current Flow.—Sometimes over a long circuit one cell will not work a telegraph sounder. In such a case, two, three, or more cells are connected so that the zinc of one is joined to the copper plate of the other. When connected in this way the cells are said to be *in series* (Fig. 251). In the figure *A* represents a voltmeter. It is found that *when cells are in series the E.M.F. of the battery is the sum of the electromotive forces of the cells.*

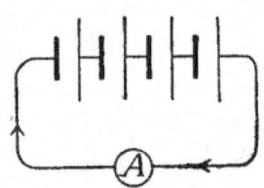

FIG. 251.—Diagram of cells connected in series.

An ammeter in the circuit shows increased current as the cells are added. Hence *if the resistance of the circuit remains unchanged, the greater the E.M.F. the greater is the current strength.* In this respect, the movement of electricity in a circuit is similar to the flow of water in a small pipe under pressure, as in the latter the flow of water increases as the pressure becomes greater. The current in a circuit may also be increased by lessening the resistance, since the current through a long wire is less than that through a short one, just as the flow of water will be greater through a short pipe than through a long one. To increase the current flowing in an electric circuit, one may therefore either increase the E.M.F. or decrease the resistance.

272. Ohm's Law.—The relation between the electromotive force applied to a circuit, its resistance, and the current produced was discovered in 1827 by George Ohm. Ohm's law, one of the most important laws of electricity, states that, in any circuit, *the current in amperes equals the electromotive force in volts divided by the resistance in ohms.*

This principle is usually expressed thus:

$$\text{Current intensity} = \frac{\text{electromotive force}}{\text{resistance}} \text{ or}$$

$$\text{Amperes} = \frac{\text{volts}}{\text{ohms}} \text{ or } I = \frac{E}{R}$$

273. Resistance of Conductors in Series.—A study of the *resistance* of conductors when alone and when grouped in various ways is of importance *since, the current flow through any circuit is dependent upon its resistance.* The two most common methods of combining several conductors in a circuit are in *series* and in *parallel*. Conductors are in *series* when all of the current passes through each of the conductors in turn (Fig. 218), thus the cell, push-button, wires, and electric bell in an electric-bell circuit are in

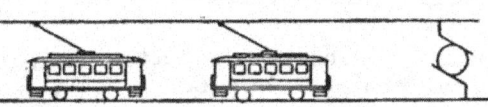

FIG. 252.—The street cars are connected in parallel with each other.

series. Conductors are in *parallel* when they are so connected that they are side by side and a part of the whole current goes through each. None of the current that passes through one conductor can go through the conductors in parallel with it. Thus the electric street cars are in *parallel* with each other. (See Fig. 252.) It is easily seen that none of the current passing through one car can go through any of the others. When the conductors are in *series* the combined resistance is the *sum* of the several resistances. Thus in an electric-bell circuit if the battery has a resistance of 1 ohm, the bell of 2 ohms, and the wire 1 ohm, the total resistance in the circuit is 4 ohms. When conductors are in *parallel* the combined resistance is always *less* than the separate resistances. Just as a crowd of people meets less resistance in leaving a building through several exits, so electricity finds less resistance in moving from one point to another along several parallel lines, than along one of the lines.

274. Resistance of Conductors in Parallel.—If three conductors of equal resistance are in parallel, the combined resistance is just one-third the resistance of each separately (Fig. 253). The rule that states the relation between the combined resistance of conductors in parallel and the separate resistances is as follows: *The combined*

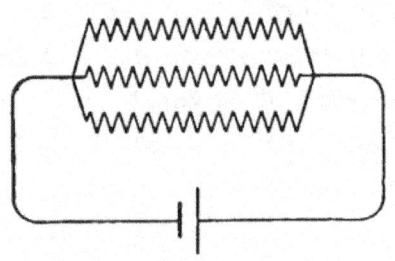

resistance of conductors in parallel is the reciprocal of the sum of the reciprocals of the several resistances. For example, find the combined resistance of three unequal resistances in parallel; the first being 4 ohms,

FIG. 253.—The three conductors are connected in parallel.

the second, 6 ohms, and third 3 ohms. The reciprocals of the three resistances are $\frac{1}{4}$, $\frac{1}{6}$, and $\frac{1}{3}$. Their sum equals $\frac{6}{24} + \frac{4}{24} + \frac{8}{24} = \frac{18}{24}$. The reciprocal of this is $\frac{24}{18}$ which equals $1\frac{1}{3}$ ohms, the combined resistance.

This rule may be understood better if we consider the *conductance* of the conductors in parallel. Since the conductance of a two ohm wire is just one-half that of a one-ohm wire, we say that the conductance of a body is inversely as the resistance, or that it is the *reciprocal of the resistance*. The conductance of the 4-, 6-, and 3-ohm coils will therefore be respectively $\frac{1}{4}$, $\frac{1}{6}$, and $\frac{1}{3}$, and since the combined conductance is the sum of the several conductances, the total conductance is $\frac{18}{24}$. Also since this is the reciprocal of the total resistance, the latter is $\frac{24}{18}$ or $1\frac{1}{3}$ ohms.

When two or more conductors are connected in parallel each one is said to be a *shunt* of the others. Many circuits are connected in *shunt* or in parallel. Fig. 254 represents four lamps in parallel. Incandescent lamps in buildings are usually connected in parallel, while arc lamps are usually connected in series. Fig. 255 represents four lamps in series.

Important Topics

1. Conditions affecting current flow, (a) E.M.F., (b) resistance.
2. Ohm's law, three forms for formula.
3. Resistance of conductors: (a) in series, (b) in parallel; how computed, illustrations.

FIG. 254.—The four lamps are connected in parallel.   FIG. 255.—The four lamps are connected in series.

Exercises

1. What current flows through a circuit if its E.M.F. is 110 volts and the resistance is 220 ohms?
2. A circuit contains four conductors in series with resistances of 10, 15, 6, and 9 ohms respectively. What current will flow through this circuit at 110 volts pressure? What will be the resistance of these four conductors in parallel?
3. What is the combined resistance of 8 conductors in parallel if each is 220 ohms? What current will flow through these 8 conductors at 110 volts pressure?
4. What is the resistance of a circuit carrying 22 amperes, if the E.M.F. is 20 volts?
5. What E.M.F. will send 8 amperes of current through a circuit of 75 ohms resistance?
6. How does the voltmeter differ from the ammeter?
7. How can one determine the resistance of a conductor?
8. The resistance of a hot incandescent lamp is 100 ohms. The current used is 1.1 amperes. Find the E.M.F. applied.
9. What is the resistance of the wires in an electric heater if the current used is 10 amperes, the voltage being 110?
10. The resistance of 1000 ft. of No. 36 copper wire is 424 ohms. How many feet should be used in winding a 200 ohms relay?
11. The resistance of No. oo trolley wire is 0.80 ohm per 1000 ft. What is the resistance of a line 1 mile long?
12. A wire has a resistance of 20 ohms. It is joined in parallel with another wire of 6 ohms, find their combined resistance.

13. The separate resistances of two incandescent lamps are 200 ohms and 70 ohms. What is their combined resistance when joined in parallel? When joined in series?

(4) METHODS OF GROUPING CELLS AND MEASURING RESISTANCE

275. Internal Resistance of a Voltaic Cell.—The current produced by a voltaic cell is affected by the resistance that the current meets in passing from one plate to another through the liquid of the cell. This is called the *internal resistance* of the cell. A Daniell cell has several (1–5) ohms internal resistance. The resistance of dry cells varies from less than 0.1 of an ohm when new to several ohms when old. If cells are joined together their combined internal resistance depends upon the method of grouping the cells.

276. Cells Grouped in Series and in Parallel.—When in *series* the copper or carbon plate of one cell is joined to the zinc of another and so on. (See Fig. 251.) The effect of connecting, say four cells, in series may be illustrated by taking four cans of water, placed one above another. (See Fig. 256.) The combined water pressure of the series is the sum of the several pressures of the cans of water, while the opposition offered to the movement of a quantity of water through the group

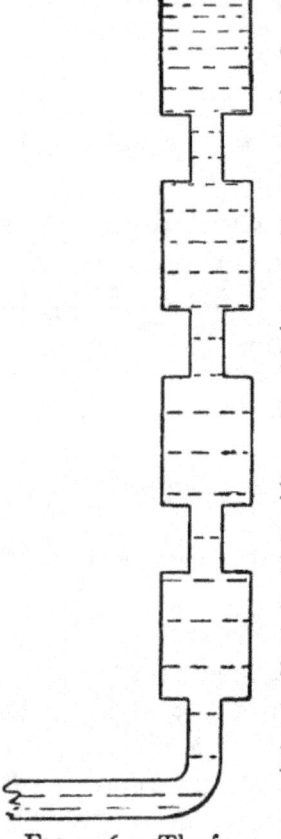

FIG. 256.—The four cans exert four times the water pressure that one can will exert.

of cans is the sum of the several resistances of the cans. In applying this illustration to the voltaic cell, we make

use of Ohm's law.   Let E represent the e.m.f. of a single
cell, r the internal resistance of the cell, and R the external
resistance or the resistance of the rest of the circuit.
Consider a group of cells in series.   If n represents the
number of cells in *series*, then Ohm's law becomes

$$I = \frac{nE}{nr + R}.$$

Cells are grouped in *series* when large E.M.F. is required
to force a current through a large external resistance such
as through a long telegraph line.   Cells
are connected in *parallel* when it is desired
to send a large current through a small
external resistance.   To connect cells in
parallel all the copper plates are joined

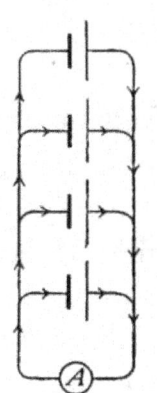

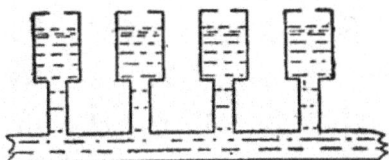

FIG.   257.—Four
cells   connected   in
parallel.

FIG. 258.—The water pressure of
the group in parallel is the same as
that of one.

and also all the zinc plates.   (See Fig. 257.)   To illustrate
the effect of this mode of grouping cells, suppose several
cans of water are placed side by side (Fig. 258).   It is
easily seen that the pressure of the group is the same as
that of a single cell, while the resistance to the flow is less
than that of a single cell.   Applying this reasoning to
the electric circuit we have by Ohm's law the formula for
the current flow of a group of

n cells arranged in parallel $I = \dfrac{E}{\dfrac{r}{n} + R}.$

277. Illustrative Problems.—Suppose that four cells are grouped in parallel, each with an E.M.F. of 1.5 volts and an internal resistance of 2 ohms. What current will flow in the circuit if the external resistance is 2.5 ohms? Substitute in the formula for cells in parallel the values given above, and we have $I = \dfrac{1.5}{0.5 + 2.5} = \dfrac{1.5}{3} = 0.5$ ampere. Suppose again that these four cells were grouped in series with the same external resistance, substituting the values in the formula for cells in series we have $I = \dfrac{4(1.5)}{4 \times 2 + 2.5} = \dfrac{6}{10.5} = 0.57$ ampere.

278. Volt-ammeter Method for Finding Resistance.— Measurements of the resistance of conductors are often made. One of these methods depends upon an application of Ohm's law. It is called the volt-ammeter method since it employs both a voltmeter and an ammeter. If the conductor whose resistance is to be measured is made a part of an electric circuit, being connected *in series with the ammeter* and *in shunt with the voltmeter*, the resistance may easily be determined, since $R = \dfrac{E}{I}$. (See Fig. 250.)

If, for example, the difference in E.M.F., or as it is often called, the *fall of potential* between the ends of the wire as read on the voltmeter is 2 volts, and the current is 0.5 ampere, then the resistance of the wire is 4 ohms. This method may be readily applied to find the resistance of any wire that is a part of an electric circuit.

279. The Wheatstone Bridge.—To find the resistance of a separate wire or of an electrical device another method devised by an Englishman named Wheatstone is commonly employed. This method requires that three known resistances, *a, b, c,* in addition to the unknown resistance

x be taken. These four resistances are arranged in the form of a parallelogram. (See Fig. 259.) A voltaic cell is joined to the parallelogram at the extremities of one diagonal while a moving-coil galvanometer is connected across the extremities of the other diagonal. The known resistances are changed until when on pressing the keys at E and K no current flows through the galvanometer. When this condition is reached, the four resistances form a true proportion, thus $a : b = c : x$.

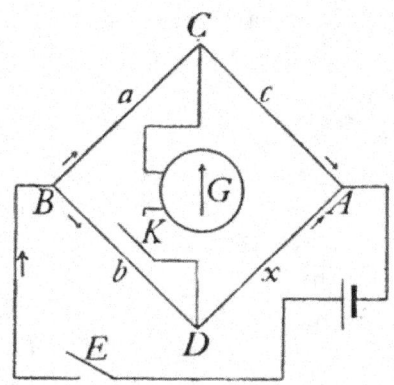

FIG. 259.—Diagram of a Wheatstone bridge.

Since the values of a, b, and c are known, x is readily computed. Thus if $a = 10$, $b = 100$, and $c = 1.8$ ohms, then x, the unknown resistance, equals 18 ohms, since $10 : 100 = 1.8 : 18$. This method devised by Wheatstone may be employed to find the resistance of a great variety of objects. It is the one most commonly employed by scientists and practical electricians.

Important Topics

1. The internal resistance of voltaic cells.
2. Ohm's law applied to groups of cells. (a) Cells in series, (b) cells in parallel.
3. Measurement of resistance: (a) volt-ammeter method, (b) Wheatstone bridge method.

Exercises

1. What is the resistance of an electric bell circuit where the E.M.F. is 3 volts and the current is 0.6 ampere?
2. A telegraph wire is broken somewhere, the ends lying upon damp ground. If an E.M.F. of 30 volts is applied from the ground

to the wire and a current of 0.1 of an ampere flows, what is the resistance of the part connected to the ammeter. (The earth which completes the circuit from the end of the wire has very small resistance.)   Why?

3. How far away is the break in the wire if the latter has a resistance of 80 ohms to the mile?   Diagram.

4. What current will flow through a bell circuit of 8 ohms resistance if it contains three cells *in series* each with an E.M.F. of 1.5 volts and an internal resistance of ⅓ ohm?

5. If the same three cells are connected in parallel on the same circuit what current flows?   Is the current in problem 4 or 5 the larger?   Why?

6. If four cells each with 1.5 volts E.M.F. and an internal resistance of 0.4 ohm are connected with a circuit having an external resistance of 0.8 ohm, what current will the parallel connection give?   The series connection?   Which gives the larger current? Why?

7. Four Daniell cells each having 1 volt E.M.F. and 3 ohms internal resistance are connected in series with 2 telegraph sounders of 4 ohms each.   The connecting wires have 6 ohms resistance. Find the current intensity.

8. A battery of 2 cells arranged in series is used to ring a door bell. The E.M.F. of each cell is 1.5 volts, internal resistance 0.3 ohm, and the resistance of the bell is 4 ohms.   What is the current in amperes?

9. In the above problem find the current if the cells are connected in parallel.

CHAPTER XIII

THE CHEMICAL AND HEAT EFFECTS OF ELECTRIC CURRENTS

(1) THE CHEMICAL EFFECT OF AN ELECTRIC CURRENT

280. Electroplating.—If two carbon rods (electric light carbons answer very well) are placed in a solution of *copper sulphate* (Fig. 260) and then connected by wires to the binding posts of an electric battery, one of the rods soon becomes covered with a coating of *metallic copper* while bubbles of gas may be seen upon the other carbon. If a solution of *lead acetate* is used in the same way a deposit of *metallic lead* is secured, while a solution of *silver nitrate* gives silver.

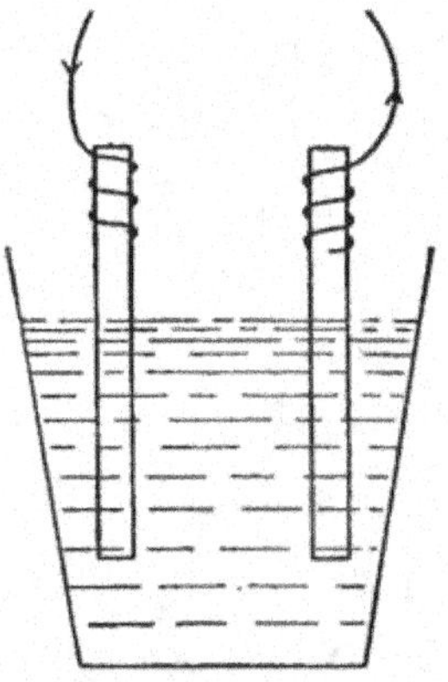

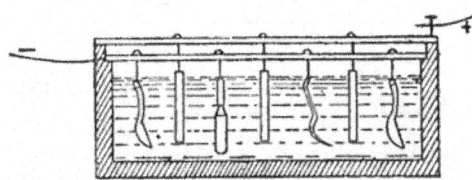

FIG. 260.—Two carbons placed in a solution of copper sulphate.

FIG. 261.—An electroplating bath.

This process of depositing metals upon the surface of solids by an electric current is called *electroplating*. Everyone has seen *electroplated* articles such as silver plated knives, forks, and spoons, and nickel-plated rods, handles, etc. *Copper electrotype* plates such as are used in printing school books are made by this process. In practical electroplating a solution of the metal to be deposited is placed in a tank; across the top of this tank are placed copper rods to act as conducting supports. From one of these rods, the cathode, objects to be plated are hung so as to be immersed in the liquid. From other rods, the anodes, are hung plates of the metal to be deposited. These are dissolved as the current deposits a coating upon the articles, thus keeping the solution up to its proper strength. (See Fig. 261.)

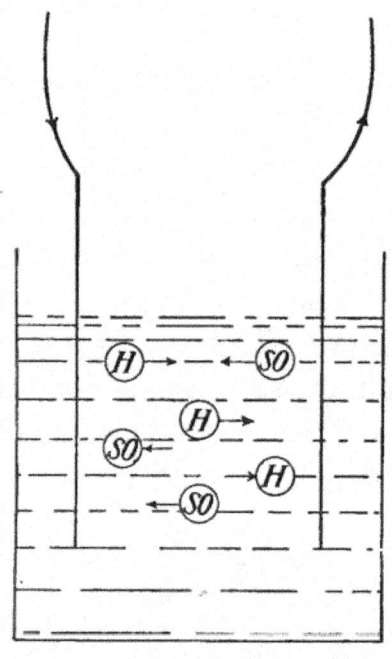

FIG. 262.—The current is carried through the solution by ions.

281. Electrolysis.—A solution from which a deposit is made by an electric current is called an *electrolyte*. The plates or other objects by which the current enters or leaves the electrolyte are called the *electrodes*. The electrode by which the current enters is called the *anode* (*an* = in) while the electrode by which it leaves is the cathode (*cath* = away). The process by which an electric current decomposes a solution and deposits a substance upon an electrode is called *electrolysis*. The *current* always flows within the cell from *anode to the cathode*. (See Fig. 262.) The metal goes with the current and is found deposited upon the cathode.

282. Theory of Electrolysis.—The action going on in an *electrolytic* cell has been carefully studied. The *theory of electrolysis*, which is supported by much experimental evidence, supposes that many of the molecules in a *dilute* solution of a substance "split up" into two parts called "ions," one ion having a positive, the other a negative charge. In a dilute solution of sulphuric acid, the *positive* ion is of hydrogen, while the *negative* ion is the (SO_4) or sulphion. These ions bearing electric charges are believed to be the *carriers of the electric current* through the electrolyte.

The positive ions move with the current from the anode to the cathode, while the negative ions apparently are repelled by the cathode and appear upon the anode. Evidence of the accumulation of the two kinds of ions at the two electrodes is furnished by the *electrolysis* of water, described below.

283. Electrolysis of Water.—Two glass tubes (Fig. 263), H and O, are attached at the bottom to a horizontal glass tube. To the latter is also connected an upright tube T. At the lower ends of H and O are inserted, fused in the glass, platinum wires, A and C. The tubes are filled with a weak solution of sulphuric acid. The tops of H and O are closed with stopcocks, T being open; a current of electricity is sent in at A and out at C. A movement of the ions at once be-

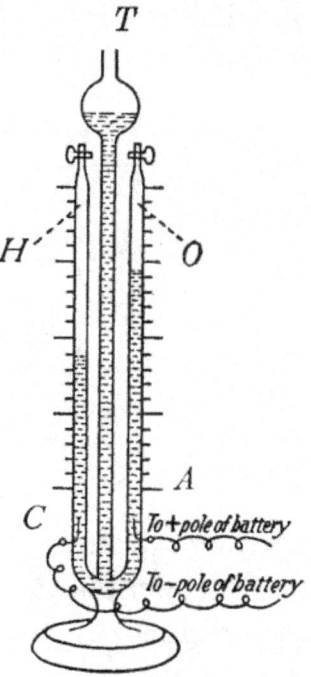

FIG. 263.—Electrolysis of water; oxygen collects in O, hydrogen in H.

gins, the positive hydrogen ions appearing at C. These accumulate as bubbles of hydrogen which rise to the top of H and displace the liquid. At the same time bubbles of oxygen appear at A. These rise in O and also displace the liquid which rises in T. After the action has continued some time it may be noticed that the volume of hydrogen is just twice that of the oxygen. This was to have been expected since the formula for water is H_2O. The nature of the gas in *H* or *O* may be tested by opening the stopcock and allowing the gas to escape slowly. The hydrogen gas can be lighted by a flame while the oxygen gas will cause a spark upon a piece of wood to glow brightly, but does not burn itself.

284. Evidence that ions are necessary to conduct a current in a liquid is furnished by the following experiment. A quart jar is carefully cleaned, and half filled with distilled water. Two pieces of zinc 5 cm. square are soldered to pieces of rubber-insulated No. 14 copper wire.

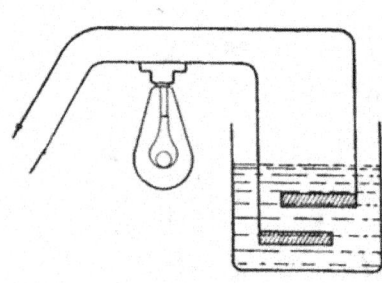

FIG. 264.—The current passes only when ions are present in the liquid.

The zincs are placed in the distilled water (Fig. 264) and the wires are connected to a 110 volt circuit with a 16 candle-power incandescent lamp in *series* with the cell, as in the figure. If the zincs have been carefully cleansed and the water is pure, no current flows as is shown by the lamp remaining dark. If a minute quantity of sulphuric acid or of common salt is placed in the water the lamp at once begins to glow. Ions are now present in the liquid and conduct the current. That some substances in solution do not form ions may be shown by adding to another jar of pure water some glycerine and some cane sugar, substances resembling the acid and salt in external appearance but which do nòt *ionize* when dissolved as is shown by the lamp remaining dark after adding the glycerine and sugar. The acid and salt are of *mineral* origin while the glycerine and sugar are *vegetable* products. This experiment illustrates the principle that the water will conduct only when it contains ions.

285. Laws of Electrolysis.—These were discovered by "Faraday in 1833, and may be stated as follows: *I. The mass of a substance deposited by an electric current from an electrolyte is proportional to the intensity of current which passes through it.*"

II. The mass of any substance deposited by a current of uniform intensity is directly proportional to the time the current flows.

These laws have been used as a basis for defining and measuring the unit of current flow, the *ampere*. (See Art. 264.)

286. Instances of Electrolysis.—(a) Medicines, especially those containing a mineral substance, are sometimes introduced into the human body by electrolysis. (b) Water and gas pipes are sometimes much weakened by the effects of electric currents in the earth, especially return currents from street railways. Such currents use the metal pipes as a conductor. At the place where the current leaves the metal and enters the ground, it removes metallic ions from the pipe. This process continuing, the pipe becomes weakened and at length breaks. (c) *Copper* is purified by the use of electric currents that remove the copper from ore or from other metals and deposit it upon electrodes. *Electrolytic* copper is the purest known. (d) *Aluminum* is obtained by the use of large currents of electricity that first heat the material used until it melts and then deposit the metal from the fluid material by electrolysis. These results are called chemical effects of the current since by the use of electric currents substances are changed chemically, that is, they are separated into different chemical substances.

Important Topics

1. Electrolysis, electroplating, anode, cathode, ion.
2. Theory of electrolysis—evidence: (a) electrolysis of water; (b) conductivity of acid and water.
3. Laws of electrolysis.
4. Practical use of electrolysis.

Exercises

1. A dynamo has an E.M.F. of 10 volts. What is the resistance in the circuit when 20 amperes are flowing?
2. How much silver will be deposited in an hour by this current?
3. Name five objects outside of the laboratory that have been acted upon by electrolysis. How in each case?
4. Why is table ware silver plated? Why are many iron objects nickel plated?
5. How is the electrolysis of water pipes prevented?
6. Two grams of silver are to be deposited on a spoon by a current of 1 ampere. Find the time required.
7. How long will it take to deposit 20 g. of silver in an electroplating bath if a current of 20 amperes is used?
8. If 1000 g. of silver are deposited on the cathode of an electrolytic reduction plant in 10 minutes, what is the current intensity employed?

(2) THE STORAGE BATTERY AND ELECTRIC POWER

287. Differences Between Voltaic and Storage Cells.
Voltaic cells in which electric currents are produced by the chemical action between metal plates and an electrolyte are often called *primary batteries*. In voltaic cells one or both plates and the electrolyte are used up or lose their chemical energy in producing the current and after a time need to be replaced by new material, the *chemical energy* of the electrolyte and of one of the plates having been *transformed* into electrical energy.

A different proceeding obtains with another type of

cell. This is called a *storage battery*, or an accumulator. In these cells, the same *plates* and electrolyte are *used* without change *for extended periods*, sometimes for a number of years. For this reason storage batteries have displaced many other types of cells, and they are now used (a) to operate many telephone, telegraph, and fire-alarm circuits, (b) to work the spark coils of gas and gasoline engines, (c) to help carry the "peak" load upon lighting and power circuits and (d) to furnish power for electric automobiles. Since a storage battery can deliver an electric current only after an electric current from an outside source has first been sent through it, they are often called *secondary batteries*.

288. Construction and Action of a Storage Cell.—The common type of storage cells consists of a *number of perforated* plates made of an alloy of lead and a little antimony. (See Figs. 265, 266, 267.) Into the perforations is pressed a paste of red lead and litharge mixed with sulphuric acid. The plates are placed in a strong solution (20 to 25 per cent.) of sulphuric acid. The plates are now ready to be charged. This is accomplished by sending a direct current from an electric generator through the cell. The hydrogen ions are moved by the current to one set of plates and change the paste to *spongy* metallic lead. The sulphions move to the other set of plates and change the paste to lead oxide. This electrolytic action causes the two plates to become quite different chemically so that when the cell is fully charged it is like a voltaic cell, in having plates that are different chemically. It has, when fully charged, an E.M.F. of about 2.2 volts. The several plates of a cell being in parallel and close together, the cell has but small internal resistance. Consequently a large current is available.

About 75 per cent. of the energy put into the storage cell

in charging can be obtained upon *discharging*. Therefore the *efficiency* of a good storage cell is about 75 per cent. Fig. 268 represents a storage battery connected to charg-

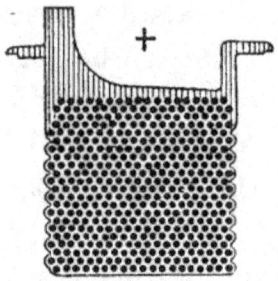

FIG. 265.—The positive plate of a storage cell.

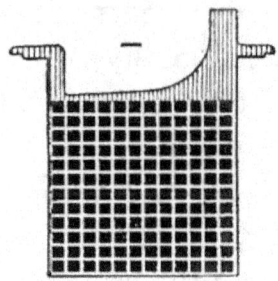

FIG. 266.—The negative plate of a storage cell.

ing and discharging circuits. The lower is the charging circuit. It contains a dynamo and a resistance (neither of which are shown in the figure) to control the current sent into the cell. The charging current enters the positive pole and leaves by the negative pole. The current produced by the cell, however, flows in the *opposite* direction through it, that is, out from the positive and in at

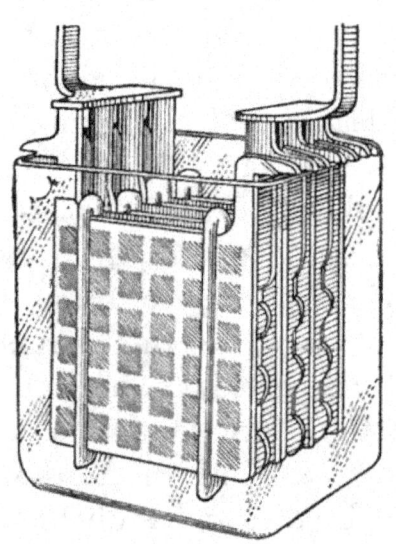

FIG. 267.—A complete storage cell.

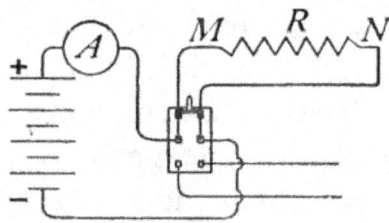

FIG. 268.

the negative pole. This current may be controlled by a suitable resistance and measured by an ammeter. Storage cells have several advantages: (a) They can be charged and discharged a great many times before the material placed in the perforations in the plates falls out.

(b) The electrical energy used in charging the plates *costs less* than the plates and electrolyte of voltaic cells. (c) Charging storage cells takes much *less labor* than replacing the electrolyte and plates of voltaic cells. (d) Storage cells produce *larger currents* than voltaic cells.

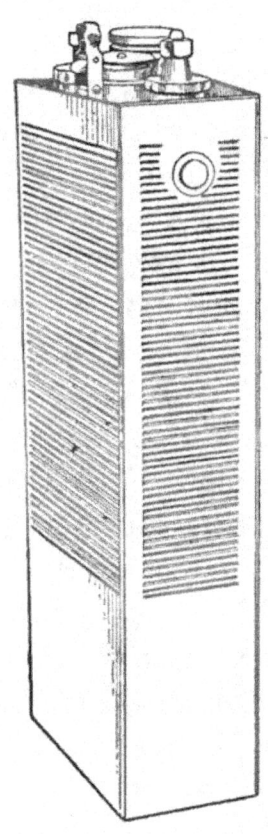

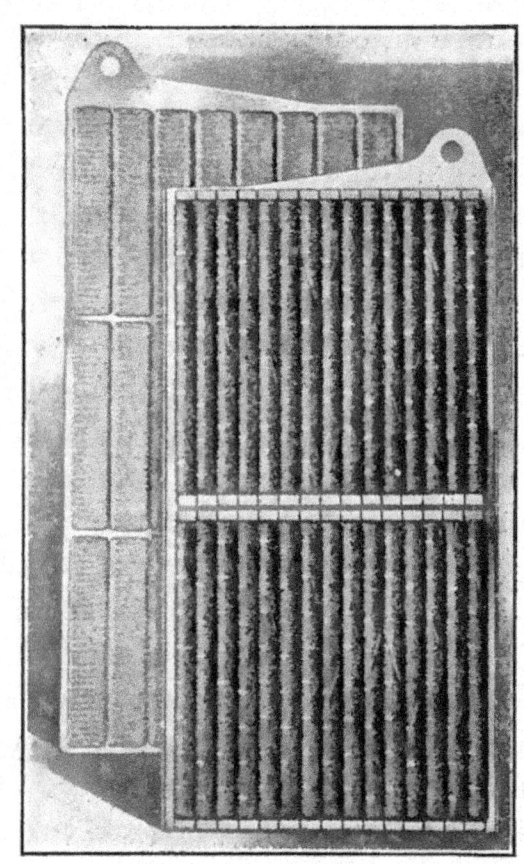

Fig. 269.—The Edison storage cell.

Fig. 270.—The plates of the Edison storage cell.

The two principal *disadvantages* of storage cells are that (a) they are *very heavy*, and (b) their initial *cost* is *considerable*.

289. The Edison storage cell (Figs. 269 and 270) has plates of iron and nickel oxide. The electrolyte is a strong solution of potassium hydroxide. These cells are lighter than lead cells of the same capacity and they are claimed to have a longer life.

290. Energy and Power of a Storage Cell.—In a storage cell, the electrical energy of the charging current is transformed into *chemical* energy by the action of electrolysis. It is this chemical energy that is transformed into the energy of the electric current when the cell is discharged. The *capacity* of storage cells is rated in "ampere hours," a 40 ampere hour cell being capable of producing a current of 1 ampere for 40 hours, or 5 amperes for 8 hours, etc. The production and extensive use of electric currents have made necessary accurate methods for measuring the *energy* and *power* of these currents. To illustrate how this is accomplished, let us imagine an electric circuit as represented in Fig. 268. Here four storage cells in series have an E.M.F. of 8 volts and in accordance with Ohm's law produce a current of 2 amperes through a resistance in the circuit of 4 ohms. Now the work done or energy expended by the current in passing through the resistance between the points M and N *depends* upon three factors (1) the E.M.F. or *potential difference;* (2) the *current intensity* and (3) the *time.* The energy is measured by their product. That is, *electrical energy = potential difference × current intensity × time.* This represents the electrical energy in *joules*, or

$$\text{Joules} = \text{volts} \times \text{amperes} \times \text{seconds, or}$$
$$j = E \times I \times t.$$

In the circuit represented in Fig. 268 the energy expended between the points M and N in 1 minute (60 seconds) is $8 \times 2 \times 60 = 960$ joules.

291. Electric Power.—Since power refers to the *time rate* at which work is done or energy expended, it may be computed by dividing the electrical energy by the time, or the *electrical power = volts × amperes*. The power of 1 joule per second is called a *watt*. Therefore,

Watts = volts × amperes, or
Watts = $E \times I$.

Other units of power are the *kilowatt* = 1000 watts and the *horse-power* = 746 watts. In the example given in Art. 290 the power of the current is $8 \times 2 = 16$ watts, or if the energy of the current expended between the joints M and N were converted into mechanical horse-power it would equal $^{16}\!/_{746}$ of a horse-power. Electrical energy is usually sold by the *kilowatt-hour*, or the amount of electrical energy that would exert a power of 1000 watts for one hour, or of 100 watts for 10 hours, or of 50 watts for 20 hours, etc.

Important Topics

1. The storage battery, its construction, electrolyte, action, uses, advantages, disadvantages.
2. Electric energy, unit value, how computed?
3. Electric power, three units, value, how computed, how sold?

Exercises

1. In what three respects are voltaic and storage cells alike? In what two ways different?
2. Name the four advantages of storage cells in the order of their importance. Give your reasons for choosing this order.
3. Why are dry cells more suitable for operating a door-bell circuit, than a storage battery? Give two reasons.
4. The current for a city telephone system is provided by a storage battery. Why is this better than dry cells at each telephone?
5. An incandescent lamp takes 0.5 ampere at 110 volts. What power is required to operate it? How much *energy* will it transform in 1 minute?
6. How long would it take for this lamp to use a kilowatt hour of energy?
7. A street car used 100 amperes at 600 volts pressure. What

power was delivered to it? Express also in kilowatts and horsepower.

8. An electric toaster takes 5 amperes at 110 volts. If it toasts a slice of bread in 2 minutes, what is the cost at 10 cents a kilowatt hour?

9. An electric flat iron takes 5 amperes at 110 volts. Find the cost of using it for 2 hours at 12 cents a kilowatt hour.

10. A ¼ kilowatt motor is used to run a washing-machine for 5 hours. What is the expense for this power at 10 cents a kilowatt hour?

11. What is the efficiency of a motor that takes 7390 watts and develops 9 horse-power?

12. How many horse-power are there in a water-fall 212 ft. high over which flows 800 cu. ft. of water per second? Express this power in kilowatts.

13. What horse-power must be applied to a dynamo having an efficiency of 90 per cent. if it is to light 20 arc lamps in series, each taking 10 amperes at 60 volts?

(3) The Heat Effect of Electric Currents

292. The Production of Heat by an Electric Current.— When no chemical or mechanical work is done by an electric current its energy is employed in overcoming the resistance of the conducting circuit and is transformed into *heat.* This effect has many practical applications and some disadvantages. Many devices employ the heating effect of electric currents, (a) the electric furnace, (b) electric lights, (c) heating coils for street cars, (d) devices about the home, as flat irons, toasters, etc. Sometimes the heat produced by an electric current in the wires of a device such as a transformer is so large in amount that especial means of cooling are employed. Unusually heavy currents have been known to melt the conducting wires of circuits and electrical devices. Hence all circuits for electric power as well as many others that ordinarily carry small currents are protected by *fuses.* An *electric* *fuse* is a short piece of wire that will melt and break the circuit if the current exceeds a determined value. The fuse wire is usually enclosed in an incombustible holder. Fuse wire is frequently made of lead or of an alloy of lead and other easily fusible metals. (See Figs. 271 and 272.)

293. Heat Developed in a Conductor.—A rule for computing the amount of heat produced in an electric circuit by a given current has been accurately determined by experiment. It has been found that 1 *calorie* of heat (Art. 142), is produced by

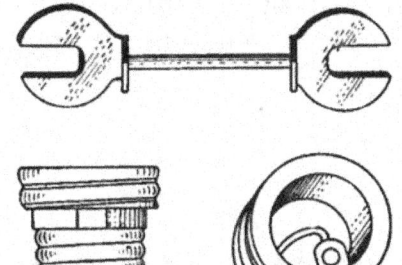

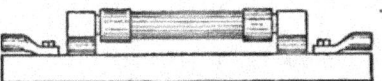

FIG. 271.—A type of enclosed fuse. FIG. 272.—A link fuse (above); plug fuses (below).

an expenditure of 4.2 joules of electrical (or other) energy. In other words, 1 joule will produce $\frac{1}{4.2}$ or 0.24 calorie. Now the number of joules of electrical energy in an electric circuit is expressed by the following formula:

Joules = volts $\times$ amperes $\times$ seconds, or since 1 joule = 0.24 calorie,

Calories = volts $\times$ amperes $\times$ seconds $\times$ 0.24 or

$$H = EI \times t \times 0.24 \qquad (1)$$

By Ohm's law, $I = \dfrac{E}{R}$ or $E = I \times R$, substituting in equation (1) IR for its equal E we have

$$H = I^2R \times t \times 0.24 \qquad (2)$$

Also since $I = \dfrac{E}{R}$ substitute $\dfrac{E}{R}$ for I in equation (I) and we have

$$H = \frac{E^2}{R} t \times 0.24 \qquad (3)$$

To illustrate the use of these formulas by a problem suppose that a current of 10 amperes is flowing in a circuit
having a resistance of 11 ohms, for 1 minute. The heat
produced will be by formula (2) = $(10)^2 \times 11 \times 60 \times 0.24$
equals 15,840 calories.

294. The Incandescent Lamp.—One of the most common devices employing the heat effect of an electric current is the *incandescent lamp*. (See Fig. 273.) In this
lamp the current is sent through a carbon filament,

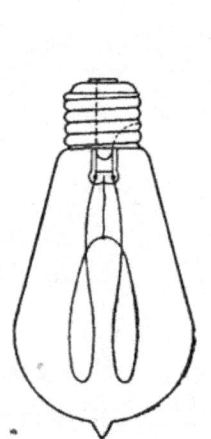

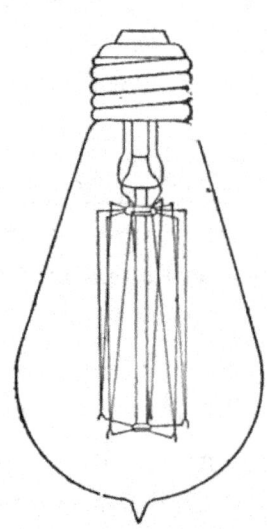

Fig. 273.—A carbon filament   Fig. 274.—A   tungsten
  incandescent lamp.   lamp.

which is heated to incandescence. In order to keep the
filament from burning as well as to prevent loss of heat
by convection, it is placed in a glass bulb from which the
air is exhausted. Two platinum wires fused in the glass
connect the carbon filament with the grooved rim and the
end piece of the base. The end piece and rim connect with
the socket so that an electric current may flow through the
filament of the lamp. The carbon incandescent lamp has
a low efficiency. It takes 0.5 ampere of current at 110
volts or in other words it requires 55 watts to cause a 16-

candle-power lamp to glow brightly, hence 1 candle power in this lamp takes $55 \div 16 = 3.43$ watts.

The *efficiency of electric lamps* is measured by the *number of watts per candle power*. This is a peculiar use of the term efficiency, as the larger the number the less efficient is the lamp. More efficient lamps have been devised with filaments of the metals *tantalum* and *tungsten* (Fig. 274). These give a whiter light than do carbon lamps, and consume but about 1.25 watts per candle power.

COMPARATIVE "EFFICIENCY" OF ELECTRIC LAMPS

Name of lamp	Watts per candle power	Name of lamp	Watts per candle power
Carbon filament............	3 to 4	Arc lamp	0.5 to 0.8
Metallized carbon..........	2.5	Mercury arc	0.6
Tantalum..................	2.0	Flaming arc	0.4
Tungsten..................	1.0 to 1.5	Nitrogen-filled tungsten	0.6 to 0.7

Incandescent lamps are connected in parallel (see Fig. 254) to wires that are kept at a constant difference of potential of 110 or 115 volts. It is customary to place not more than twelve lamps upon one circuit, each circuit being protected by a fuse and controlled by one or more switches.

295. The Arc Light.—The electric *arc* light (see Fig. 275) is extensively used for lighting large rooms, also in stereopticons and motion picture machines. The light is intense, varying from 500 to 1700 candle power. The so-called mean spherical candle power of the arc light is about 510. The candle power in the direction of greatest intensity is about 1200. It is produced at an expenditure of

21

about 500 watts. It is therefore more efficient than the incandescent lamp, often taking less than 0.5 watt per candle power produced. The arc light was first devised by Sir Humphrey Davy in 1809, who used two pieces of charcoal connected to 2000 voltaic cells. The arc light requires so much power that its production by voltaic cells is very expensive. Consequently it did not come into common use until the dynamo had been perfected. Fig. 276

FIG. 275.—An electric arc light.

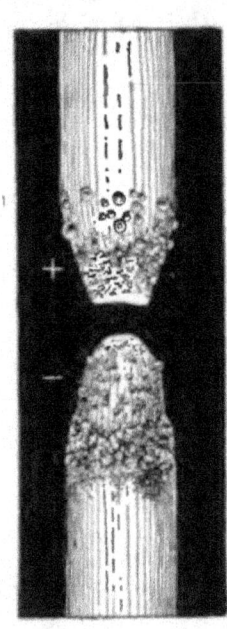

FIG. 276.—The appearance of a pair of used carbons.

shows the appearance of the two carbons in an arc light. If a direct current is used the positive carbon is heated more intensely, and gives out the greater part of the light. The positive carbon is consumed about twice as fast as the negative and its end is concave, the negative remaining pointed.

With alternating currents, the rods are equally consumed and produce equal amounts of light. In the stereopticon, the carbons are usually placed at right angles as in Fig. 277. In the stereopticon as well as in outdoor lighting the direct current is more effective, although the alternating current is often used, since the latter can be produced and distributed more cheaply than can direct currents. In arc lamps, placing an inner glass globe (Fig. 278) about the carbons, decreases the consumption of the carbons materially. The carbon rods of *enclosed* arc lamps often last 60 to 100 hours.

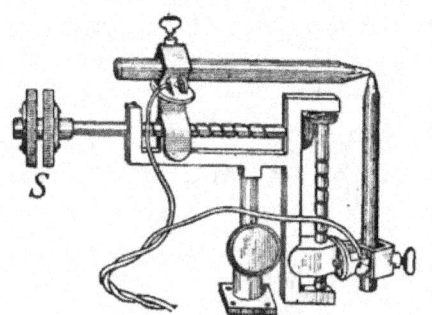
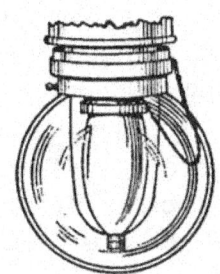

FIG. 277.—A right-angle electric   FIG. 278.—An enclosed
arc lamp for a stereopticon.   arc lamp.

The reason why an *open* arc lamp needs to be "retrimmed" oftener than the *enclosed* lamp, that is, have new carbons placed in it, is because the carbons "burn" freely, that is unite with the oxygen of the air. In the enclosed arc lamp, the supply of oxygen in the inner globe is limited and is soon consumed, therefore the carbons last many times longer in such lamps.

Some carbon rods have soft cores containing calcium salts. These vaporize in the arc producing the *flaming arc light* of a bright yellow color, and give more light than the ordinary lamp.

Important Topics

1. Heat effects of electric currents, uses and applications.
2. Computation of the heat developed in a circuit. Three formulas.
3. Electric lamps; incandescent and arc; construction, uses, efficiency.

Exercises

1. Sketch a circuit containing 10 incandescent lamps in parallel. If each lamp when hot has a resistance of 220 ohms, and the E.M.F. is 100 volts, what current will flow?
2. What will it cost to use these lights for 3 hours a day for 30 days at 10 cents a kilowatt hour?
3. How much heat will these lamps produce per minute?
4. How could you connect 110-volt lamps to a street car circuit of 660 volts? Explain this arrangement and draw a diagram.
5. A certain arc lamp required 10 amperes of current at 45 volts pressure. What would it cost at 10 cents per kilowatt hour if used 3 hours a day for 30 days?
6. Show a diagram of 3 arc lamps in series. If each takes 45 volts and 10 amperes, how much E.M.F. and current will they require?
7. If an electric toaster uses 5 amperes at 115 volts, how much heat will this develop in half an hour?
9. How much heat is developed in an electric toaster in 2 minutes, if it uses 5 amperes at 100 volts?
10. How many B.t.u.'s are given off in an electric oven that takes 10 amperes at 110 volts for 1 hour? (1 B.t.u. equals 252 calories.)
11. An electric heater supplies heat at the rate of 700 B.t.u.'s an hour. How much power does it require?
12. How many watts are required to operate 120 incandescent lamps in parallel if each takes 0.5 amperes at 110 volts?
13. An electric lamp takes 12 amperes at a P.D. of 110 volts. How many B.t.u.'s are radiated from it each second? How many calories?
14. If a 110-volt incandescent lamp is submerged for 10 minutes in 400 gr. of cold water while a current of 0.5 amperes is flowing, how many degrees centigrade will the water be warmed?
15. In an electric furnace a current of 3000 amperes is used át a P.D. of 10 volts. Find the heat developed in 1 minute.
16. How many candle power should a 20-watt tungsten lamp give if its efficiency is one watt per candle power?
17. What is the "efficiency" of a 40-watt tungsten lamp if it gives 34 candle power?

Review Outline: Current Electricity

Produced by—Chemical action; voltaic and storage cells.

Three { Magnetic, electromagnet, uses and applications.
Principal { Chemical, electrolysis, applications.
Effects: { Heat, lighting and heating devices.

Theories: (a) of voltaic cells, (b) of electrolysis.

Units: Ampere, ohm, volt, watt, joule, kilowatt, horse power.

Measurement—(a) magnetic effect; galvanometer, ammeter, voltmeter, wattmeter, Wheatstone bridge, construction and use.

(b) chemical effect; voltameter.

Laws: (a) Right hand rules, for conductor and helix.

(b) Resistance, Conductors in series and parallel.

(c) Ohm's law, heat law, power law, 3 forms for
each.

(d) Cells in parallel and series.

Problems: Upon applications of the laws and formulas studied.

Devices { Voltaic cells; wet, dry, and Daniell.
and { Electrolysis and the storage battery.
Instruments: { Measuring instruments, electric bell, sounder.
{ heating and lighting devices.

Terms: Anode, cathode, electrolyte, ion, circuit switch, current,
e.m.f., resistance, potential.

CHAPTER XIV

INDUCED CURRENTS

(1) Electromagnetic Induction

296. Current Induced by a Magnet.—The discovery in 1819 that a current in a conductor can deflect a magnetic needle or that it has a magnetic effect, led to many attempts *to produce an electric current by means of a magnet.* It was not until about 1831, however, that *Joseph Henry* in America and *Michael Faraday* in England, independently discovered how to accomplish this important result.

At the present time, voltaic cells produce but a very small part of the current electricity used. Practically all that is employed for *power, light, heat, and electrolysis is produced by the use of magnetic fields, or by electromagnetic induction.*

297. Laws of Induced Currents. [1]—To illustrate how a current can be produced by electromagnetic induction:

Connect a coil of 400 or more turns of No. 22 insulated copper wire to a sensitive galvanometer. (See Fig. 279.) Now insert a bar magnet in the coil. A sudden movement of the galvanometer will be noticed, indicating the *production of a current.* When the magnet stops moving, however, the current stops, and the coil of the galvanometer returns to its first position. If now the magnet is removed, a movement of the galvanometer coil *in the opposite direction is* noticed. This action may be repeated as often as desired with similar results.

Careful experiments have shown that it is the *magnetic field* of the magnet that produces the action, and that

[1] An induced current is one produced by changing the number of magnetic lines of force passing through a coil.

only when the *number of lines of force in the coil is changing* do we find a current produced in the coil. These facts lead to *Law I. Any change in the number of magnetic lines of force passing through or cut by a coil will produce an electromotive force in the coil.* In the account of the experiment just given, *electric currents* are produced, while in Law I, *electromotive forces* are mentioned. This difference is due to the fact that an E.M.F. is *always* produced in a

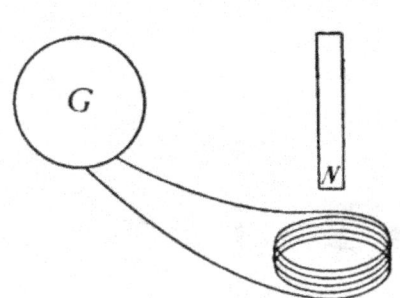

FIG. 279.—The moving magnet induces a current in the coil.

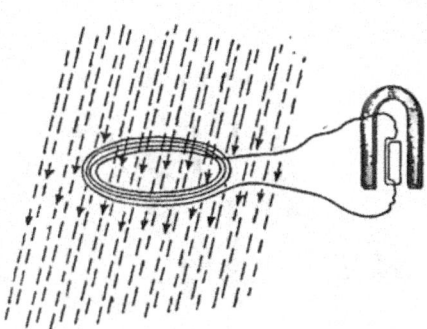

FIG. 280.—A current may be induced by turning the coil in the earth's magnetic field.

coil when the magnetic field within it is changed, while a current is found only when the coil is part of a *closed circuit.* The inductive action of the earth's magnetic field (see Fig. 280), may be shown by means of a coil of 400 to 500 turns a foot in diameter.

Connect its ends to a sensitive galvanometer and hold it at right angles to the earth's field. Then quickly revolve the coil through 180 degrees and note the movement of the galvanometer. Reverse the coil and the galvanometer swings in the opposite direction.

If the magnet in Fig. 279 is moved *in* and *out* of the coil at first *slowly* and *later swiftly, small and large* deflections of the galvanometer coil are noticed. The quicker the movement of the magnetic field the greater are the galvanometer deflections produced. This leads to *Law II.*

The electromotive forces produced are proportional to the number of lines of force cut per second.

298. The magneto is a device that illustrates the laws of induced currents stated in Art. 297. The magneto (see Fig. 281), consists of several permanent, "U"-shaped magnets placed side by side. Between the poles of these magnets is placed a slotted iron cylinder having a coil of many turns of fine insulated copper wire wound in the slot as in Fig. 282. The cylinder and coil form what is

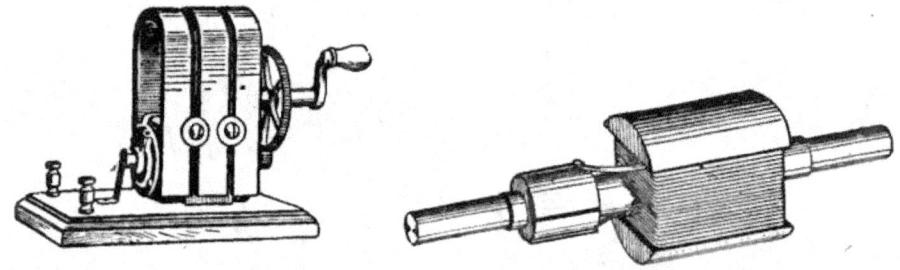

Fig. 281.—A magneto.　　　Fig. 282.—A shuttle armature.

called an *armature*. The armature is mounted so as to be revolved between the poles of the "U"-shaped magnets by means of a handle. As the armature revolves, the lines of force from the magnets pass through the coil first in one direction and then in the other. This repeated change in the lines of force passing through the coil produces an E.M.F. which may be felt by holding in the hands the two wires leading from the armature coil. On turning the armature *faster* the current is felt *much stronger*, showing that the E.M.F. in the coil increases as the rate of cutting the magnetic lines of force by the coils increases.

299. Lenz's Law.—While one is turning the armature of a magneto if the two wires leading from its coil are connected, forming what is called a "short circuit," the difficulty of turning the armature is at once increased. If now the circuit is broken, the armature turns as easily as at first. The increased difficulty in turning the armature is due to the *current* produced in the coil. This current sets up a magnetic field of its own that opposes the field from the steel magnets. This opposition makes it necessary for *work* to be done to keep up the motion of the coil when a current is passing through it. This fact is called *Lenz's Law.* It may be expressed as follows: *Whenever a current is induced by the relative motion of a magnetic field and a conductor, the direction of the induced current is always such as to set up a magnetic field that*

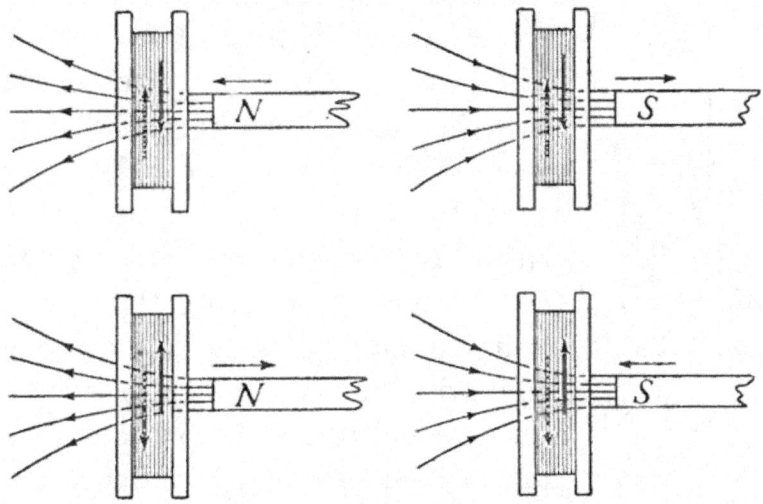

Fig. 283.—The induced current has a field which opposes the motion of the magnet. The heavy line represents the direction of the induced current.

opposes the motion. Lenz's Law follows from the principle of conservation of energy, that energy can be produced only from an expenditure of other energy. Now since an electric current possesses energy, such a current can be produced only by doing mechanical work or by expending some other form of energy. To illustrate Lenz's Law, suppose that the north-seeking pole of a bar magnet be inserted in a closed coil of wire. (See Fig. 283.) The current induced in the coil has a direction such that its lines

of force will pass within the coil so as to *oppose* the field
of the bar magnet, when the north pole of the magnet is
inserted so as to point to the left.   That is, the north pole
of the helix is at the right.   Applying the right-hand rule
to the coil, its current will then be *counter clockwise*.   On
withdrawing the magnet, the current reverses, becoming
clockwise with its field passing to the left within the coil.

A striking illustration of the opposition offered by the
field of the induced current to that of the inducing field
is afforded by taking a strong electro magnet (see Fig.

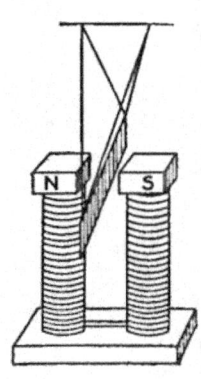

284) and suspending a sheet of copper so
as to swing freely between the poles.
When no current flows through the magnet the sheet swings easily for some
time. When, however, the coils are
magnetized, the copper sheet has induced
within it, currents that set up magnetic
fields strongly opposing the motion, the
swinging being stopped almost instantly.

FIG. 284.—The
magnetic field
stops the swinging
of the sheet of
copper.

The principle is applied in good ammeters and voltmeters to prevent the swinging of the needle when deflected.   The
current induced in the metal form on
which is wound the galvanometer coil is sufficient to make
the needle practically "dead beat."

300. The Magneto and the Dynamo.—Magnetos are
used to *develop small* currents, such as are used for telephone signals, and for operating the *sparking* devices of
gasoline *engines*. They are therefore found in automobiles containing gasoline motors. The most important
device for producing electric currents by electromagnetic
induction, however, is the *dynamo*. It is employed whenever large currents are desired. The principle of this
device is similar to that of the magneto except that it con-

Lord Kelvin (Sir William Thomson), (1824–1907). Professor of Physics, Glasgow University. Invented the absolute scale of temperature: also many practical electrical measuring instruments. The foremost physicist of the latter part of the nineteenth century.

LORD KELVIN
"By Permission of the Berlin Photographic Co., New York."

Michael Faraday (1791–1867). Famous English Physicist. Made many discoveries in electricity and magnetism; "Greatest experimentalist of the nineteenth century."

MICHAEL FARADAY
"By Permission of the Berlin Photographic Co., New York."

tains an *electromagnet* for producing the magnetic field. Since the electromagnet can develop a much stronger field than a permanent magnet, the dynamo can produce a higher E.M.F. and a much larger current than the magneto.

301. The Magnetic Fields of Generators.—In the magneto, the magnetic field is produced by *permanent* steel magnets. In dynamos powerful *electromagnets* are used.

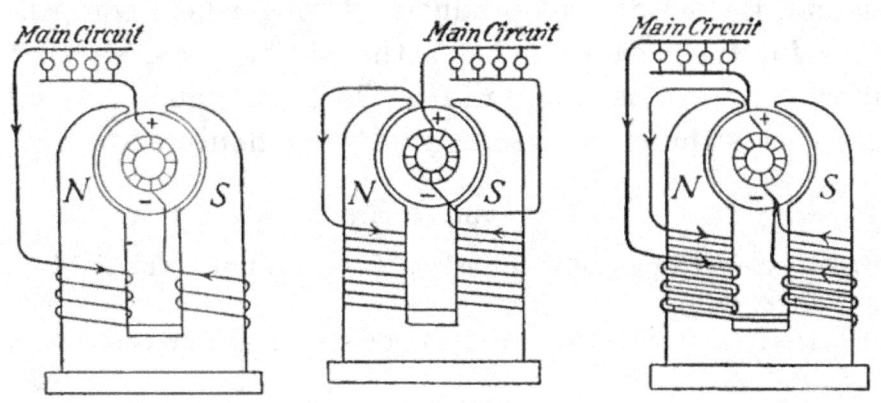

FIG. 285.—A series-wound dynamo.   FIG. 286.—A shunt-wound dynamo.   FIG. 287.—A compound-wound dynamo.

The latter are sometimes excited by currents from some other source, but usually current from the armature is sent around the field coils to produce the magnetic fields. Dynamos are classified according to the manner in which the current is sent to their field coils.

A. The *series wound dynamo* (see Fig. 285) is arranged so that *all* of the current produced by the armature is sent through coils of coarse wire upon the fields, after flowing through the external circuit.

B. The *shunt wound dynamo* (see Fig. 286) sends a part only of the current produced through the field coils. The latter are of many turns of fine wire so as to use as little current as possible. The greater part of the current goes to the main circuit. If the number of lamps or motors connected to the main circuit is increased, the voltage is lessened which weakens the current in the field coils, causing a weaker field and still lower voltage, producing a fluctuating E.M.F. which is unsatisfactory for many purposes. This fault is overcome by

C. the *compound wound dynamo*. This dynamo has both shunt and series coils upon its fields. (See Fig. 287.) If more current is drawn into the main circuit with this dynamo, the series coils produce a stronger field compensating for the weaker field of the shunt coils, so that uniform voltage is maintained. The compound wound generator is therefore the one most commonly employed.

Important Topics

1. Laws of electromagnetic induction (a) conditions, (b) E.M.F., (c) direction.
2. Devices, (a) magneto, (b) dynamo: series, shunt, compound.
3. Illustrations of the laws.

Exercises

1. Under what conditions may an electric current be produced by a magnet?
2. Show how Lenz's Law, follows from the principle of conservation of energy.
3. A bar magnet is fixed upright with its north-seeking pole upward. A coil is thrust down over the magnet. What is the direction of the current induced in the coil? Explain.
4. In what two ways may a current be induced in a closed coil?
5. What method is employed in the magneto? In the dynamo?
6. What is the nature of the current produced in the armature coil of a magneto, that is, is it direct or alternating? Why?
7. What is the resistance of a 20-watt tungsten lamp if the E.M.F. is 115 volts?
8. Find the resistance of a 40-watt tungsten lamp when the voltage is 115? How much heat will it produce per minute?
9. An Edison storage battery cell on a test gave a discharge of 30 amperes. The average voltage was 1.19. What was the resistance of the cell?

10. Eight storage cells are connected in series.   Each has an E.M.F. of 1.2 volts and an internal resistance of 0.03 ohms.   What will be the current flowing through a voltmeter having 500 ohms resistance in circuit with them?

(2) THE DYNAMO AND THE MOTOR

302. The Dynamo may be defined as a machine for transforming mechanical energy into the energy of electric currents by electromagnetic induction.   Although elec-

 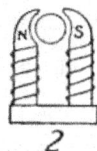 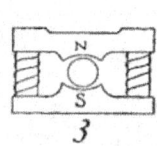 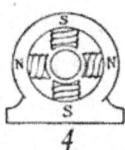 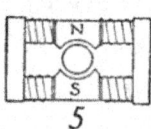

1 *2* *3* *4* *5*

FIG. 288.—Several methods of arranging the field coils and the armature of a dynamo.

tromagnetic induction was discovered in 1821, practical dynamos were not built for about 40 years or until between 1860 and 1870.   The great development in the production and use of electric currents has come since the latter date. The principle parts of the dynamo are (a) the *field magnet*, (b) the *armature*, (c) the *commutator* or *collecting rings*, (d) the *brushes*. Fig. 288 shows several common methods of arranging the field coils and the armature.

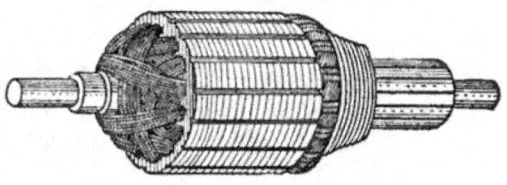

FIG. 289.—A drum armature.

The field coils vary in number and position.   The purpose of their construction is always to send the largest possible number of lines of force through the armature. Some dynamos are *bipolar*, or have *two* poles, others are multipolar or have more than two.   In Fig. 288 No. 4 has four poles.   The *armature* of a dynamo differs from a magneto armature in that it consists of a series of coils of insulated copper wire wound in numerous slots cut in the surface of a cylindrical piece of iron. Fig. 289 shows a *side* view of the iron core of such an armature. Iron is used to form the body of the armature since the magnetic lines of force flow easily through the iron. The iron by its permeability also concentrates and increases the magnetic flux. The best armatures are made of many thin sheets of soft iron. These are called *laminated* armatures. An armature made of a solid piece of iron becomes hot when revolving in a magnetic field. This is due to electric currents induced in the iron itself. This heating is largely reduced by *laminating* the armature. Why?

303. Methods of Collecting Current from the Armature. —The electric currents produced in the armature are conducted away by *special sliding contacts*. The stationary part of the sliding contact is called a *brush*. The

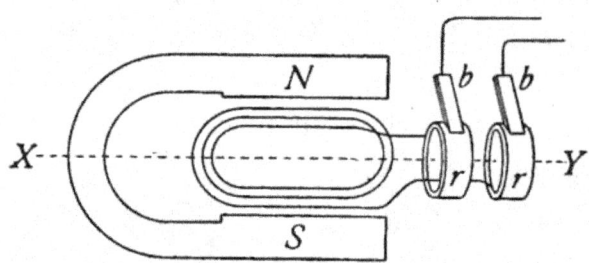

FIG. 290.—Armature connected to slip rings producing an alternating current.

moving part is a *slip ring* or a *commutator*. Fig. 290 shows an armature coil connected to slip rings. As the armature revolves, the coils and slip rings revolve with it. The two ends of the armature coils are connected to the two rings respectively. Now as the armature revolves it cuts the lines of force first in one direction and then in the other. This produces in the coils an E.M.F. first one way and then the other. This E.M.F. sets up a current which is conducted to the outside circuits through the slip rings and brushes. Such a current which repeatedly reverses its direction is called an *alternating current*. Fig. 291 (1) indicates graphically how the current moves alternately one way and then the other. Alternating currents are extensively used for electric *light, heat, and power*. *Direct currents* or those going continuously in one direction are however in much demand especially for *street car service, for electrolysis*, and for *charging storage batteries*.

304. The Commutator.—For a dynamo to deliver a *direct current* it must carry upon the shaft of the armature a *commutator*. The commutator is used to *reverse* the connections of the ends of the armature coils at the instant that the current changes its direction in the

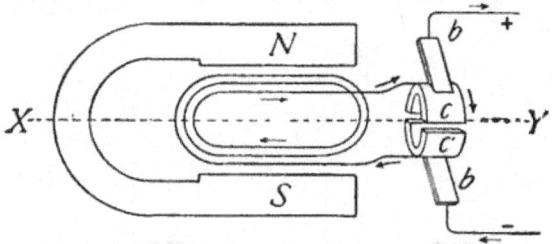

FIG. 291.—The armature coils are connected to a commutator producing a direct current.

armature. This reversal of connection when the direction of current changes, keeps the current in the outside circuit flowing in the same direction. Fig. 291 is a diagram of an armature with a commutator. The commutator is a *split ring*, having as many parts or *segments* as there are coils upon the armature. The brushes touch opposite points upon the commutator as they slide over the surface of the latter. Suppose that the armature viewed from the commutator end rotates in a counter-clockwise direction, also that the currents from the upper part move toward the commutator and out the top brush.

As the armature revolves, its coils soon begin to cut the force lines in the opposite direction. This change in the direction of cutting the lines of force causes the current to reverse in the coils of the armature. At the instant the current changes in direction, what was the upper segment

22

of the commutator slips over into contact with the lower brush, and the other segment swings over to touch the upper brush. Since the current has reversed in the coils

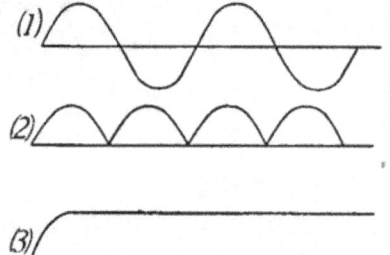

it continues to flow out of the upper brush. This change in connection at the brushes takes place at each half turn of the armature, just as the current changes in direction in the coils. This is the manner in which the commutator of a dynamo changes the alternating current produced in the armature coils, into a direct current in the external circuit.

FIG. 292.—Graphic representation of (1) an alternating current; (2) a pulsating current; (3) a continuous current.

Fig. 292 (1) represents graphically an alternating current, (2) of the

FIG. 293.—DeLaval multi-stage turbine and gear driving 750-kw., 750-r.p.m., 600-volt direct-current generator.

same figure shows current taken from the brushes of the commutator of a dynamo with one coil on the armature.

A practical dynamo, however, has many coils upon its

armature with a corresponding number of segments upon
the commutator. (See Figs. 289 and 293.) As each coil
and commutator segment passes a brush, it contributes
an impulse to the current with the result that armatures
with many coils produce currents that flow quite evenly.
(See Fig. 292, 3.)

The current represented in Fig. 292 (2) is called a *pulsating* current.

305. The electric motor is a machine which transforms
the energy of an electric current into mechanical energy
or motion. The *direct current motor* consists of the same

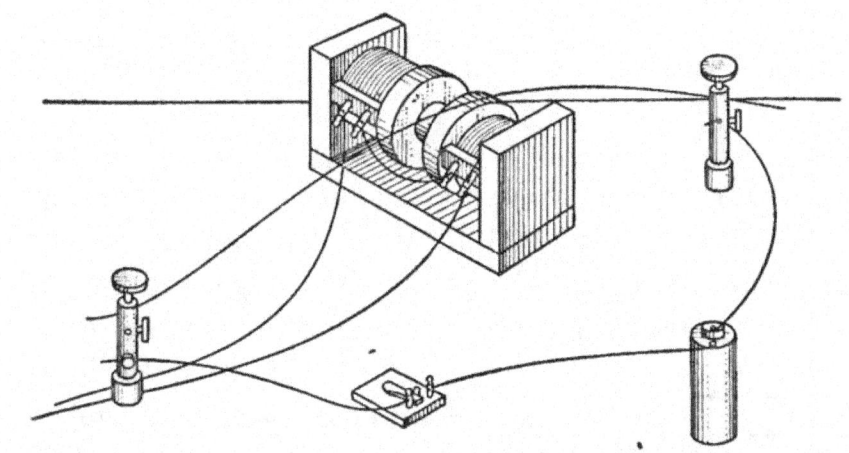

FIG. 294.—A wire carrying a current across a magnetic field is pushed
sideways by the field.

essential parts as a direct current dynamo, viz., the field
magnet, armature, commutator and brushes. Its operation is readily comprehended after one understands the
following experiment:

Set up two bar electromagnets with unlike poles facing
each other about an inch apart. A wire connected to a
source of current is hung loosely between the poles as in
Fig. 294. The circuit through the wire should contain a
key or switch. If a current is sent through the electromagnets and then another is sent through the wire, the

latter will be found to be pushed either up or down, while
if the current is reversed through the wire it is pushed in
the opposite direction.   These results may be explained
as follows:

Consider the magnetic field about a wire carrying a cur-

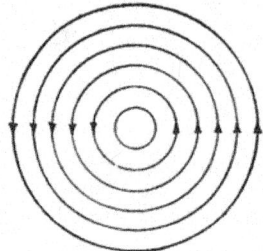

FIG. 295.—The mag-
netic field about a wire
carrying a current.

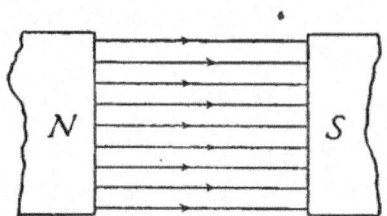

FIG. 296.—The magnetic field between two unlike poles.

rent.   (See Fig. 295.)   If such a wire is placed in the magnetic field between two opposite poles of an electromagnet
(Fig. 296), the wire will be moved either up or down.
The reason for this is shown by the diagram in Fig. 297.
Here a wire carrying a current and therefore surrounded
by a magnetic field passes
across another magnetic
field.   The two fields
affect each other causing
a crowding of the force
lines either above or below
the wire.   The wire at
once tends to move side-
ways across the field away

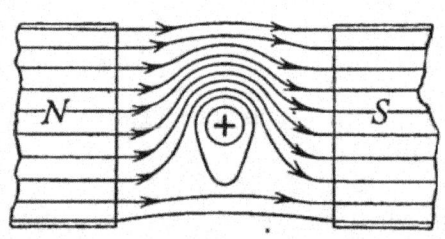

FIG. 297.—The crowding of the lines
of force above the wire, pushes it
downward.

from the crowded side.   In the figure, the wire tends to
move downward.

In a practical motor, the wires upon the armature are
so connected that those upon one side (see Fig. 298), carry
currents that pass in, while on the other side they pass out
To represent the direction of the current in the wires, the

following device is employed; a circle with a cross (to represent the feather in the tail of an arrow) indicates a current going away from the observer, while a circle with a dot at its center (to represent the tip of an arrow) indicates a eurrent coming toward the observer.

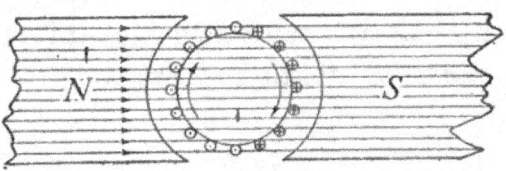

In Fig. 298 the north pole is at the left and the south pole at the

FIG. 298.—The crowding of the lines of force causes the armature to revolve in a clockwise direction.

right. The field of the magnets therefore passes from left to right as indicated in the figure. Now in the armature the currents in the wires on the left half of the armature are coming toward the observer while those on the right move away. Applying the right-hand rule, the magnetic lines will crowd *under* the wires on the left side of the armature while they will crowd *over* the wires on the right side. This will cause a rotation up on the left side and down on the right, or in a *clockwise* direction.

FIG. 299.—View of a one-half horse-power motor.

If the current in the armature is reversed (in on the left and out on the right), the lines of force will crowd the armature around in the opposite direction or *counter clockwise*. The rotation of the armature will also be reversed if, while the current in the armature is unchanged in direction, the poles of the magnet are changed thus reversing the magnetic field.

The motorman of a street car reverses the motion of his car by reversing the direction of the current in the *armature* of the motor.

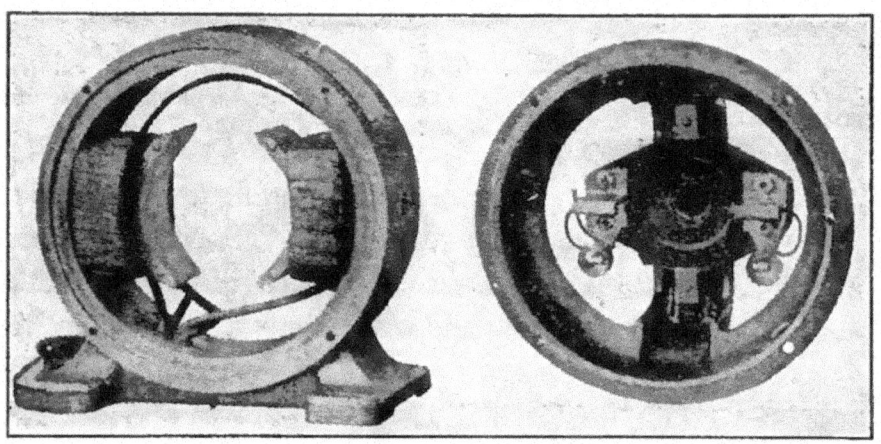

FIG. 300.—The frame and electromagnet (at left), front bracket and brush holder (at right) of the motor shown in Fig. 299.

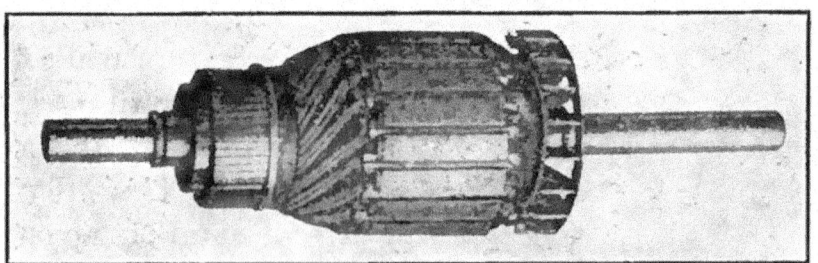

FIG. 301.—The armature of a motor.

306. Practical motors have many coils upon the armature with a corresponding number of segments upon the commutator. A large number of coils and commutator segments enables some one of the coils to exert its greatest efficiency at each instant, hence a steady force is provided for turning the armature which causes it to run smoothly.

Fig. 299 represents a ½ horse-power motor ready for use while Fig. 300 shows the frame and poles and the front bracket and brush holder, and Fig. 301 represents the armature.

Important Topics

1. The dynamo, four essential parts, action (a) for alternating currents, (b) for direct currents.
2. The electric motor: (a) essential parts, (b) action.

Exercises

1. *Why* is an alternating current produced in the armature of a dynamo?
2. *How* is this current produced?   Give careful explanations.
3. What is the result of Lenz's law as applied to the dynamo?
4. Apply the first two laws of electromagnetic induction to the dynamo.
5. What is the power of a dynamo if it produces 40 amperes of current at 110 volts?
6. How much power must be applied to this dynamo if its efficiency is 90 per cent.?
7. A motor takes 10 amperes of current at 220 volts; what is the *power* of the current in *watts?*   If this motor has an efficiency of 95 per cent., how many horse-power of mechanical energy can it develop?
8. Explain why reversing the current in the armature of a motor reverses the direction of rotation.
9. Find the cost of running a washing machine using a ½-horse-power motor 2 hours if the cost of the electricity is 10 cents a kilowatt hour.
10. A ⅛-horse-power motor is used to run a sewing machine. If used for 3 hours what will be the cost at 11 cents a kilowatt hour?

(3) THE INDUCTION COIL AND THE TRANSFORMER

307. The Induction Coil.—Practically all electric currents are produced either by voltaic cells or by dynamos. It is frequently found, however, that it is desirable to

change the E.M.F. of the current used, either for purposes
of *effectiveness, convenience, or economy.* The *induction
coil* and the *transformer*, devices for changing the E.M.F.
of electric currents, are therefore in common use. *The
induction coil* (see Fig. 302) consists of a *primary* coil of
coarse wire *P* (Fig. 303) wound upon a core of soft iron

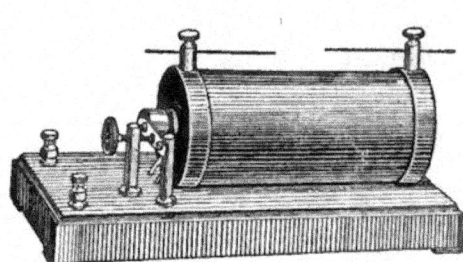

wire, and a *secondary* coil,
S, of several thousand
turns of fine wire. In
circuit with the primary
coil is a battery, *B*, and a
current interrupter, *K*,
which works iike the in-
terrupter upon an electric

FIG. 302.—An induction coil.

bell. The ends of the secondary coil are 'brought to
binding posts or spark points as at *D*.

The current from the battery flows through the primary
coil magnetizing the iron core. The magnetism in the
core attracts the soft-iron end of the interrupter, drawing
the latter over and breaking the circuit at the screw contact, *K*. This abruptly
stops the current and at
once the core loses its
magnetism. The spring
support of the inter-
rupter now draws the
latter back to the con-
tact, *T*, again complet-
ing the circuit. The
whole operation is re-

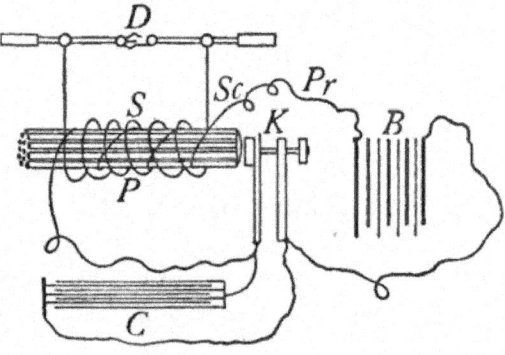

FIG. 303.—Diagram showing the parts of
an induction coil.

peated, the interrupter vibrating rapidly continually opening and closing the circuit.

**308. The Production of Induced Currents in the
Secondary Coil.**—When the current flows through the

primary it sets up a magnetic field in the *core*. When the current is interrupted, the field disappears. The increase and decrease in the field of the core induces an E.M.F. in the secondary coil, in accordance with the first law of electromagnetic induction. The E.M.F. produced depends upon (a) the number of turns in the secondary, (b) the strength of the magnetic field and (c) the rate of change of the field. The rate of change in the field is more rapid at the break than at the make. When the circuit is closed it takes perhaps $\frac{1}{10}$ of a second for the current to build up to its full strength while at a break the current stops in perhaps 0.00001 of a second, so that the induced E.M.F. is perhaps 10,000 times as great at "break" as at make. To increase the suddenness of the "make" and "break," a condenser is often connected in the primary circuit, in parallel, with the interrupter. (See Fig. 303, *C*.) This condenser provides a place to hold the rush of current at the instant that the interrupter breaks the circuit. This stored up charge reinforces the current at the make producing a much more sudden change in the magnetic field with a corresponding increase in the E.M.F. The induced currents from induction coils are sometimes called *faradic currents* in honor of Faraday who discovered electromagnetic induction. They are used to operate sparking devices upon gas and gasoline engines and in many devices and experiments in which high-tension electricity is employed.

309. The Transformer.—This is like the induction coil in that it uses a *primary* and a *secondary* coil, and an iron core to carry the magnetic field. (See Fig. 304.) They differ in that the transformer has a *closed* core or one forming a continuous iron circuit, while the induction coil has an *open* core, or one in which the magnetic field must travel in air from the north to the south poles of the core. The transformer must always be used with an *alternating* current while the induction coil may use either a direct or an alternating current. Further, the *induction* coil always produces a higher E.M.F. while the transformer may pro-

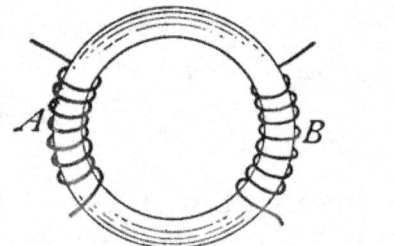 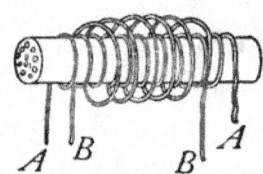

FIG. 304.—The transformer has a closed core; the induction coil, an open core.

duce an E.M.F. in its secondary coil that is either higher or lower than the one in the primary. The former is called "*step-up*" while the latter is a "*step-down*" transformer. The alternating current in the primary coil of the transformer produces an *alternating magnetic flux* in the iron

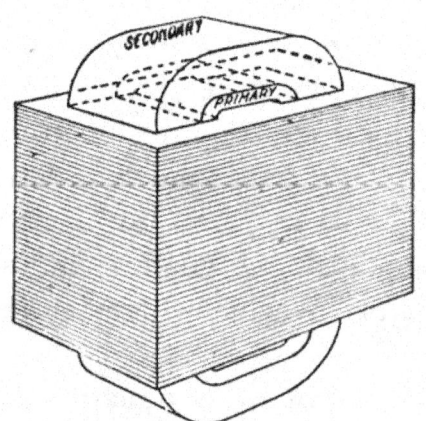

FIG. 305.—The laminated iron core of a transformer.

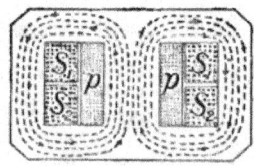

FIG. 306.—Cross-section of the transformer shown in Fig. 305 showing the magnetic field around the primary and secondary coils.

core. This iron core is *laminated* (see Fig. 305) to prevent the heating that would result if a solid core were used. The alternating magnetic flux induces in the secondary coil an E.M.F. in accordance with the following rule. The ratio of the *number* of *turns* in the *primary* to the *number* of the *turns* in the *secondary* coil equals the ratio of the electromotive forces in these respective coils. If the secondary coil has 8 turns while the primary has 4, the E.M.F. of the secondary will be just twice that of the primary. Or, if in the primary coil of the transformer Fig. 306 is an E.M.F. of 110 volts, in the secondary will be found an E.M.F. of 220 volts.

310. Uses of Transformers.—In electric lighting systems, dynamos often produce alternating currents at 1000 to 12,000 volts pressure. It is very dangerous to admit currents at this pressure into dwellings and business houses, so that transformers are installed just outside of buildings to "step-down" the high voltage currents to 110 or 220 volts. The lighting current that enters a house does not come directly from a dynamo. It is an induced current produced by a transformer placed near the house. (See Fig. 307.) In a perfect transformer the efficiency would be 100 per cent. This signifies that the energy that is sent into the primary coil of the transformer exactly equals the energy in the secondary coil. The best transformers actually show efficiencies better than 97 per cent. The lost energy appears as heat in the transformer. "The transfer of great power in a large transformer from one circuit to another circuit entirely separate and distinct, without any motion or noise and almost without loss, is one of the most wonderful phenomena under the control of man."

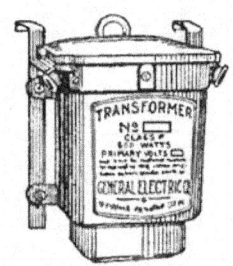

FIG. 307.—A commercial transformer.

311. The mercury arc rectifier is a device for changing an alternating current into a direct current. It is frequently used for charging storage batteries where only alternating current is supplied by the electric power company. It consists of an exhausted bulb containing two carbon or graphite electrodes marked *G* in Fig. 308 and a mercury electrode marked *M*. It is found that current will pass through such a bulb only from the graphite to the mercury but not in the reverse direction. In operating the device, the secondary terminals of an alternating current transformer *T* are connected to the graphite terminals of the rectifier. A wire connected to the center of the secondary of the transformer at *C* is attached to the *negative* terminal of the storage battery *SB*. The *positive* terminal of the battery is connected to the mercury electrode of the rectifier tube through a reactance or choke coil *R*. This coil serves to sustain the arc between the alternations. *Sw* is a starting switch, used only in striking the arc. It is opened immediately after the tube begins to glow.

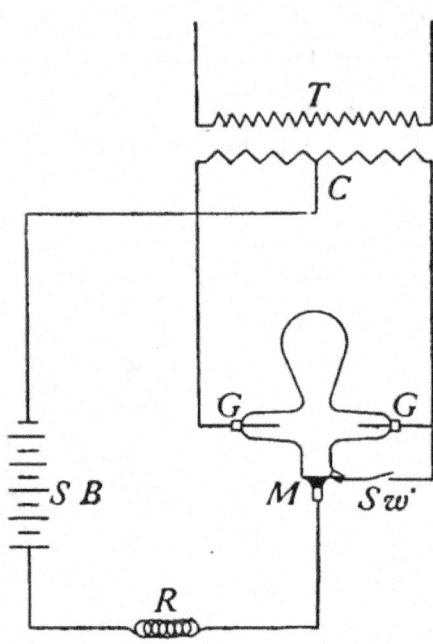

FIG. 308.—Diagram of a mercury arc rectifier.

Important Topics

Transformer, induction coil, mercury are rectifier, construction, action; uses of each.

Exercises

1. Does the spark of an induction coil occur at "make" or a "break?" Why?

2. What must be the relative number of turns upon the primary and secondary coils of a transformer if it receives current at

220 volts and delivers current at 110? Also show by diagram.

3. Would the transformer work upon a direct current? Why?

4. Explain why the interrupter is a necessary part of the induction coil and not of the transformer.

5. If a building used eighty 110-volt incandescent lamps, what E.M.F. would be necessary to light them if they were joined in series? Why would this not be practical?

6. If a 16-candle-power lamp requires 0.5 ampere upon a 110-volt circuit what current and voltage will be needed to operate 12 such lamps in parallel?

7. What will it cost to run these lamps 4 hours a night for 30 days at 10 cents per kilowatt hour?

8. If a mercury arc rectifier uses 5 amperes of current at 110 volts alternating current to produce 5 amperes of direct current at 70 volts, what is the efficiency of the rectifier?

9. Compute the heat produced in a 40 watt tungsten lamp in 1 minute.

10. Compute the heat produced in a 60 watt carbon incandescent lamp in 1 hour.

(4) THE TELEPHONE

312. The Electric Telephone.—This is an instrument for reproducing the human voice at a distance by an electric current. The modern electric telephone consists of at least four distinct parts (see Fig. 312); viz., a *transmitter*, an induction coil, an electric battery, and a *receiver*. The first three of these are concerned in sending, or *transmitting* over the connecting wires a fluctuating electric current, which has been modified by the waves of a human voice. The receiver, is affected by the fluctuating current and reproduces the voice. It will be considered first, in our study.

313. The telephone receiver was invented in 1876 by Alexander Graham Bell. It consists of a permanent steel magnet, U shaped, with a coil of fine insulated copper wire about each pole. (See Fig. 310.) A disc of thin sheet iron is supported so that its center does not quite touch the poles of the magnet. A hard rubber cap or ear piece with an opening at its center is screwed on so as to hold the iron disc firmly in place.

The action of the receiver may be understood from the following explanation: The electric current sent to the

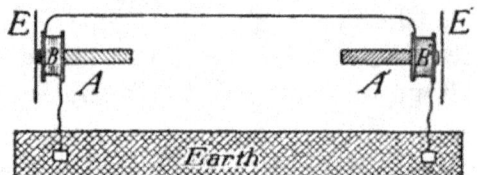

FIG. 309.—The simplest telephone system. It consists of two telephone receivers connected in series on a circuit. It will work, but not satisfactorily.

receiver, comes from the secondary coil of the induction coil; it is an alternating current, fluctuating back and forth just in time with the waves of the voice affecting it at the transmitter.

This alternating current flows around the coils on the poles of the permanent magnet. When this current flows in one direction, its magnetic field assists the field of the permanent magnet, strengthening it. This stronger magnetic field draws the thin iron disc in front of the poles of the magnet a little closer to them. When the current in the coils flows the other way, its magnetic field weakens the field of the steel magnet, and the disc is drawn back by the force of its own elasticity. Thus the disc of the receiver vibrates with the alternations of the current, and reproduces the same sounds that were spoken into the transmitter.

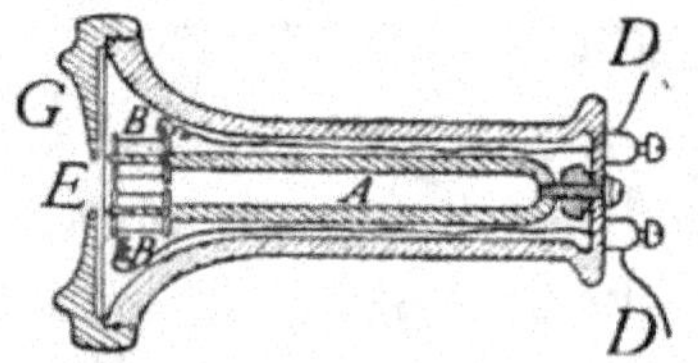

314. The Telephone Transmitter.—The telephone receiver just described has great sensitiveness in reproducing sound, but it is not satisfactory as a transmitter or sending apparatus. The *transmitter* commonly used is represented in cross-section in Fig. 311. In this figure, back of the mouthpiece, is a thin carbon disc, *D*. Back of this

FIG. 310.—A telephone receiver. This receiver has a permanent horseshoe magnet with a coil about each pole.

disc is a circular compartment containing granular carbon, *g*. The wires of the circuit are connected to the carbon disc and to the back of the case containing granular carbon. The circuit through the transmitter also includes a voltaic or storage cell and the primary coil of an induction coil. (See Fig. 312.)

315. The action of the transmitter is explained as follows: When the sound waves of the voice strike upon the carbon disc, the latter vibrates, alternately increasing and decreasing the pressure upon the granular carbon. When the pressure *increases*, the electrical resistance of the

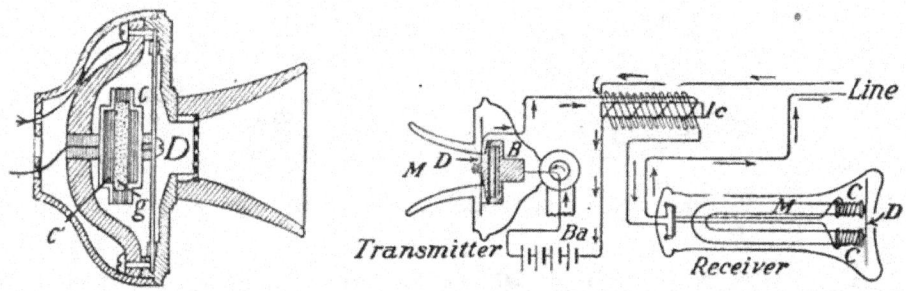

FIG. 311.—A telephone   FIG. 312.—Telephone instruments at one
    transmitter.   end of a talking circuit.

granular carbon is *lessened*, and when the pressure upon it is *decreased*, its resistance *increases*. This changing resistance causes fluctuations in the electric current *that* correspond exactly with the sound waves of the voice affecting it.

316. A complete telephone system operating with a local battery is shown in Fig. 312. A person speaking into the transmitter causes a fluctuation in the electric current in the transmitter as described in Art. 315. This fluctuating current passes through the primary coil of the induction coil *Ic*. This fluctuating current produces a fluctuating magnetic field in its core. This fluctuating field induces an *alternating* current in the secondary coil which alternates just as the primary current fluctuates, but with a much higher E.M.F. than the latter. The alternating current passes to the receiver which reproduces the speech as described in Art. 313. The line circuit includes the secondary of the induction coil, the receiving instrument and the receiver of the sending instrument so that the voice is reproduced in both receivers. An electric bell is placed at each station to call the attention of parties wanted. The movement of the receiver hook when the receiver is lifted, disconnects the bell and closes the talking circuit. The latter is opened and the bell connected when the receiver is hung up again.

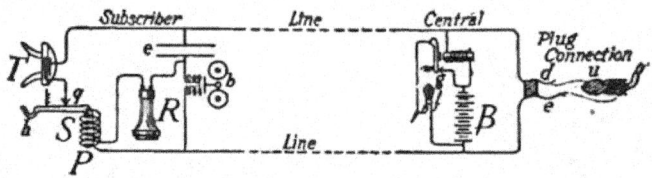

Fig. 313.—Diagram of a telephone system as used in a large exchange.

In cities and towns, the telephone system in use differs from the one described in usually having one large battery placed in the central exchange, instead of dry cells at each instrument. (See Fig. 313.) Also the operator at *central* is called by simply taking the receiver from the hook instead of being "rung up" by the subscriber. The operations of the transmitter, induction coil and receiver, however, are the same in all telephones.

Important Topics

1. Receiver: parts, action.
2. Transmitter: parts, action.
3. Induction coil, bell, line wires, etc.
4. Action of the whole device.

Exercises

State three important electrical laws or principles that are employed in the operation of the telephone. What is the application of each?

2. Connect the binding posts of a telephone receiver with a sensitive galvanometer and press on the diaphragm of the receiver; a deflection of the galvanometer will be noticed. Release the diaphragm and a reflection in the opposite direction is seen. Explain.

3. Is the current passing through the transmitter the one going to the receiver of the instrument? Explain.

4. Does the receiver at the telephone used by a person repeat the speech of the person? Explain.

5. How many 0.5 ampere lamps can be used with a 6 ampere fuse?

6. Why is it necessary to have a rheostat connected in series with a stereopticon or moving picture machine while a rheostat is not used with arc lights out doors?

7. How many candle power should a 60 watt carbon incandescent lamp give, if its efficiency is 3.4 watts per candle power?

8. Three incandescent lamps having resistances of 100, 150, and 240 ohms, respectively, are connected in parallel. What is their combined resistance?

Review Outline: Induced Currents

Induced currents; 3 laws, illustrations.

Construction, action, and uses of—magneto, dynamo, induction coil, transformer, motor, telephone. Mercury arc rectifier.

Terms—primary, secondary, for coils and currents, armature, commutator, slip ring, brush, rectifier, open core, series, shunt, and compound connections for dynamos.

CHAPTER XV

SOUND

(1) SOUND AND WAVE MOTION

317. What is a Sound?—This question has two answers, which may be illustrated as follows: Suppose that an alarm clock is set so that it will strike in one week and that it is placed upon a barren rock in the Pacific Ocean by

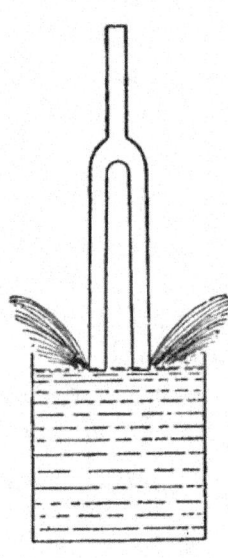

sailors who immediately sail away. If when the tapper strikes the bell at the end of the week no ear is within a hundred miles, is any sound produced? The two view-points are now made evident, for some will answer "no" others "yes." Those answering "no" hold that sound is a *sensation* which would not be produced if no ear were at hand to be affected. Those answering "yes" understand, by the term sound, *a mode of motion capable of affecting the auditory nerves*, and that sound exists wherever such motions are present. This latter point of view is called the *physical* and is the one we are to use in this study.

FIG. 314.—The tuning fork is vibrating.

318. Source of Sound.—If we trace any sound to its source, it will be found to originate in a body in rapid motion usually in what is called a state of *vibration*. To illustrate, take a tuning fork, strike it to set it in vibration and place its stem firmly against a thin piece of wood; the sound will be strengthened materially by the vibration of the wood. If now the vibrating fork is placed with the tips of the prongs in water, the vibration is plainly shown by the spattering of the water (Fig. 314). When one *speaks*, the vibrating body is in the *larynx* at the top of the windpipe. Its vibration may be plainly felt by the hand placed upon the throat while speaking.

319. Sound Media.—Usually sounds reach the ear through the air. The air is then said to be a *medium for sound*. Other substances may serve as a sound medium, for if the head is under water and two stones, also under water, are struck together a sharp sound is heard. Also if one end of a wooden rod is held at the ear and the other end of the rod is scratched by a pin, the sound is more plainly perceived through the wood than through the air. Think of some illustration from your own experience of a solid acting as a medium for sound. If an electric bell is

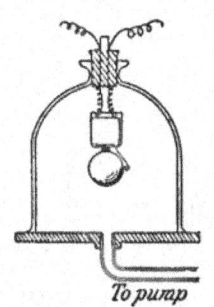

To pump

FIG. 315.— Sound does not travel in a vacuum.

placed in a bell jar attached to an air pump, as in Fig. 315, on exhausting the air the loudness of the sound is found to diminish, indicating that in a perfect vacuum no sound would be transmitted. This effect of a vacuum upon the transmission of sound is very different from its effect upon radiation of heat and light. Both heat and light are known to pass through a vacuum since both come to the earth from the sun through space that so far as we know contains no air or other matter. Sound differs from this in that it is always transmitted by some material body and cannot exist in a vacuum.

320. Speed of Sound.—Everyone has noticed that it takes time for sound to travel from one place to another. If we see a gun fired at a distance, the report is heard a few seconds after the smoke or flash is seen. The time elapsing between a flash of lightning and the thunder shows that sound takes time to move from one place to another. Careful experiments to determine the speed of sound have been made. One method measures accurately the time required for the sound of a gun to pass between two stations several miles apart. A gun or cannon is placed at each station. These are fired alternately, first the one at one station and then the one at the other so as to avoid an error in computation due to the motion of wind. This mode of determining the speed of sound is not accurate. Other methods, more refined than the one just described have given accurate values for the speed of sound. The results of a number of experiments show that at the freezing temperature, 0°C., the speed of sound in air is 332 meters or 1090 ft. a second. The speed of sound in air is affected by the temperature, increasing 2 ft. or 0.6 meter per second for each degree that the temperature rises above 0°C. The speed decreases the same amount for each degree C. that the air is cooled below the freezing point. The speed of sound in various substances has been carefully determined. It is greater in most of them than in air. In water the speed is about 1400 meters a second; in wood, while its speed varies with different kinds, it averages about 4000 meters a second; in brass the speed is about 3500 meters; while in iron it is about 5100 meters a second.

321. The Nature of Sound.—We have observed that sound originates at a vibrating body, that it requires a medium in order to be transmitted from one place to another, and that it travels at a definite speed in a given substance. Nothing has been said, however, of the *mode* of transmission, or of the *nature* of *sound*. Sounds continue to come from an alarm clock even though it is placed under a bell jar. It is certain that nothing material can pass through the glass of the jar. If, however, we consider that *sound is transmitted by waves through substances* the whole matter can be given a simple explanation. In order to better understand the nature of sound a study of waves and wave motion will be taken up in the next section.

Important Topics

Sound: two definitions, source, medium, speed, nature.

Exercises

1. Give two illustrations from outside the laboratory of the fact that sound is transmitted by other materials than air.

2. Name the vibrating part that is the source of the sound in three different musical instruments.

3. Is sound transmitted more strongly in solids, liquids or gases? How do you explain this?

4. How far away is a steamboat if the sound of its whistle is heard 10 seconds after the steam is seen, the temperature being 20°C.? Compute in feet and in meters.

5. How many miles away is lightning if the thunder is heard 12 seconds after the flash in seen, the temperature being 25°C.?

6. Four seconds after a flash of lightning is seen the thunder clap is heard. The temperature is 90°F. How far away was the discharge?

7. The report of a gun is heard 3 seconds after the puff of smoke is seen. How far away is the gun if the temperature is 20°C.?

8. An explosion takes place 10 miles away. How long will it take the sound to reach you, the temperature being 80°F?. How long at 0°F.?

9. How long after a whistle is sounded will it be heard if the distance away is ¼ mile, the temperature being 90°F.?

10. The report of an explosion of dynamite is heard 2 minutes after the puff of smoke is seen. How far away is the explosion the temperature being 77°F.?

(2) WAVES[1] AND WAVE MOTION

322. Visible Waves.—It is best to begin the study of wave motion by considering some waves which are familiar

---

[1] A wave is a disturbance in a substance or medium that is transmitted through it.

to most persons. Take for example the waves that move over the surface of water (Fig. 316). These have an onward motion, yet boards or chips upon the surface simply rise and fall as the waves pass them. They are not carried onward by the waves. The water surface simply rises and falls as the waves pass by. Consider also the waves that may be seen to move across a field of tall

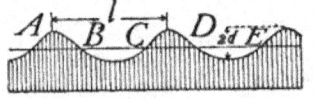

FIG. 316.—Water waves.

grass or grain. Such waves are produced by the bending and rising of the stalks as the wind passes over them. Again, waves may be produced in a rope fastened at one end, by suddenly moving the other end up and down. These waves move to the end of the rope where they are *reflected* and return. The three types of waves just mentioned are illustrations of *transverse* waves, the ideal case being that in which the particles move at *right angles* to the path or course of the wave. Such waves are therefore called *transverse* waves.

323. Longitudinal Waves.—Another kind of wave is found in bodies that are elastic and compressible and have inertia, such as gases and coiled wire springs. Such waves may be studied by considering a wire spring

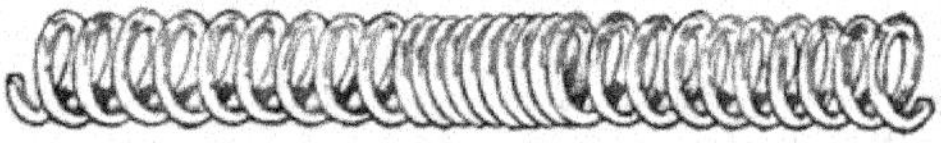

FIG. 317.—The compression wave travels through the spring.

as the medium through which the waves pass. (See Fig. 318.)

If the end of the wire spring shown in Fig. 317 is struck the first few turns of the spring will be compressed. Since the spring possesses elasticity, the turns will move forward a little and compress those ahead, these will press the next in turn and so on. Thus a *compression* wave will move to the end of the spring, where it will be reflected and return. Consider the turns of the spring as they move toward the end. On account of their *inertia* they will continue moving until they have separated from each other *more* than at first, before returning to their usual position. This condition of a greater separation of the turns of the spring than usual is called a *rarefaction*. It moves along the spring following the wave of compression. The condensation and rarefaction are considered as together forming a complete wave. Since the turns of wire move back and forth in a direction parallel to that in which the wave is traveling, these waves are called *longitudinal*.

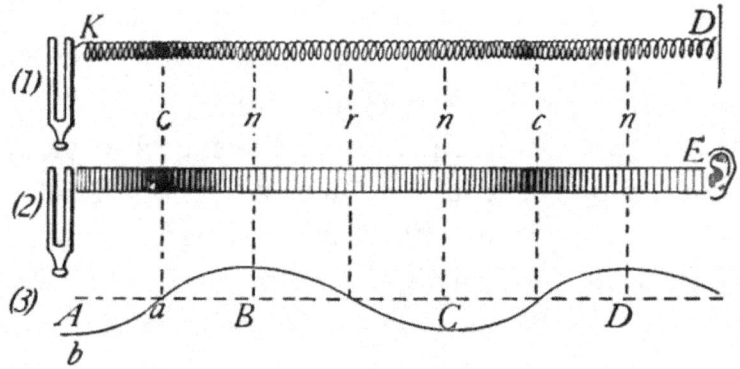

Fig. 318.—Longitudinal waves (1) in a spring, (2) in air, and (3) graphic representation showing wave length, condensations, and rarefactions.

324. The transmission of a sound by the air may be understood by comparing it with the process by which a *wave is transmitted by a wire spring*. Consider a light spring (Fig. 318, 1) attached at the end of a vibrating tuning fork, K, and also to a diaphragm, D. Each vibration of the fork will first compress and then separate the coils of the spring. These impulses will be transmitted by the spring as described in Art. 315, and cause the diaphragm to vibrate *at the same rate* as the tuning fork. The diaphragm will then give out a sound similar to that of the tuning fork. Suppose that the spring is replaced by air, and the diaphragm, by the ear of a person, E, (Fig.

318, 2.) When the prong of the fork moves toward the ear it starts a compression and when it moves back a rarefaction. The fork continues vibrating and these impulses move onward like those in the spring at a speed of about 1120 ft. in a second. They strike the diaphragm of the ear causing it to move back and forth or to vibrate at the same rate as the tuning fork, just as in the case of the diaphragm attached to the spring.

325. Graphic Representation of Sound Waves.—It is frequently desirable to represent sound waves graphically. The usual method is to use a curve like that in (Fig. 318, 3). This curve is considered as representing a train of waves moving in the same direction as those in Fig. 318 1 and 2, and also having the same length. The part of the wave *A–B* represents a condensation of the sound wave and the part *B–C* represents a rarefaction. A complete wave consisting of a condensation and a rarefaction is represented by that portion of the curve *A–C*. The portion of the curve *B–D* also represents a *full wave length* as the latter is defined as *the distance between two corresponding parts of the adjacent waves*. The curve, Fig. (318, 3) represents not only the wave length, but also the height of the wave or the amount of movement of the particles along the wave. This is called the *amplitude* and is indicated by the distance *A–b*. Since the *loudness* or intensity of a sound is found to depend upon the amount of movement of the particles along the wave, the *amplitude* of the curve is used to indicate the loudness of the sound represented. All of the characteristics of a sound wave may be graphically represented by curves. Such curves will be used frequently as an aid in explaining the phenomena of wave motion both in sound and in light.

326. Reflections of Sound.—It is found that a wave moving along a wire spring is reflected when it reaches the end and returns along the spring. Similarly a sound wave in air is reflected upon striking the surface of a body. If the wave strikes perpendicularly it returns along the line from which it comes, if, however, it strikes at some other angle it does not return along the same line, but as in other cases of reflected motion, the *direction* of the *reflected* wave is described by the *Law of Reflected Motion* as follows: *The angle of reflection is always equal to the angle of incidence.* This law is illustrated in Fig. 319. Suppose that a series of waves coming from a source of sound move from H to O. After striking the surface IJ the waves are reflected and move toward L along the

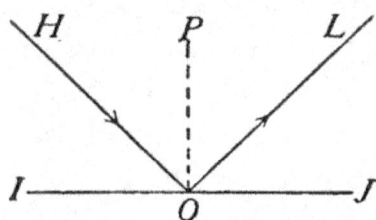
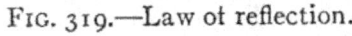

FIG. 319.—Law of reflection.

FIG. 320.—An ear trumpet.

line *OL*. Let *PO* be perpendicular to the surface *IJ* at *O*. Then *HOP* is *the angle* of incidence and *LOP* is the *angle of reflection*. By the law of reflected motion these angles are equal. In an ordinary room when a person speaks the sound waves reflected from the smooth walls reinforce the sound waves moving directly to the hearers. It is for this reason that it is usually easier to speak in a room than in the open air. Other illustrations of the reinforcement of sound by reflection are often seen. Thus an *ear trumpet* (Fig. 320), uses the principle of reflection and concentration of sound. So-called *sounding boards* are sometimes placed back of speakers in large halls to reflect sound waves to the audience.

327. Echoes.—*An echo is the repetition of a sound caused by its reflection from some distant surface* such as that of a building, cliff, clouds, trees, etc. The interval of time between the production of a sound and the perception of its echo is the time that the sound takes to travel from its source to the reflecting body and back to the listener. Experiments have shown that the sensation of a sound persists about one-tenth of a second. Since the velocity of sound at 20°C. is about 1130 ft., during one-tenth of a second the sound wave will travel some 113 ft. If the reflecting surface is about 56 ft. distant a *short* sound will be followed immediately by its echo as it is heard one-tenth of a second after the original sound. The reflected sound tends to strengthen the original one if the reflecting surface is less than 56 ft. away. If the distance of the reflecting surface is much more than 56 ft. however, the reflected sound does not blend with the original one but forms a distinct echo. The echoes in large halls especially those with large smooth walls may very seriously affect the clear perception of the sound. Such rooms are said to have poor *acoustic* properties. Furniture, drapery, and carpets help to deaden the echo because of diffused reflection. The Mormon Tabernacle at Salt Lake City, Utah, is a fine example of a building in which the reflecting surfaces of the walls and ceiling are of such shape and material that its acoustic properties are remarkable, a pin dropped at one end being plainly heard at the other end about 200 ft. away.

Important Topics

1. Waves: transverse, longitudinal; wave length, condensation, rarefaction.
2. Wave motion: in coiled spring, in air, on water.
3. Reflection of waves: law, echoes.

Exercises

1. A hunter hears an echo in 8 seconds after firing his gun. How far is the reflecting surface if the temperature is 20°C.?
2. How far is the reflecting surface of a bulding if the echo of one's footsteps returns in 1 second at 10°C.?
3. Why is it easier to speak or sing in a room than out of doors?
4. Draw a curve that represents wave motion. Make it exactly three full wave lengths, and state why your curve shows this length. Indicate the parts of the curve that correspond to a condensation and to a rarefaction.
5. How long does it take the sound of the "pin drop" to reach a person at the farther end of the building mentioned at the end of Art. 327?
6. An echo is heard after 6 seconds. How far away is the reflecting surface, the temperature being 70°F.?
7. Why are outdoor band-stands generally made with the back curving over the band?
8. A man near a forest calls to a friend. In 4 seconds the echo comes back. How far away is he from the forest?
9. Would it be possible for us ever to hear a great explosion upon the moon? Explain.
10. If a sunset gun was fired exactly at 6:00 P. M. at a fort, at what time was the report heard by a man 25 miles away, if the temperature was 10°C.?

(3) Intensity and Pitch of Sounds

328. Musical Sounds and Noises Distinguished.—The question is sometimes raised, what is the difference

Fig. 321.—Graphic representations of (a) a noise, (b) a musical sound.

between a *noise* and a *musical sound?* The latter has been found to be produced by an even and regular vibration such as that of a tuning fork or of a piano string. A noise on the other hand is characterized by sudden or irregular vibrations such as those produced by a wagon bumping over a stony street. These differences may be represented graphically as in Fig. 321, (a) represents a noise, (b) a musical tone.

329. Characteristics of Musical Sounds.—Musical tones differ from one another in three ways or are said to have *three characteristics*, viz., *intensity*, *pitch*, and *quality*. Thus two sounds may differ only in intensity or *loudness*; that is, be alike in all other respects except this one, as when a string of a piano is struck at first gently, and again harder. The second sound is recognized as being louder. The difference is due to the greater *amplitude* of vi-

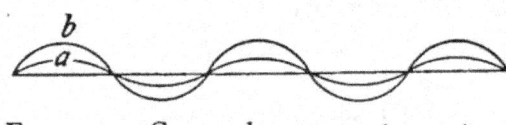

FIG. 322.—Curve *b* represents a tone of greater intensity.

bration caused by more energy being used. Fig. 322 shows these differences graphically. Curve *b* represents the tone of greater intensity or loudness, since its amplitude of vibration is represented as being greater.

330. Conditions Affecting the Intensity of Sound.— The intensity of sounds is also affected by the *area* of the vibrating body. This is shown by setting a tuning fork in vibration. The area of the vibrating part being small, the sound is heard but a short distance from the fork. If, however, the stem of the vibrating fork is pressed against the panel of a door or the top of a box, the sound may be heard throughout a room. The stem of the fork has communicated its vibrations to the wood. The vibrating area, being greater, the sound is thereby much increased in intensity, producing a wave of greater amplitude. The same principle is employed in the sounding boards of musical instruments as in the piano, violin, etc. It is a common observation that sounds decrease in loudness as the distance from the source increases. This is due to the increase of the surface of the spherical sound waves spreading in all directions from the source. Careful experiments have shown that in a uniform medium *the intensity of a sound is inversely proportional to the square of the distance from its source.* If a sound is confined so that it cannot spread, such as the sound moving through a speaking tube, it maintains its intensity for a considerable distance. An *ear trumpet* (see Fig. 320) also applies this principle. It is constructed so that sound from a given area is *concentrated* by reflection to a much smaller area with a corresponding increase in intensity. The *megaphone* (Fig.

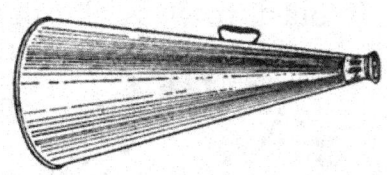

FIG. 323.—The megaphone.

323), and the *speaking trumpet* start the sound waves of the voice in one direction so that they are kept from spreading widely, consequently by its use the voice may be heard several times the usual distance. The intensity of a sound is also affected by the *density* of the transmitting medium. Thus a sound produced on a mountain top is fainter and thinner than one produced in a valley. The sound of a bell in the receiver of an air pump becomes weaker as the air is exhausted from the latter. *Four* factors thus influence the intensity of a sound, the *area* of the vibrating body, its *amplitude* of vibration, the *distance* of the source and the *density* of the transmitting medium. It is well to fix in mind the precise effect of each of these factors.

331. Pitch.—The most characteristic difference between musical sounds is that of *pitch.* Some sounds have a high pitch, such as those produced by many insects and birds. Others have a low pitch as the notes of a bass drum or the sound of thunder. How notes of different pitch are produced may be shown by the siren (Fig. 324). This is a disc mounted so as to be rotated on an axis. Several rows of holes are drilled in it in concentric circles. The number of holes in successive rows increases from within outward. If when the siren is rapidly rotated air is blown through a tube against a row of holes a clear musical tone is heard   The tone is due to the succession of pulses in the air produced by the row of holes in the rotating disc alternately cutting off and permitting the air blast to pass through at very short intervals. If the

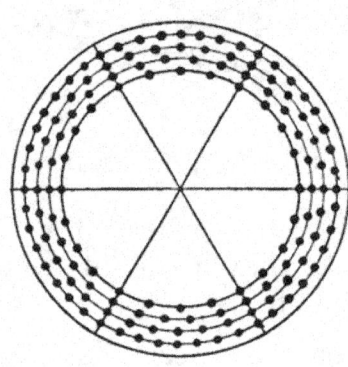

blast is directed against a row of holes nearer the circumference the pitch is higher, if against a row nearer the center the pitch is lower. Or if the blast is sent against the same row of holes the pitch rises when the speed increases and lowers when the speed lessens. These facts indicate that the pitch of a tone is due to the

FIG. 324.—A siren.

number of pulses or vibrations that strike the ear each second; also that *the greater the rate of vibration, the higher the pitch.*

332. The Major Scale.—If a siren is made with eight rows of holes, it may indicate the relation between the notes of a *major scale.* To accomplish this, the number of holes in the successive rows should be 24, 27, 30, 32, 36, 40, 45, 48. If a disc so constructed is rapidly rotated at a uniform rate, a blast of air sent against all of the rows in succession produces the tones of the scale   These facts indicate that the relative vibration numbers of the notes of any *major scale* have the same relation as the numbers 24, 27, 30, 32, 36, 40, 45, 48.

The note called middle C is considered by physicists as having 256 vibrations a second. This would give the following *actual vibration* numbers to the remaining notes of the major scale that begins with "Middle C" D.–288, E.–320, F.–341.3, G.–384, A.–426.6, B.–480, C'.–512. Musicians, however, usually make use of a scale of slightly higher pitch. The *international* standard of pitch in this country and in Europe is that in which "A" has 435 vibrations per second. This corresponds to 261 vibrations for middle C.

333. The Relation between Speed, Wave Length, and Number of Vibrations per Second.—Since the notes from the various musical instruments of an orchestra are noticed to harmonize as well at a distance as at the place produced, it is evident that notes of all pitches travel at the same rate, or have the *same speed*. Notes of high pitch, having a high vibration rate produce more waves in a second than notes of low pitch, consequently the former are shorter than the latter. The following formula gives the relation between the speed (v), wave length (l), and number of vibrations per sec. (n):

$$v = l \times n, \text{ or } l = v/n$$

that is, *the speed of a sound wave is equal to the number of vibrations per second times the wave length, or the wave length is equal to the speed divided by the number of vibrations per second.* This formula may also be employed to find the *number* of vibrations when the wave length and speed are given.

Important Topics

1. Difference between noise and music.
2. Factors affecting intensity: area, amplitude, density, distance.
3. Pitch, major scale, relative vibration numbers.
4. Relation between speed, wave length and vibration rate.

Exercises

1. Give an illustration from your own experience of each of the factors affecting intensity.
2. Write the relative vibration-numbers of a major scale in which *do* has 120 vibrations.
3. What is the wave length of the "A" of international concert pitch at 25°C.? Compute in feet and centimeters.
4. At what temperature will sound waves in air in unison with "Middle C" be exactly 4 ft. long?
5. Explain the use of a megaphone.
6. What tone has waves 3 ft. long at 25°C.?
7. What is the purpose of the "sounding board" of a piano?
8. Two men are distant 1000 and 3000 ft. respectively from a fog horn. What is the relative intensity of the sounds heard by the two men?
9. The speaking tone of the average man's voice has 160 vibrations per second. How long are the waves produced by him at 20°C.?

(4) Musical Scales and Resonance

334. A musical interval *refers to the ratio between the pitches*[1] *of two notes* as indicated by the results of the siren experiment. The simplest interval, or ratio between two notes is the *octave, C′:C,* or 2:1 (48:24). Other important intervals with the corresponding ratios are the *fifth,* G:C, or 3:2 (36:24); the *sixth,* A:C, or 5:3 (40:24); the *fourth,* F:C, 4:3 (32:24); the *major third,* E:C, or 5:4(30:24); and the *minor third,* G:E, 6:5. The interval between any two notes may be determined by finding the ratio between the vibration numbers of the two notes. Thus, if one note is produced by 600 vibrations a second and another by 400, the interval is 3:2, or a *fifth,* and this would be recognized by a musician who heard the notes sounded together or one after the other. Below is a table of musical nomenclatures, showing various relations between the notes of the major scale.

[1] Pitch as used here, means *vibration rate.*

TABLE OF MUSICAL NOMENCLATURES

Name of note	C	D	E	F	G	A	B	C'
Frequency in terms of "do"	n	$\frac{9}{8}n$	$\frac{5}{4}n$	$\frac{4}{3}n$	$\frac{3}{2}n$	$\frac{5}{3}n$	$\frac{15}{8}n$	$2n$
Intervals.............		$\frac{9}{8}$	$\frac{10}{9}$	$\frac{16}{15}$	$\frac{9}{8}$	$\frac{10}{9}$	$\frac{9}{8}$	$\frac{16}{15}$
Name of note in vocal music	do	re	mi	fa	sol	la	ti	do
Treble clef.								
Bass clef.								
International pitch of treble clef..........	261	293.6	326.3	348.	391.5	435	489.4	522
Scientific scale........	256	288	320	341.3	384	426.6	480	512
Relative vibration numbers.	24	27	30	32	36	40	45	48

335. Major and Minor Triads.—The notes C, E, G (*do, mi, sol*) form what is called a *major triad*. The *relative vibration numbers* corresponding are 24, 30, 36. These in simplest terms have ratios of 4:5:6. Any three other tones with vibration ratios of 4:5:6 will also form a major triad. If the octave of the lower tone is added, the four make a major chord. Thus: F, A, C' (*fa, la, do*), 32:40:48, or 4:5:6, also form a major triad as do G, B, D' (*sol, ti, re*), 36:45:54, or 4:5:6. Inspection will show that these three major triads comprise all of the tones of the major scale D' being the octave of D. It is, therefore, said that the major scale is based, or built, upon these three major triads. The examples just given indicate the mathemat-

24

ical basis for harmony in music. Three notes having vibration ratios of 10:12:15 are called *minor triads*. These produce a less pleasing effect than those having ratios of 4:5:6.

336. The Need for Sharps and Flats.—We have considered the key of C. This is represented upon the piano or organ by white keys only (Fig. 325). Now in order (a) to give variety to instrumental selections, and (b) to accommodate instruments to the range of the human voice, it has been necessary to introduce other notes in musical instruments. These are represented by the *black keys* upon the piano and organ and are known as *sharps* and

FIG. 325.—Section of a piano keyboard.

flats. To illustrate the necessity for these additional notes take the major scale starting with B. This will give vibration frequencies of 240, 270, 300, 320, 360, 400, 450, and 480. The only white keys that may be used with this scale are E 320 and B 480 vibrations. Since the second note on this scale requires 270 vibrations about halfway between C and D the black key C sharp is inserted. Other notes must be inserted between D and E (D sharp), between F and G (F sharp), also G and A sharps.

337. Tempered Scales.—In musical instruments with fixed notes, such as the harp, organ, or piano, complications were early recognized when an attempt was made to adapt these instruments so that they could be played in all keys. For the vibration numbers that would give a perfect major scale starting at C are not the same as will give a perfect major scale beginning with any other key. In using the various notes as the keynote for a major scale, 72 different notes in the octave would be required. This would ake it more difficult for such instruments as the piano to be played. To avoid these complications as much as possible, it has been found necessary to abandon the simple ratios between successive notes and to substitute another ratio in order that the vibration ratio between any two successive notes will be equal in every case. The differences between semitones are abolished so that; for example, C sharp and D flat become the same tone instead of two different tones. Such a scale is called a *tempered scale*. The tempered scale has 13 notes to the octave, wi h 12 equal intervals, the ratio between two successive notes being the $\sqrt[12]{2}$ or 1.059. That is, any vibration rate on the tempered scale may be computed by multiplying the vibration rate of the preceding note by 1.059 While this is a necessary arrangement, there is some loss in perfect harmony. It is for this reason that a quartette or chorus of voices singing without accompaniment is often more harmonious and satisfactory than when accompanied with an instrument of fixed notes as the piano, since the simple harmonious ratios may be employed when the voices are alone. The imperfection introduced by *equal temperament tuning* is illustrated by the following table:

	C	D	E	F	G	A	B	C
Perfect Scale of C.	256.0	288.0	320.0	341.3	384.0	426.6	480.0	512.0
Tempered Scale...	256.0	287.3	322.5	341.7	383.6	430.5	483.3	512.0

338. Resonance.—If two tuning forks of the same pitch are placed near each other, and one is set vibrating, the other will soon be found to be in vibration. This result is said to be due to *sympathetic vibration*, and is an example of *resonance* (Fig. 326). If water is poured into a glass tube while a vibrating tuning fork is held over its top, when the air column has a certain length it will start vibrating, reinforcing strongly the sound of the tuning fork. (See Fig. 327.) This is also an example of resonance. These and other similar facts indicate that *sound waves started by a vibrating body will cause another body near it to start vibrating if the two have the same rate of vibration.* Most persons will recall illustrations of this effect from their own experience.

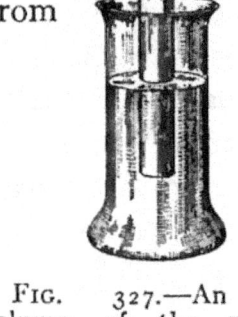

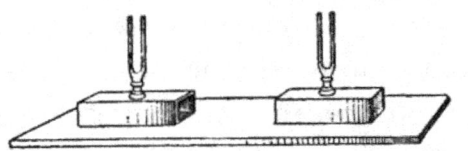

FIG. 326.—One tuning fork will vibrate in sympathy with the other, if they have exactly equal rates of vibration.

FIG. 327.—An air column of the proper length reinforces the sound of the tuning fork.

339. Sympathetic vibration is explained as follows: Sound waves produce very slight motions in objects affected by them; if the vibration of a given body is exactly in time with the vibrations of a given sound each impulse of the sound wave will strike the body so as to increase the vibratory motion of the latter. This action continuing, the body soon acquires a motion sufficient to produce audible waves. A good illustration of sympathetic vibration is furnished by the bell ringer, who times his pulls upon the bell rope with the vibration rate of the swing of the bell. In the case of the resonant air column over which is held a vibrating tuning fork (see Fig. 328), when the prong of the fork starts downward from 1 to 2, a condensation wave moves down to the water surface and back just in time to join the condensation wave *above* the fork as the prong begins to move from 2 to 1; also when the prong starts upward from 2 to 1, the rarefaction, produced under it moves to the bottom of the air column and back so as to join the rarefaction *above* the fork as the prong returns. While the prong is making a *single* movement, up or down, it is plain that the air wave moves twice the length of the open tube. During a *complete* vibration of the fork, therefore, the sound wave moves four times the length of the air column. In free air, the sound progresses a wave length during a complete vibration, hence the resonant air column is one-fourth the length of the sound wave to which it responds. Experiments with tubes of different lengths show that the diameter of the air column has some effect upon the length giving best resonance. About 25 per cent. of the diameter of the tube must be added to the length of the air column to make it just one-fourth the wave length. The sound heard in seashells and

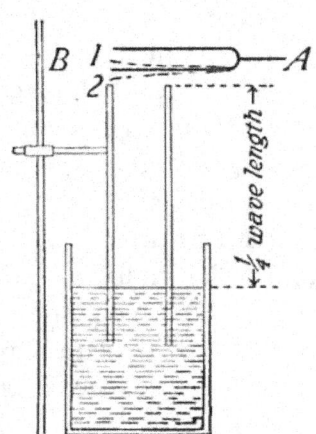

Fig. 328.—Explanation of resonance.

in other hollow bodies is due to resonance. Vibrations in the air too feeble to affect the ear are intensified by sympathetic vibration until they can be heard. A tuning fork is often mounted upon a box called a *resonator*, which contains an air column of such dimensions that it reinforces the sound of the fork sympathetic vibration.

Important Topics

1. Musical intervals: octave, sixth, fifth, fourth, third.
2. Major chord, 4 : 5 : 6.
3. Use of sharps and flats.   Tempered scale.
4. Resonance, sympathetic vibration; explanation, examples.

Exercises

1. What is a major scale?   Why is a major scale said to be built upon three triads?
2. Why are sharps and flats necessary in music?
3. What is the tempered scale and why is it used?   What instruments need not use it?   Why?
4. Mention two examples of resonance or sympathetic vibration from your own experience out of school.   .
5. An air column 2 ft. long closed at one end is resonant to what wave length?   What number of vibrations will this sound have per second at 25°C.?
6. At 24°C. what length of air column closed at one end will be resonant to a sound having 27 vibrations a second?
7. A given note has 300 vibrations a second.   What will be the number of vibrations of its (a) octave, (b) fifth, (c) sixth, (d) major third?
8. In the violin or guitar what takes the place of the sounding board of the piano?
9. Can you explain why the pitch of the bell on a locomotive rises as you rapidly approach it and falls as you recede from it?
10. Do notes of high or low pitch travel faster?   Explain.
11. An "A" tuning fork on the "international" scale makes 435 vibrations per second.   What is the length of the sound waves produced?

(5) Wave Interference, Beats, Vibration of Strings

340. Interference of Waves.—The possibility of two trains of waves combining so as to produce a reduced motion or a *complete destruction* of motion may be shown graphically.   Suppose two trains of waves of equal wave length and amplitude as in Fig. 329 meet in *opposite phases*.   That is, the parts corresponding to the *crests* of A coincide with the *troughs* of B, also the troughs of A with the crests of B; when this condition obtains, the result is that shown at C, the union of the two waves resulting in complete destruction of motion.   *The more or less complete destruc-*

tion of one train of waves by another similar train is an illustration of **interference.** If two sets of water waves so unite as to entirely destroy each other the result is a level water surface. If two trains of sound waves combine they may so interfere that silence results. The conditions for securing interference of sound waves may

FIG. 329.—Interference of sound waves.

readily be secured by using a tuning fork and a resonating air column. If the tuning fork is set vibrating and placed over the open end of the resonating air column (see Fig. 328), an increase in the sound through resonance may be heard. If the fork is rotated about its axis, in some positions no sound is heard while in other positions the sound

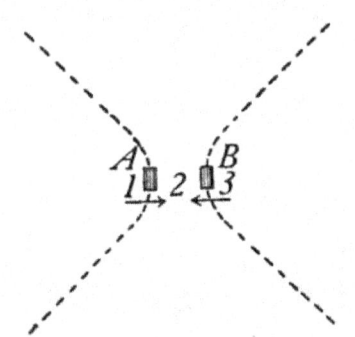

FIG. 330.—At 2 is a condensation; at 1 and 3 are rarefactions.

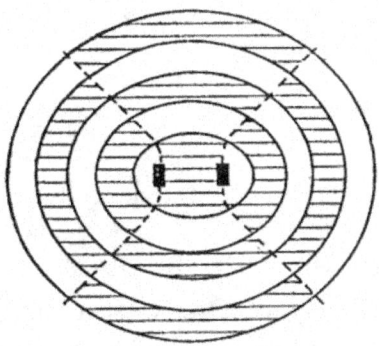

FIG. 331.—The condensations and rarefactions meet along the dotted lines producing silence.

is strongly reinforced. Similar effects may be perceived by holding a vibrating fork near the ear and slowly rotating as before. In some positions interference results while in other positions the sound is plainly heard. The explanation of interference may be made clear by the use of a diagram. (See Fig. 330.) Let us imagine that we are looking at the two square ends of a tuning fork. When the fork is vibrating the two prongs approach each other and then recede. As they approach, a condensation is produced at 2 and rarefactions at 1 and 3. As they separate, a rarefaction is produced at 2 and condensations at 1 and 3. Now along the lines at which the simultaneously produced rarefactions and condensations meet there is more or less complete interference. (See Fig. 331.) These positions have been indicated by dotted lines extending through the ends of the prongs. As indicated above, these positions may be easily found by rotating a vibrating fork over a resonant air column, or near the ear.

341. Beats.—If two tuning forks of slightly different pitch are set vibrating and placed over resonating air

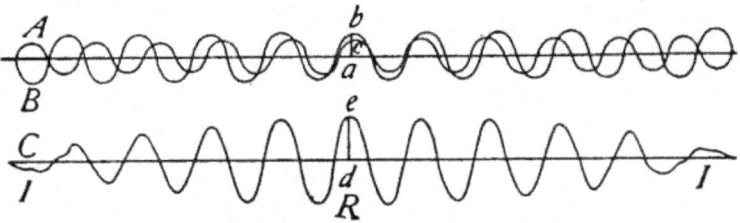

FIG. 332.—Diagram illustrating the formations of beats.

columns or with the stem of each fork upon a sounding board, so that the sounds may be intensified, a peculiar pulsation of the sound may be noticed. This phenomenon is known as *beats*. Its production may be easily understood by considering a diagram (Fig. 332). Let the curve A represent the sound wave sent out by one tuning fork and B, that sent out by the other. C represents the effect produced by the combination of these waves. At R the two sound waves meet in the *same phase* and reinforce each other. This results in a louder sound than either produces alone. Now since the sounds are of slightly

different pitch, one fork sends out a few more vibrations
per second than the other. The waves from the first
fork are therefore a little shorter than those from the other.
Consequently, although the two waves are at one time in
the *same* phase, they must soon be in opposite phases as at
I. Here interference occurs, and silence results. Immediately the waves reinforce, producing a louder sound and
so on alternately. The resulting rise and fall of the sound
are known as *beats*. The number of beats per second
must, of course, be the same as the difference between the
numbers of vibrations per second of the two sounds. One
effect of beats is *discord*. This is especially noticeable
when the number of beats per second is between 30 and
120. Strike the two lowest notes on a piano at the same
time. The beats are very noticeable.

342. Three Classes of Musical Instruments.—There
are three classes or groups of musical instruments, if

FIG. 333.—Turkish
    cymbals.

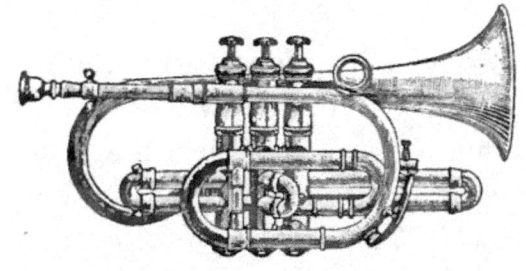

FIG. 334.—The cornet.

we consider the vibrating body that produces the sound
in each: (A) Those in which the sound is produced by a
vibrating *plate* or *membrane*, as the drum, cymbals (Fig.
333), etc.; (B) those with vibrating *air columns*, as the
flute, pipe organ, and cornet (Fig. 334), and (C) with
vibrating *wires* or *strings*, as the piano, violin, and guitar.
It is worth while to consider some of these carefully.

We will begin with a consideration of vibrating wires and strings, these often producing tones of rich quality.

Let us consider the strings of a piano. (If possible, look at the strings in some instrument.) The range of the piano is $7\frac{1}{3}$ octaves. Its lowest note, A_4, has about 27 vibrations per second. Its highest, C^4, about 4176. This great range in vibration rate is secured by varying the length, the tension, and the diameter of the strings.

343. The Laws of Vibrating Strings.—The relations between the vibration rate, the length, the tension and the diameter of vibrating strings have been carefully studied with an instrument called a *sonometer* (Fig. 335). By

FIG. 335.—A sonometer.

this device it is found that the pitch of a vibrating string is raised one octave when its vibrating length is reduced to one-half. By determining the vibration rate of many lengths, the following law has been derived:

(Law I) *The rate of vibration of a string is inversely proportional to its length.*

Careful tests upon the change of vibration rate produced by a change of *tension* or pull upon the strings show that if the pull is increased four times its vibrations rate is *doubled*, and if it is increased nine times its rate is tripled, that is: (Law II) *The vibration rates of strings are directly proportional to the square roots of their tensions.*

Tests of the effects of diameter are made by taking wires of equal length and tension and of the same material but of different diameter. Suppose one is twice as thick as the other. This string has a tone an octave lower or vibrates one-half as fast as the first. Therefore:

(Law III) *The vibration rates of strings are inversely proportional to the diameters.* These laws may be expressed by a formula $n \propto \dfrac{\sqrt{t}}{dl}$.

The vibration of a string is rarely a simple matter. It usually vibrates in parts at the same time that it is vibrating as a whole. The tone produced by a string vibrating as a whole is called its *fundamental*. The vibrating parts of a string are called *loops* or *segments* (see Fig. 336), while the points of least or no vibration are *nodes*. Segments are often called *antinodes*.

344. Overtones.—The *quality* of the tone produced by a vibrating string is affected by its vibration in parts when

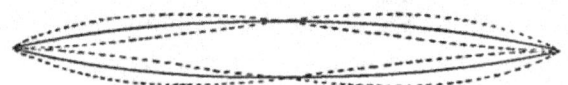

Fig. 336.—A string yielding its fundamental and its first overtone.

it is also vibrating as a whole. (See Fig. 336.) The tones produced by the vibration in parts are called *overtones* or *partial* tones. The presence of these overtones may often be detected by the sympathetic vibration of other wires near-by. What is called the *first* overtone is produced by a string vibrating in *two* parts, the *second* overtone by a string vibrating in *three parts*, the *third* overtone by its vibration in *four* parts and so on. In any overtone, the number of the parts or vibrating segments of the string is one more than the number of the overtone. For example, gently press down the key of middle C of a piano. This will leave the string free to vibrate. Now strongly strike the C an octave lower and then remove the finger from this key. The middle C string will be heard giving its tone.   In like manner try E^1 and G^1, with C.   This experiment shows that the sound of the C string contains these tones as overtones.   It also illustrates sympathetic vibration.

Important Topics

1. Interference, beats, production, effects.
2. Vibration of strings, three laws.
3. Three classes of musical instruments.
4. Fundamental and overtones, nodes, segments, how produced? Results.

Exercises

1. What different means are employed to produce variation of the pitch of piano strings?   For violin strings?
2. How many beats per second will be produced by two tuning forks having 512 and 509 vibrations per second respectively?
3. A wire 180 cm. long produces middle C.   Show by a diagram, using numbers, where a bridge would have to be placed to cause the string to emit each tone of the major scale.
4. How can a violinist play a tune on a single string?
5. What are the frequencies of the first 5 overtones of a string whose fundamental gives 256 vibrations per second?
6. One person takes 112 steps a minute and another 116.   How many times a minute will the two walkers be in step?   How many times a minute will one be advancing the left foot just when the other advances the right?
7. Why is it necessary to have a standard pitch?
8. How can the pitch of the sounds given by a phonograph be lowered?
9. How many beats per second will occur when two tuning forks having frequencies of 512 and 515 respectively, are sounded together?
10. Which wires of a piano give the highest pitch?   Why?

(6) Tone Quality, Vibrating Air Columns, Plates

345. Quality.—The reason for the *differences in tone quality* between notes of the same pitch and intensity as produced, *e.g.*, by a violin and a piano, was long a matter of conjecture. Helmholtz, a German physicist (see p. 397) first definitely proved that tone quality is due to the *various overtones* present along with the fundamental and *their relative intensities*. If a tuning fork is first set vibrating by drawing a bow across it and then by striking it with a hard object, a difference in the *quality* of the tones produced is noticeable. It is thus evident that the manner of setting a body in vibration affects the overtones produced and thus the quality. Piano strings are struck by felt hammers at a point about one-seventh of the length of the string from one end. This point has been selected by experiment, it having been found to yield the

Fig. 337.—Chladni's plate.

best combination of overtones as shown by the quality of the tone resulting.

346. Chladni's Plate.—The fact that vibrating bodies are capable of many modes of vibration is well illustrated

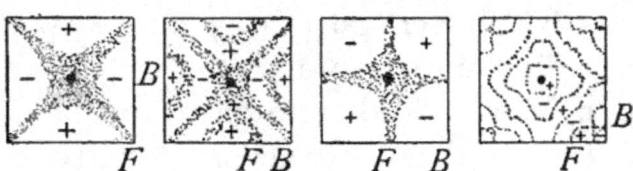

Fig. 338—Chladni's figures.

by what is known as Chladni's plate. This consists of a circular or square sheet of brass attached to a stand at its center so as to be held horizontally. (See Fig. 337.) Fine sand is sprinkled over its surface and the disc is set vibrating by drawing a violin bow across its edge. The mode of vibration of the disc is indicated by the sand accumulating along the lines of least vibration, called *nodal lines*. A variety of nodal lines each accompanied by its characteristic tone may be obtained by changing the position of the bow and by touching the fingers at different points at the edge of the disc. They are known as Chladni's figures. (See Fig. 338.)

347. Manometric Flames.—The actual presence of overtones along with the fundamental may be made *visible*

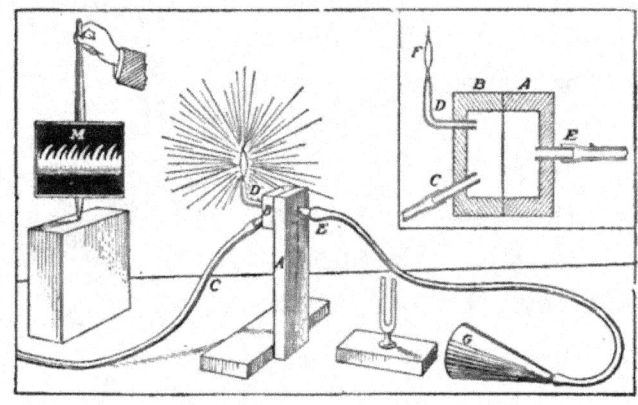

Fig. 339.—Manometric flame apparatus.

by the *manometric flame apparatus*. This consists of a wooden box, C, mounted upon a stand. (See Fig. 339.) The box is divided vertically by a flexible partition or diaphragm. Two outlets are provided on one side of the partition, one, C, leads to a gas pipe, the other is a glass tube, D. On the other side of the partition a tube, E, leads to a mouthpiece. A mirror, M, is mounted so as to be rotated upon a vertical axis in front of F and near it. Gas is now turned on and lighted at F. The sound of the voice produced at the mouthpiece sends sound waves through the tube and against the diaphragm which vibrates back and forth as the sound waves strike it.

This action affects the flame which rises and falls. If now the mirror is rotated, the image of the flame seen in the mirror rises and falls, showing not only the fundamental or principal vibrations but also the overtones. If the different vowel sounds are uttered in succession in the mouthpiece, each is found to be accompanied by its characteristic wave form (Fig. 340). In some, the fundamental is strongly prominent, while in others, the overtones produce marked modifications. Other devices have been invented which make possible the accurate analysis of sounds into their component vibrations, while still others unite simple tones to produce any complex tone desired.

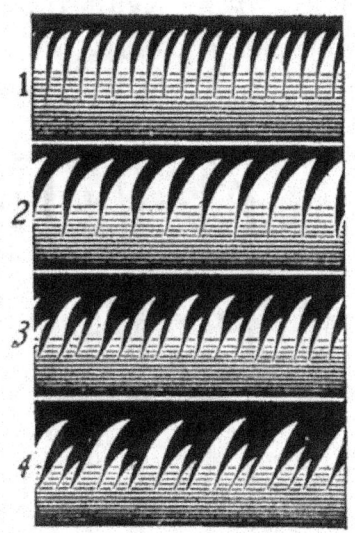

FIG. 340. — Characteristic forms of manometric flames.

348. The Phonograph.—The *graphophone* or *phonograph* provides a mechanism for cutting upon a disc or cylinder a groove that reproduces, in the varying form or depth of the tracing, every peculiarity of the sound waves affecting it. The reproducer consists of a sensitive diaphragm to which is attached a needle. The disc or cylinder is rotated under the reproducing needle. The irregularities of the bottom of the tracing cause corresponding movements of the needle and the attached diaphragm, which start waves that reproduce the sounds that previously affected the recorder. The construction of the phonograph has reached such perfection that very accurate reproduction of a great variety of sounds is secured.

349. Wind Instruments.—In many musical instruments as the *cornet*, *pipe-organ*, *flute*, etc., and also in *whistles*, the vibrating body that serves as a source of sound is *a column of air*, usually enclosed in a tube. Unlike vibrating strings, this vibrating source of sound changes but little in tension or density, hence changes in the pitch of air columns is secured by changing their length. The law being similar to that with strings, *the vibration rates of air columns are inversely proportional to their lengths.*

If an *open* organ pipe be sounded by blowing gently through it, a tone of definite pitch is heard. Now if one end is closed, on being sounded again the pitch is found

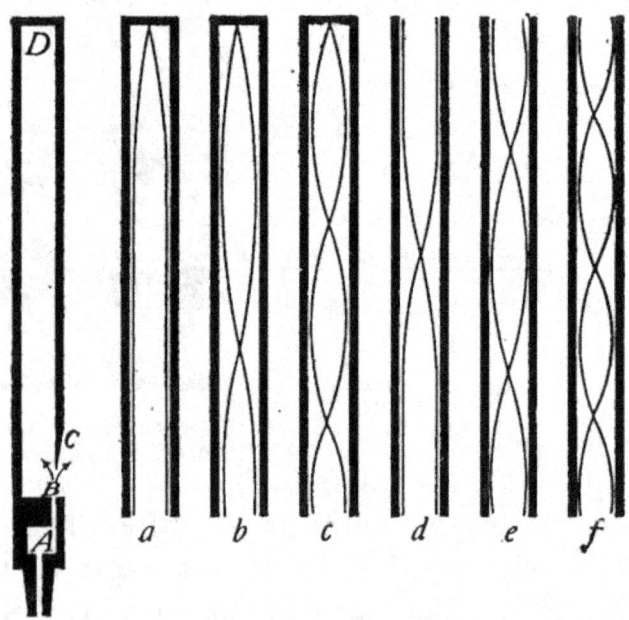

FIG. 341.—(R) Cross-section of an organ pipe showing action of tongue at C. (a) The fundamental tone in a closed pipe has a wave length four times the length of the pipe; (b) and (c) how the first and second overtones are formed in a closed pipe; (d) the fundamental tone of an open pipe has a wave length equal to twice the length of the pipe; (e) and (f) first and second overtones of open pipe.

to be an octave lower. Therefore, *the pitch of a closed pipe is an octave lower than that of an open one of the same length.*

350. Nodes in Organ Pipes.—Fig. 341, R represents a cross-section of a wooden organ pipe. Air is blown through A, and strikes against a thin tongue of wood C. This

starts the jet of air vibrating thus setting the column of
air in vibration so that the sound is kept up as long as air
is blown through A. To understand the mode of vibration of the air column a study of the curve that represents
wave motion (Fig. 342) is helpful Let AB represent such
a curve, in this 2, 4 and 6 represent nodes or points of
least vibration, while 1, 3 and
5 are antinodes or places of
greatest motion. A full wave
length extends from 1–5, or
2–6. Now in the open organ
pipe (Fig. 341d), the end of

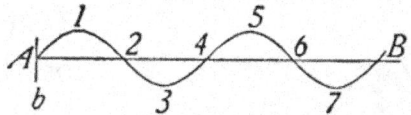

FIG. 342.—Graphic representation of sound waves.

the air column d is a place of great vibration or is an antinode. At the other end also occurs another place of
great vibration or an antinode; between these two antinodes must be a place of least vibration or a node. The
open air column therefore extends from antinode to antinode (or from 1–3) or is *one-half* a wave length. *The
closed air column* (Fig. 341a) extends from a place of *great*
vibration at a to a place of *no* vibration at the closed end.

FIG. 343.—A clarinet.

The distance from an antinode to a node is that from 1–2
on the curve and is *one-fourth* a wave length.

When a pipe is blown strongly it yields overtones.
The *bugle* is a musical instrument in which notes of different pitch are produced by differences in blowing. (See
Fig. 341 (d), (e), (f). In playing the *cornet* different
pitches are produced by differences in blowing, and by
valves which change the length of the vibrating air column.
(See Fig. 334.)   The *clarinet* has a mouthpiece containing

25

a reed. similar to that made by cutting a tongue on a straw or quill. The length of the vibrating air column in the clarinet is changed by opening holes in the sides of the tube. (See Fig. 343.)

351. How We Hear.—Our hearing apparatus is arranged in three parts. (See Fig. 344.) *The external ear* leads to the *tympanum*. *The middle ear* contains three bones that convey the vibrations of the tympanum to the *internal ear*. The latter is filled with a liquid which conveys the vibrations to a part having a coiled shell-like structure called the *Cochlea*. Stretched across within the cochlea are some 3000 fibers or strings. It is believed that each is sensitive to a particular vibration rate and that each is also attached to a nerve fiber. The sound waves of the air transmitted by the tympanum, the ear bones and the liquid of the internal ear start sympathetic vibrations in the strings of the cochlea which affect the auditory nerve and we hear. The highest tones perceptible by the human ear are produced by from 24,000 to 40,000 vibrations per second. The average person cannot hear sounds produced by more than about 28,000 vibrations. The usual range of hearing is about 11 octaves. The tones produced by higher vibrations than about 4100 per second are shrill and displeasing. In music the range is 7⅓ octaves, the lowest tone being produced by 27.5 vibrations, the highest by about 4100 per second.

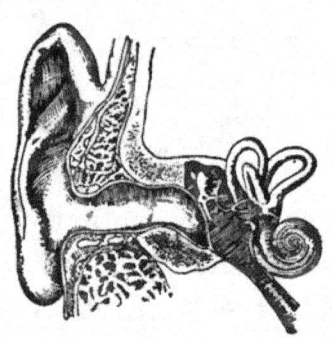

FIG. 344.—The human ear.

The tones produced by men are lower than those of women and boys. In men the vocal cords are about 18 mm. long; in women they are 12 mm. long

The compass of the human voice is about two octaves,

although some noted singers have a range of two and one-half octaves. In ordinary conversation the wave length of sounds produced by a man's voice is from 8 to 12 ft. and that of a woman's voice is from 2 to 4 ft.

Important Topics

1. Tone quality. Fundamental and overtones. Chladni's plate.
2. Manometric flame apparatus.
3. Phonograph recorder and reproducer.
4. Air columns and wind instruments.
5. How we hear.

Exercises

1. What determines the pitch of the note of a toy whistle?
2. The lowest note of the organ has a wave length of about 64 ft. What is the length of a closed pipe giving this note? Of an open pipe?
3. What is the first overtone of C? What are the second and third overtones? Give vibration numbers and pitch names or letters.
4. Why is the music of a band just as harmonious at a distance of 400 ft. as at 100 ft.?
5. A resonant air column 60 cm. long closed at one end will respond to what rate of vibration at 10°C.?
6. Can you find out how the valves on a cornet operate to change the pitch of the tone?
7. How is the trombone operated to produce tones of different pitch?
8. The lowest note on an organ has a wave length of about 64 ft. What must be the length of a closed pipe giving this note?
9. What is the approximate length of an open organ pipe which sends out waves 4 ft. long?

Review Outline: Sound

Sound—definition, source, medium, speed, nature.
Waves—longitudinal, transverse, illustrations.
Characteristics of Musical Sounds: $\begin{cases} \text{intensity—area, amplitude, density, distance.} \\ \text{pitch—scales; major, tempered, triads, } N = \\ V/L \text{ quality—fundamental and overtones.} \end{cases}$
Sympathetic Vibrations—resonance, interference, beats, discord.
Musical Instruments—string, air column, membrane or plate.
Laws of; (a) vibrating strings (3), (b) vibrating air columns (2).

CHAPTER XVI

LIGHT

(1) Light, Its Rectilinear Propagation, Shadows

352. A Comparison of Sound and Light.—Light from the standpoint of physics is considered much as is *sound*, as a *mode* of *motion;* one affecting the ear, the other producing the result called *vision*. There are other differences also worth considering. (a) While sound travels as vibrations of some *material* medium, light travels only as vibrations of the *ether;* solids, liquids, and gases act so as to hinder rather than to assist in its movement. That is, light travels best in a vacuum or in a space devoid of ordinary matter. (b) The *speed* of light is so great that at ordinary distances on the earth its motion is practically instantaneous. Experiments have shown that its speed is about 186,000 miles to 300,000 kilometers a second.

353. Luminous and Illuminated Bodies.—If we consider the objects within a room, some of them, as books and furniture, would be invisible if all light from external sources were excluded. On the other hand, some other objects, such as a lighted lamp, a burning coal, or a red hot iron, would be seen if no outside light were present. Such bodies are said to be luminous. Most luminous bodies are hot and become non-luminous on cooling. There are, however, some bodies that are luminous at ordinary room temperatures, as the firefly and some phosphorescent paints. When light emitted by a luminous body strikes an object, a portion of it is always *reflected*.

It is this reflected light that makes the illuminated object *visible.* If the object is a sheet of glass, some of the light is *transmitted.* If a substance is so clear that objects can be seen through it, the substance is *transparent,* but if objects cannot be seen through it, the substance is said to be *translucent.* Objects transmitting no light are *opaque.* Some of the light falling upon a body is neither reflected nor transmitted, but is *absorbed* and tends to warm the body. The light falling upon a body is therefore either *reflected, transmitted,* or *absorbed.* Thus Fig. 345 represents light coming from *S* to a piece of glass *GL.* A portion of the light represented by *R* is reflected Another part *A* is absorbed and disappears, while stil another part *T* is transmitted and passes on.

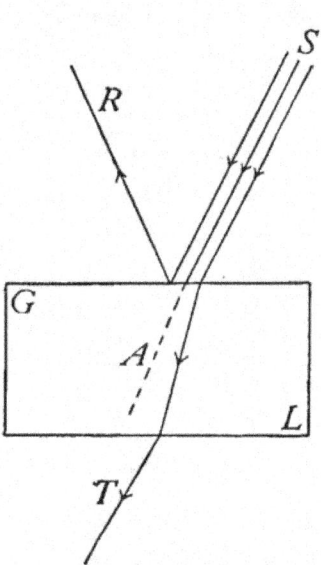

FIG. 345.—The light is transmitted (*T*), reflected (*R*), or absorbed (*A*).

There is no sharply drawn line between transparent and opaque bodies. Very thin sheets of gold transmit a greenish light, and experiments have shown that substances as transparent as clear water absorb enough light so that at considerable depths in an ocean or lake little or no light is ever found. All light whether from luminous bodies or reflected from non-luminous objects shows certain properties which will now be considered.

354. The Rectilinear Propagation of Light.—If a beam of light passes through a hole in a window shade into a darkened room, it is seen to follow a perfectly straight course. If a person while coughing holds a book before the face, the sound passes around the book and is heard

at any point in the room while the face is hidden by the
book. In other words, light ordinarily does not pass
around corners as sound does, but travels in *straight lines*.
This fact is made use of when one aims a gun or merely
looks at an object. So well established in our minds is the
idea that an *object* is in the direction from which we see the
light coming to us from it, that we are sometimes deceived
as to the real position of an object, when the course of the
light from it has been changed by a mirror or some other
reflecting surface. Many *illusions* are produced in this
way, of which the *mirage* of the desert is one example.
(See Art. 381.)

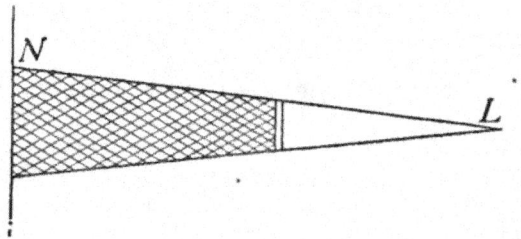

FIG. 346.—Shadow from a small source of light.

355. Shadows.—*A shadow is the space from which light
is cut off by an opaque body.* Thus if a book (see Fig. 346)
is held between a screen, *N*, and a *small* source of light, *L*,

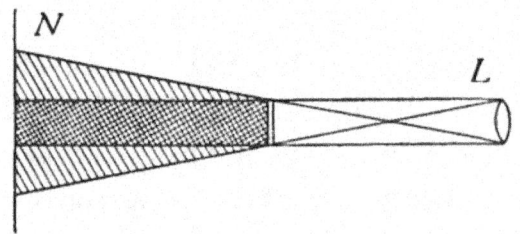

FIG. 347.—Shadow when source of light is large.

a shadow is produced which extends from the book to the
screen Notice that the shadow is a *space* and not an
area. If a *large* gas flame (see Fig. 347) is used as the
source of light, the shadow of the book is no longer clear

cut at the edges as before, but has a darker central part with a lighter fringe of partial shadow at the edges. The dark portion within the shadow has all the light excluded from it and is called the *umbra*. The lighter portion of the shadow at the edges has only a part of the light from the flame cut off. This portion is called the *penumbra*. When one stands in sunlight his shadow extends from his body to the ground or object on which the shadow falls. At night we are in the earth's shadow, which extends out into space beyond the earth.

356. Eclipses.—Since the sun is a very large object the shadow cast by the earth contains both umbra and penumbra. (See Fig. 348.) When the moon passes into the

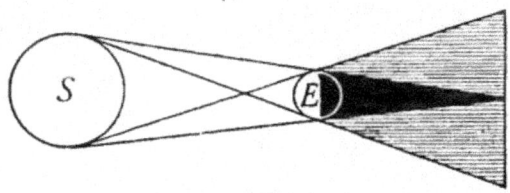

FIG. 348—Character of the earth's shadow.

shadow of the earth, there is said to be an eclipse of the moon, while if the moon's shadow falls upon the earth, the portion of the earth cut off from the sun's light has an eclipse of the sun.

357. Images by Small Apertures.—The straight line movement of light makes possible the *pin-hole* camera, by which satisfactory photographs have been made. The action of this device may be illustrated by placing a luminous body, a lighted candle, an incandescent lamp, or a gas flame, in front of a piece of cardboard, *S*, which has a small opening in it   Light from the object (see Fig. 349) falls upon a screen, S_2, so as to produce an *inverted image*. Other applications of this principle will be given later.

In Fig. 349 let *PQ* represent a gas flame, then light from point *P* at the *top* of the flame will pass in a straight line through the opening or aperture of the cardboard and strike at P_2 at the *bottom* of the illuminated spot upon the screen. Light from *Q* passing in straight lines through the aperture will strike at Q_2 at the top of the lighted space. This spot of light will have the same outlines as the luminous body *PQ* and being formed as just described will be *inverted*.

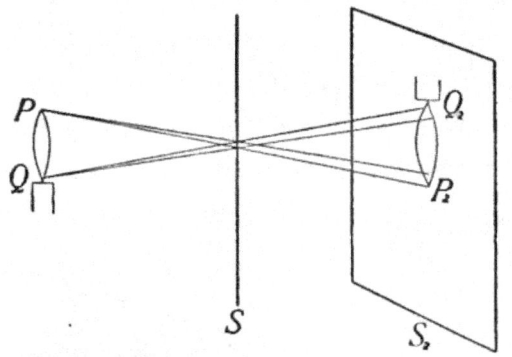

·FIG. 349.—Image formed by a small aperture is inverted.

This spot of light, resembling in its outlines the flame, is called an *image*. *An image is defined as an optical counterpart of an object.* Images are formed in a variety of devices, such as *apertures, mirrors*, and *lenses*. The *pinhole camera* is simply a light-tight box with a small aperture in one side Light passing through this aperture forms an image upon the opposite side of the interior of the box, of whatever object is in front of the camera. Light entering a room through a *large* aperture such as a window produces a multitude of overlapping images which blend to form a somewhat evenly illuminated surface.

Important Topics

1. Light contrasted with sound (three differences).
2. Bodies: transparent, translucent, opaque.
3. Light: reflected, transmitted, absorbed.

4. Light travels in straight lines, evidence, shadows, umbra, penumbra.

5. Formation of images by small apertures.

Exercises

1. Consider the circumference of the earth as 25,000 miles. How many times would the speed of light cover this distance in a second?

2. How soon after any great disturbance takes place on the sun, 93,000,000 miles distant, can it be seen upon the earth?

3. Construct a diagram of the moon's shadow. How much of the sun can one see when in the moon's umbra? When in its penumbra? Have you ever been in either? When? Have you ever been in the earth's umbra? In its penumbra?

4. Explain, using a diagram, the formation of an inverted image by a small aperture.

5. If the sun is 45 degrees above the horizon, what is the height of a pole casting a shadow 60 ft. long?

6. If a shadow 6 ft. long is cast by a 10-ft. pole standing vertically upon a walk, how tall is the tree whose shadow is 42 ft. long, both measurements being made at the same time?

7. Why are the shadows caused by an electric arc lamp so sharply defined?

8. Why should schoolroom windows be all on one side and reach to the ceiling?

9. What is the relation between the size of an image and its distance from the aperture forming it? Can you prove this by geometry?

10. What are silhouettes and how are they produced?

(2) PHOTOMETRY AND THE LAW OF REFLECTION

358. Photometry.—It is desirable at times to compare the intensities of illumination produced by light from different sources. This is done to determine the *relative cost or effectiveness* of various illuminants such as candles, kerosene and gas lamps, and electric lights The process of determining the relative intensity of lights or lamps is called photometry. (*Photos* = light.)

The unit for measuring the power of light is called a *candle power*. It is the light produced by a sperm candle burning 120 grains per hour. An ordinary gas light burns 5 or more cubic feet of gas per hour and yields from 15 to 25 candle power. A Welsbach gas lamp, consuming 3 cu. ft. per hour, produces 50 to 100 candle power.

Instead of using candles, for practical photometry, incandescent lamps standardized by the Bureau of Standards are used for testing or calibration purposes.

It is necessary to distinguish between the intensity of a *luminous* body, *i.e.*, as a source of light, and the *intensity* of *illumination* upon some surface produced by a light. It is considered that two sources of light are of *equal intensity* if they produce equal illumination at equal distances.

359. Law of Intensity of Light.—A device for measuring the candle power of a light is called a *photometer*. Its use is based upon the *law of intensity of light*. *The intensity of illumination of a surface is inversely proportional to the square of its distance from the source of light.* This relation is similar to that existing between the intensity of a sound and the distance from its source. The following device illustrates the truth of this law in a simple manner.

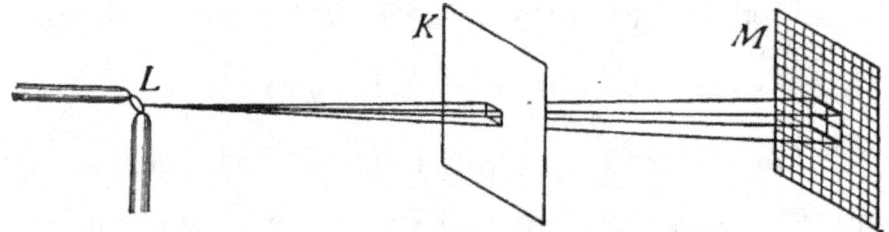

FIG. 350.—The light spreads over four times the area at twice the distance.

Cut a hole 1 in. square in a large sheet of cardboard (*K*) and place the card in an upright position 1 meter from an arc light or other *point source* of light (*L*). Now rule inch squares upon another card (*M*) and place it parallel to the first card and 2 meters from it. (See Fig. 350.) The light that passed through the hole of 1 sq. in.

at a distance of 1 meter is spread over 4 sq. in. at a distance of 2 meters. Therefore, the intensity of illumination on each square inch of M is one-fourth that upon the surface of K. If M is placed 3 meters from the light, 9 sq. in. are illuminated, or the intensity is one-ninth that at 1 meter distance.

These relations show that the intensity of illumination is inversely proportional to the square of the distance from the source of light. An application of the law of intensity is made in using a simple (Bunsen) photometer. This consists of a card containing a spot soaked with oil or melted wax. (See Fig. 351.) The lights whose intensities

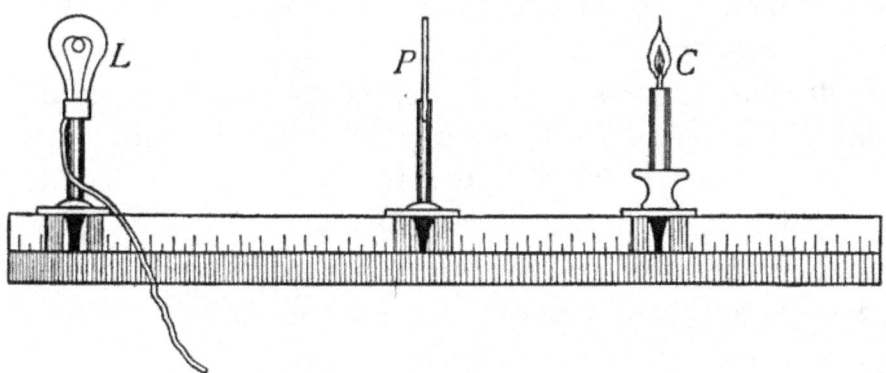

FIG. 351.—The Bunsen photometer.

are to be compared are placed upon opposite sides of the card. The card is then adjusted so that the spot appears the same on both sides. The illumination is now equal on both sides of the card and the *candle powers of the two lights are proportional to the squares of their distances from the card.* The simple device just described will give approximate results only. For accurate results more elaborate apparatus is required.

360. Measurement of the Intensity of Illumination.— A standard candle (Art. 358) produces when lighted 1 candle power. The illumination caused by this upon a surface 1 ft. away and at right angles to the light rays is called a **foot-candle.** It is the unit of intensity of illumination. A 4-candle-power lamp, at a distance of 1 ft., produces 4 foot-candles. A 16-candle-power lamp at a distance of 2 ft. also produces 4 foot-candles-(16 ÷ 2²).

The intensity of illumination required for a good light for seeing varies with the conditions. Thus, for stage and store lighting about 4 foot-candles are needed, while homes and churches may require but 1 foot-candle.

Too great an intensity of illumination is as harmful as not enough. Exposed lights having an intensity of more than 5 candle power per square inch are often a cause of eye trouble. Such lights should be protected by frosted globes.

A pleasing form of lighting for large halls and public buildings is the *indirect system.* In this, the lamps are hidden by reflectors which throw the light upon the ceiling from which it is diffused over the room. This form of lighting is more expensive than other systems since but a part of the light is reflected. Its cost therefore is an important factor when considering its use.

361. The Reflection of Light.— The light reflected from the surfaces of bodies about us gives us information concerning our surroundings. A knowledge of the behavior of light undergoing reflection is not usually gained from ordinary observation. The law of reflection of light may be shown, however, by an experiment.

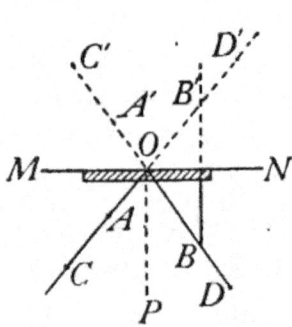

Fig. 352.—*B'* is as far back of the mirror as *B* is in front of it.

A plane mirror, *M*, is held in a vertical position resting upon a sheet of paper. (See Fig. 352.) Pins are set upright in the paper at *A* and *B*. On placing the eye along the line *AC* and looking toward the mirror an image of *B* may be seen in the mirror due to

Christian Huygens (1629–1695).
Dutch physicist; invented the pendulum clock (1656); developed the wave theory of light; discovered polarization of light (1690).

CHRISTIAN HUYGENS
(Popular Science Monthly)

Hermann von Helmholtz (1821–1894) Germany. Established the doctrine of conservation of energy; made many discoveries in sound; invented the ophthalmoscope; established the physical basis of tone quality.

H. V. HELMHOLTZ
"By Permission of the Berlin Photographic Co., New York."

the light reflected from its surface. Pins C and D are now set in the paper so that when one looks along the line BD toward the mirror one may see all four pins apparently in one line. This indicates that the light from A and C passing along CA toward O is reflected back along the light CBD. By means of a ruler, draw lines through BD and AC till they intersect at O. Also draw PO perpendicular to the mirror at O.

Then the angles AOP and BOP will be found equal. These are called the angles of *incidence* and *reflection* respectively. *The law of reflection* is therefore stated: *The angle of reflection is equal to the angle of incidence.* These angles are in the same plane, that of the paper. This law applies in all cases of reflection of light. It is similar to the law of reflection of sound (Art. 326.)

Important Topics

1. Photometry, law of intensity, candle power, foot-candle.
2. Intensity of illumination.
3. Reflected light and law of reflection.

Exercises

1. Both sides of a card are equally illuminated when two lights are on opposite sides of it and 10 and 30 cm. respectively from it. What are their relative intensities?

2. What are the relative intensities of illumination from a gas light upon a book 6 ft. and 2 ft. respectively from the light?

3. Which is more expensive per candle power? How many times as expensive? A 50-watt 16-candle-power incandescent lamp at 10 cents per kilowatt-hour or a 100-candle-power Welsbach light burning 5 cu. ft. of gas per hour at 80 cents per 1000 cu. ft. of gas. (Find cost of each per hour, and then the cost of 1 candle power hour for each.)

4. Why are not ordinary shadows perfectly dark?

5. At what distance will a 16-candle-power lamp give the same illumination as single candle at 10 in.?

6. If the sun is at an elevation of 30 degrees what is the angle of incidence at which it strikes the surface of water. What is the angle between the incident and the reflected rays?

7. What is the difference between the phenomena of reflection of light from a white sheet of writing paper and from a piece of clear window glass?

8. A horizontal ray of light, traveling due east, strikes a vertical mirror so that after reflection it is traveling due north. If the mirror be now turned 10 degrees about a vertical axis, the north edge moving east, what will be the direction of the reflected ray?

9. The necessary illumination for reading is about 2 foot-candles. How far away may an 8-candle-power lamp be placed?

10. What is the illumination in foot-candles upon a surface 20 ft. from an arc lamp having an intensity of 1000 candle power?

11. How far from a 100-candle-power Welsbach light would the illumination be 2 foot-candles?

(3) Mirrors and the Formation of Images

362. Mirrors.—The many purposes served by mirrors in our everyday life has made their use familiar to everyone. Yet without study and experiment few understand

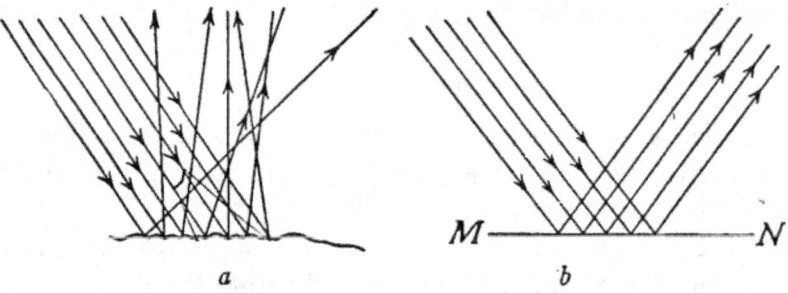

FIG. 353.—Reflection of light, (a) diffused, (b) regular.

their properties and action. *Any smooth* surface may serve as a mirror, as that of glass, water, polished wood, or metal. Most objects, unlike mirrors, have irregular surfaces; these scatter or diffuse the light that falls upon them. (See Fig. 353a.) This is called *diffused or irregular reflection*. The reflection of light from the smooth surface of a mirror is *regular*. (See Fig. 353b.) In every case of reflected light, however, the angle of reflection equals the angle of incidence, diffusion being due to the irregularity of the surface. It is by means of the light "diffused" from the surface of illuminated bodies, such as plants, animals, food, and manufactured articles, that we "see" the various objects about us, and it is this light that enables us to judge of their distance, size, form, color, etc. The moon is seen by the sunlight reflected from its surface. Moonlight is therefore sunlight diffused by reflection. The *new moon* is that phase or condition of the moon when only a narrow strip of the moon's illuminated surface is turned toward the earth. At the time of the *full moon* the whole illuminated surface is seen.

363. Images Formed by a Plane Mirror.—The most common use of mirrors is in the formation of images. The way in which images are formed by a plane mirror may be illustrated by diagrams. Thus in Fig. 354, let L represent a luminous body and E and E' two positions of the observer's eye. Take any line or ray as LO along which the light from L strikes the mirror O–O'. It will be reflected so that angle LOP equals angle POE. Similarly with any other ray, as LO', the reflected ray O'E' has

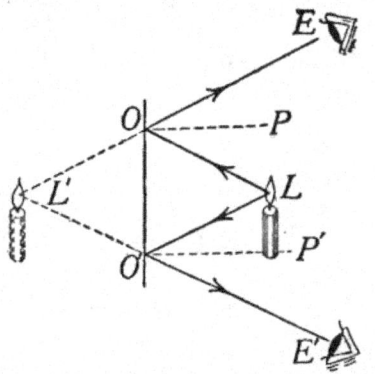

FIG. 354.—The virtual image of a fixed object as seen in a plane mirror, has the same location from every position of the observer's eye.

a direction such as that angle L'O'E' equals angle P'O'E'. Any other rays will be reflected in a similar manner, each of the reflected rays appearing to the eye to come from a point L' behind the mirror.

364. Light Waves and Wave Diagrams.—Just as a stick continually moved at the surface of a body of water sets up a series of waves spreading in all directions, so one may imagine a train of waves sent out by a luminous

26

body L (as in Fig. 355) to the mirror MN. These waves will be reflected from the mirror as if the source of light were at L'. It is much simpler and more convenient to locate the position of the image of a point by the use of lines or "rays" (as in Fig. 354) than by the wave diagram (as in Fig. 355). In all *ray diagrams*, however, it should be kept in mind that the *so-called* ray is a symbol used to represent the direction taken by a part of a light wave.

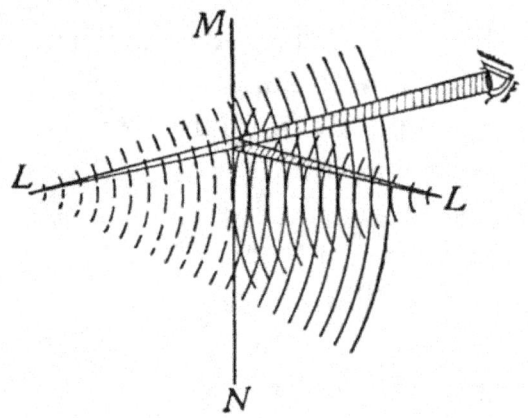

FIG. 355—Wave diagram of image formed in a plane mirror.

Thus in Fig. 354, the light from L moving toward O is reflected to E along the line OE, the heavy lines repsenting rays.

365. To locate the image of an object formed by a plane mirror *requires* simply an application of the law of reflection. Thus in Fig. 356 let AB represent an object and MN a plane mirror. Let AA' be a ray from A striking the mirror *perpendicularly*. It is therefore reflected back along the same line toward A. Let AO represent any other ray from A. It will be reflected along OE so that angle r equals i. The intersection of AC and OE at A' behind the mirror locates the image of the point A, as seen by reflection from the mirror. The triangles ACO and $A'CO$ may be proved equal by geometry. Therefore

$A'C$ equals AC.   This indicates that *the image of a point
formed by a plane mirror is the same distance back of the
mirror as the point itself is in front of it.*   This principle
may be used in locating the image of point B at B'.   Locating the position of the *end points* of an image determines
the position of the whole image as $A'B'$.

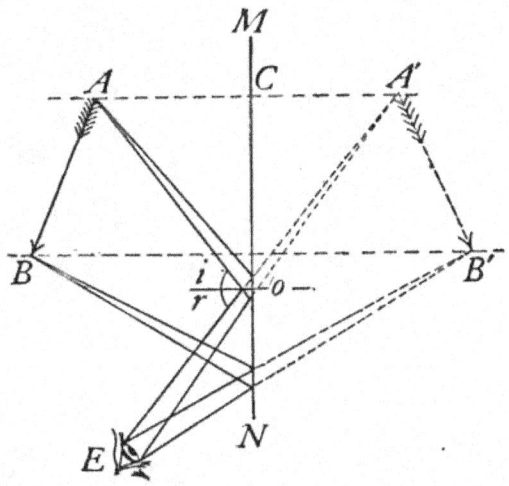

Fig. 356.—The image A' B' is as far back of the mirror M N as the
object A B is in front of the mirror.

366. How the Image is Seen.—Suppose the eye to be
placed at E.   It will receive light from A by reflection
as if it came from A'.   Similarly light starting from B
reaches the eye from the direction of B'.   There is nothing
back of the mirror *in reality* that affects our sight, the light
traveling only in the space in front of the mirror.   Yet
the action of the reflected light is such that it produces
the same effect as if it came from behind the mirror.
Images such as are seen in plane mirrors are called *virtual*
to distinguish them from *real* images, in which light
actually comes to the eye from the various parts of the
visible image, as from the real image formed by a projecting lantern upon a screen, or by an aperture as in the pin-

hole camera. Real images therefore are those that can be obtained upon a screen while virtual images cannot.

367. Multiple Reflection.—If the light from an object is reflected by two or more mirrors various effects may be produced, as may be illustrated by the *kaleidoscope*. This consists of three plane mirrors so arranged that a cross-section of the three forms an equilateral triangle. The

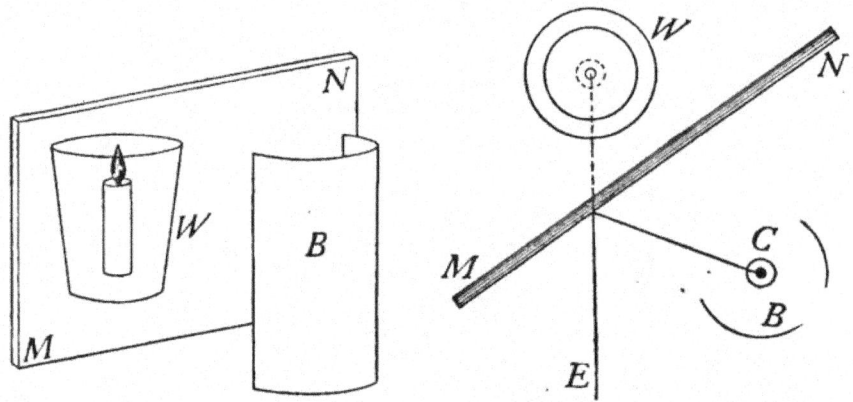

FIG. 357.—Perspective view of "Pepper's ghost."

FIG. 358.—Diagram of the "Pepper Ghost" illusion.

mirrors are placed in a tube across the end of which is a compartment with a translucent cover containing pieces of colored glass. On looking through the tube, the reflections from the several surfaces produce beautiful hexagonal designs.

368. Optical Illusions by a Plane Mirror.—The *illusion* called *Pepper's Ghost* is typical of many illusions produced by reflection. It may be illustrated by taking a piece of plate glass, M–N, a tumbler of water, W, and a lighted candle, C, placed in a box, B, having one side open and arranged as shown in perspective in Fig. 357, and in section in Fig. 358. If the effect is produced in a darkened room, the observer at E sees a virtual image of the lighted candle as if it were in the glass of water, the water being seen by transmitted light *through* the plate glass, the latter forming a virtual image of the candle by reflection. Some of the illusions produced by this means are: (a) the figure suspended in mid air; (b) the bust of a person without a trunk; (c) the stage ghost; (d) the disappearing bouquet.

369. Concave Mirrors.—Another useful piece of physical apparatus is the concave spherical mirror. It is frequently made from plano-convex lenses by silvering the convex surface of the lens, thus making a

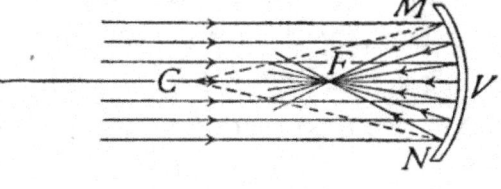

FIG. 359.—Action of a concave mirror on parallel rays of light.

concave reflecting surface from the inner surface of the silvered part; they are also made by polishing the inner surfaces of metallic spherical shells. The concave mirror is represented in section in Fig. 359 by the curve MN; C is the *center of curvature* or the center of the surface of which this mirror MN is a part; the line VC through the center V of the mirror is called the *principal axis;* while any other line passing through C is called a *secondary axis.* The point midway between the vertex V and center of curvature C is called

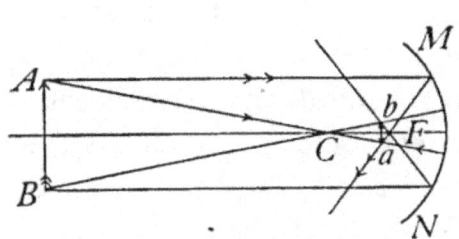

FIG. 360.—Real image formed by a concave mirror.

the *principal focus, F.* It is the point through which parallel incident rays pass after reflection. The angle MCN which the curve of the mirror subtends at the center is called the aperture of the mirror. We learned in Art. 361, the angle of reflection of a ray of light is always equal to the angle of incidence no matter what the nature of the reflecting surface may be. If the reflecting surface is a regular concave surface, like the inner surface of a sphere, the rays of light coming from a point source may after reflection come to a focus, forming a real image. The two extreme points of an object should be selected¯ for locating its image; Fig. 360 shows the construction. The real images formed by concave mirrors are always inverted. The principal focus of a concave mirror may be observed by holding the mirror in a beam of sunlight entering a darkened room. The sun's rays after reflection converge to form a small, round, intense spot of light, which is a real image of the sun, located at the principal focus of the mirror. The distance of the principal focus from the mirror is the least distance that a real image can be formed in front of a concave mirror. ·

370. Virtual Images by Concave Mirrors.—When light comes from a small point situated between a concave mirror and its principal focus, the reflected rays are divergent and hence no real image of the object can be found in front of the mirror. But if the rays are extended behind the mirror they will meet in a point called the *virtual focus*. This is the point from which they appear to come.

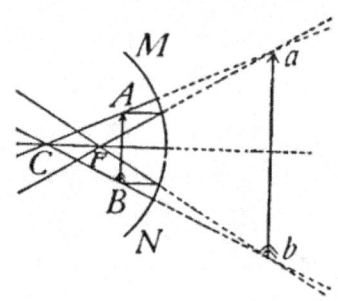

FIG. 361.—Virtual image formed by a concave mirror.

Any image of an object situated between the principal focus and a concave mirror is therefore a virtual image, erect and larger than the object. (See Fig. 361.)

371. Construction of Real Images.—There are five positions at which an object may be situated in front of a concave mirror, namely: (1) *beyond C*; (2) *at C*; (3) *between C and F*; (4) *at F* and (5) *between F and V*. There are two ways by means of which the image formed at each of these positions may be located, namely; (1) *experimentally*, by allowing the rays of light from a luminous body to focus on a screen and (2) *diagrammatically*. By the latter method the two rays of light are considered the course of each of which may easily be determined; first, the ray which strikes the mirror parallel to its principal axis and which after reflection passes through the principal focus; second, the ray which passing through the center of curvature strikes the mirror at right angles and therefore after reflection must pass directly back along its incident path. Where these two reflected rays intersect is located the real image of the object. Whenever these two rays of light do actually intersect, as in Fig. 360, a real image (*ab*) is formed of the object *AB*.

The points *A* and *a*, *B* and *b* and others similarly situated on an axis extending through the center of curvature *C* are called *conjugate foci*, for they are so related that an object being at either one, its image will be found at the other.

372. The Convex Mirror.—There are few practical uses to which convex mirrors can be put. They are sometimes used to give the chauffeur of an automobile a view of the road behind him. It is then attached to the wind shield by a short rod. The reflected rays coming from a convex mirror are always divergent (see Fig. 362), hence the image is always virtual and

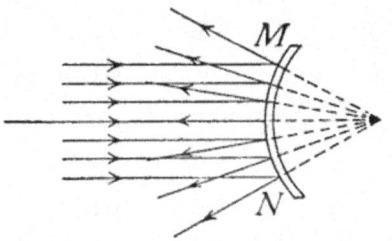

FIG. 362.—Action of a convex mirror upon parallel rays of light.

located behind the reflecting surface. The method of construction for images formed by a convex mirror is similar to that for concave mirrors. (See Fig. 363.) The center of curvature and principal focus are behind the mirror and consequently the reflected rays have to be produced backward until they meet. The images are always *virtual, erect* and *smaller* than the object.

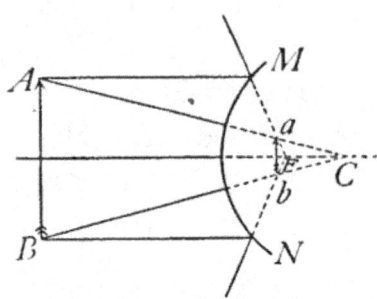

FIG. 363.—Construction of an image by a convex mirror.

373. Spherical Aberration.

Sometimes in a concave mirror when the aperture MCN (Fig. 364) is large the images are blurred or indistinct. This is due to the fact that the incident rays near the outer edge of the mirror do not focus after reflection at the same point as those which pass into the mirror near the vertex, but cross the principal axis at points between the mirror and principal focus as is shown in Fig. 364; this result is called *spherical aberration*. The larger the aperture of the mirror the more the image is blurred. Concave mirrors in practical use do not have an aperture much greater than 10 degrees. This non-focusing of the rays of light by curved reflecting surfaces may be noticed in many places, as when light is reflected from the inside of a cup that contains

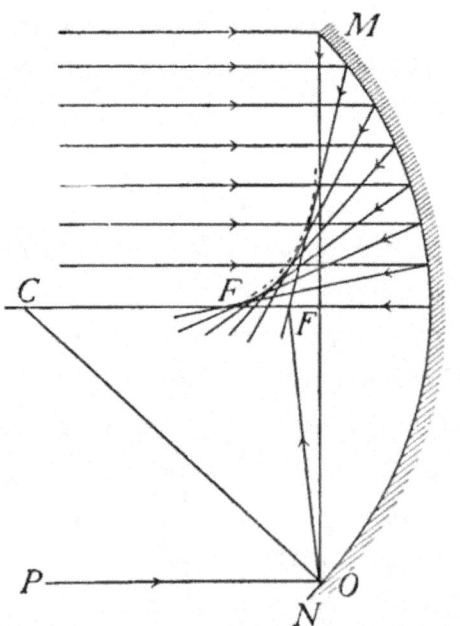

FIG. 364.—Illustrations of Spherical Aberration.

milk or from the inside of a wide gold ring placed on top of a piece of white paper. The pupil will note other instances.   This curve of light observed is called the *caustic by reflection*.

374. Parabolic Mirrors.—The best possible surface to give to concave mirrors is parabolic.   This is a curve which may be generated by moving a point so that its distance from a fixed point and a fixed line are always equal. If a source of light is placed at F the rays after reflection are rendered parallel. See Fig. 365.   This reflector is used in automobile lamps, head-lights of locomotives, search-lights, etc.   It is also used in large reflecting astronomical telescopes to collect as large an amount of light as possible from distant stars and bring it to a focus.   Such mirrors may be made exceedingly accurate.

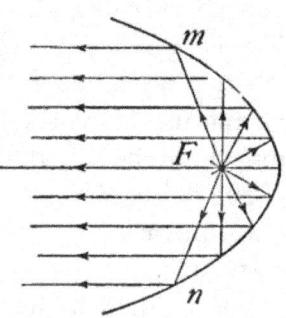

FIG. 365.—Parabolic mirror.

Important Topics

1. Reflection: regular, diffused; plane mirrors; laws of reflection.
2. Formation and location of images by plane mirrors.   Wave and ray diagrams.
3. Multiple reflection, illusions.
4. Curved mirrors, uses; concave, convex, parabolic.

Exercises

1. Distinguish between regular and diffused reflection.   By means of which do we see non-luminous bodies?
2. Could a perfect reflecting surface be seen?   Explain.
3. A pencil is stood upright in front of a plane mirror set at an angle of 45 degrees to the vertical.   Shown by a diagram the location and position of the image.
4. Show by diagrams the position and location of the images of a pencil (a) when standing erect and in front of a *vertical* mirror. (b) When standing upon a horizontal mirror.
5. What is the difference between a real and a virtual image?

6. A standard candle and a lamp give equal illuminations to a screen that is 1 ft. from the candle and 6 ft. from the lamp. What is the candle power of the lamp? Explain.

7. Why are walls finished in rough plaster or painted with soft tones without gloss better for schoolrooms than glossy paints or smooth white plaster?

8. Try to read a printed page by looking at its image in a mirror. Write your name backward on a sheet of paper, and then look at the image of the writing in a mirror. What effect is produced by the mirror in each case?

9. If the point of a pencil is held to the surface of a piece of plate-glass mirror two or more images may be seen in the mirror. Explain.

10. Given a small lighted candle, a concave mirror, a meter stick, and a white screen, how would you prove the statements made in Arts. 369 and 370 concerning the location of images formed by concave mirrors? Make the diagram in each case.

11. Why do images seen in a quiet pond of water appear inverted? Explain by a diagram.

(4) REFRACTION OF LIGHT

375. Common Examples of Refraction.—Everyone has noticed the apparent bending of an oar, of a stick, or of a

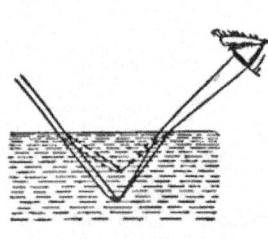

FIG. 366.—The stick appears to be bent on account of refraction.

spoon when placed in water (see Fig. 366), while many have observed that the bottom of a pond or stream looks nearer to the surface than it really is. These and similar illusions are due to the *refraction* or bending of light rays as they pass from one medium to another. The principles of refraction are among the most useful found in the study of light since application is made of them in the construction and use of important optical instruments, such as the camera, microscope, telescope, and the eye.

376. Action of Light Undergoing Refraction.—If a beam of sunlight be admitted to a darkened room and reflected by a mirror so that it strikes the surface of water in a glass jar, a part of the beam may be seen to be reflected while another portion is transmitted through the water (Fig. 367). The reflected beam follows the law of reflection while the transmitted beam is seen to be *refracted*, or to have its courses slightly changed in direction upon entering the water. If the mirror is turned so that the angle at which the light strikes the water is changed, the amount

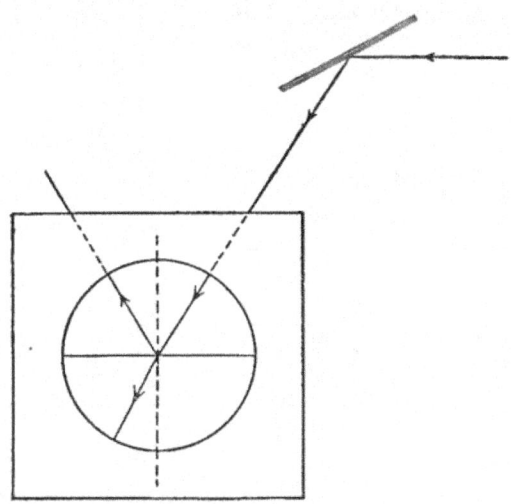

FIG. 367.—Part of the ray is reflected and part passes into the water and is refracted.

of refraction or change of course of the light is varied. When the light strikes the water perpendicularly there is no refraction. On the other hand, the greater the angle at which the light strikes the water the greater the bending.

377. Laws of Refraction. The action of light on entering, passing through, and leaving a great variety of substances has been carefully studied. A summary of

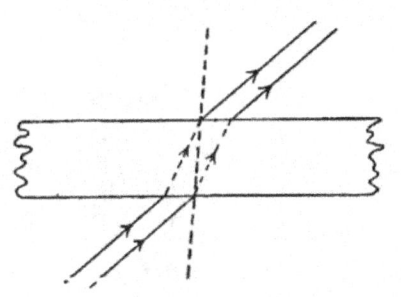

FIG. 368.—Illustrating the laws of refraction of light.

the results of these observations is given in the following *laws of refraction:* I. *When light enters a transparent body, perpendicularly, it passes on without changing its direction.*

II. *When light enters a denser transparent body obliquely, it is bent toward the perpendicular; when light enters a less dense body obliquely, it is bent away from the perpendicular.* (See Fig. 368.)

378. The cause of refraction may be illustrated by considering a line of men moving across a field and occupy-

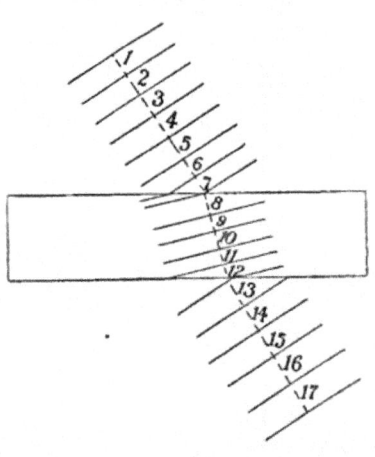

FIG. 369.—Diagram illustrating the cause of refraction.

ing at equal time intervals the successive positions 1, 2, 3, etc., indicated in Fig. 369. Suppose that the upper and lower parts of the field have a smooth hard surface, while at the center is a strip of newly ploughed ground. The line will move more slowly over the ploughed field than over the hard field. This will result in a retardation of the end of the line first striking the soft ground with a resulting change of direction of the line, *toward* the perpendicular to the edge of the field (*on entering the place of more difficult travel*), and *away* from the perpendicular on moving into a place where *increased speed results.*

379. Index of Refraction.—By studying the change of direction of the marching men as shown in Fig. 369 it is evident *first* that it is due to a difference in speed in the two media. It is not easy to measure the speed of light in a medium. However, the amount of refraction may be determined easily and from this the *relative* speed may be computed. The *number that expresses the ratio of the speed of light in air to its speed in another medium is called the index of refraction of that medium.* The relative speeds of light, or the indices of refraction for some substances, are:

water, 1.33, crown glass, 1.51, flint glass, 1.61, diamond, 2.47, carbon bisulphide, 1.64.

380. Plates, Prisms, Lenses.—The refraction of light is usually observed when it is passing through a plate, a prism, or a lens. The important differences between the effects of each in refracting light are illustrated in Figs. 370, 371 and 372. In Fig. 370 it is seen that the refraction of the ray on entering the glass is counteracted by the refraction away from the perpendicular upon leaving it. So that the entering and emergent rays are *parallel.*

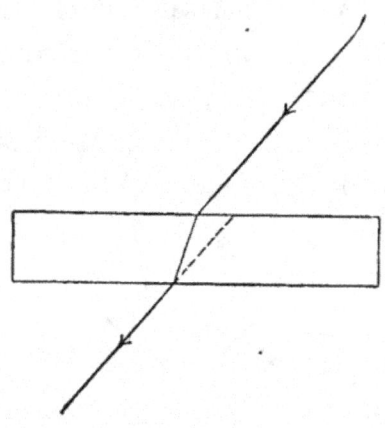

FIG. 370.—The incident ray and the emergent ray are parallel.

In Fig. 371 the refraction at the two surfaces of the prism results in a change of direction of the ray, the course being bent toward the *thicker* part of the *prism.* In

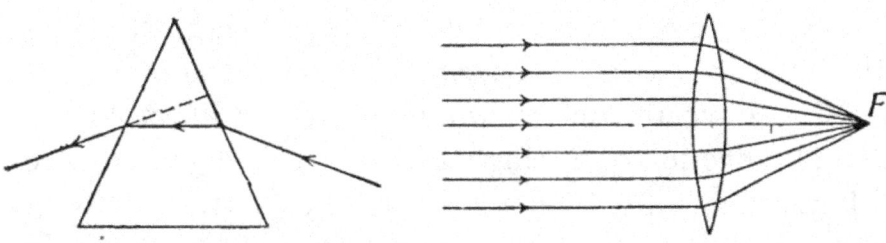

FIG. 371.—Effect of a prism upon a ray of light.

FIG. 372.—The convex lens brings the rays of light to a focus.

Fig. 372 it may be noticed that the convex lens resembles two prisms with their bases together. Since all parts of the lens refract light toward the thicker part, the center, the effect of the convex lens is to bring the rays of light to a focus, at *F*.

381. Total Reflection.—It has been shown that when light passes from a *denser to a lighter* medium, as from glass or water to air, that the beam is refracted *away* from the perpendicular. This is illustrated in Fig. 373. The diagram represents the change in the course of a ray of light that passes through water to a surface with air above it. A ray striking perpendicularly passes through without refraction. Other rays show increasing refraction with increasing angle of incidence. For one ray the angle of refraction is so large that the refracted ray is parallel to the surface When this condition is reached, the

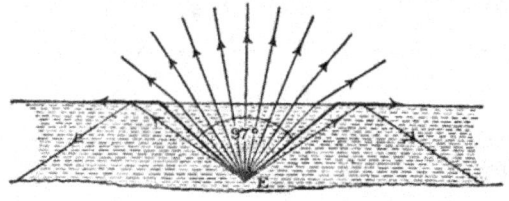

FIG. 373.—An example of total reflection.   FIG. 374.—Total reflection in a right-angle prism.

angle of incidence is called the *critical angle*. Any increase in the angle of incidence causes all of the light to be reflected as is the beam *E*. This action is called *total reflection*, the course of the reflected ray being according to the law of reflection. *A right-angle prism* (see Fig. 374) is often used where a mirror would ordinarily be employed, the total reflection occurring within the prism giving more satisfactory results than a mirror. See Art. 398 for a description of the Zeiss binocular field-glass for an example of this use of total reflection.

The mirage (see Fig. 375) is an optical illusion by which distant objects, below the horizon, are sometimes plainly seen. This phenomenon is most frequently observed in hot, desert regions, when the air conditions are such that the lower strata near the ground are very much hotter than those above. These lower strata, having expanded the most, are less dense than the cooler ones above. Hence a ray of light traveling obliquely downward is refracted more and more until total reflection takes place. The images seen are inverted giving a representation of trees or other objects reflected on the surface of still water. The mirage is also frequently seen at sea, ships being observed, sometimes erect, sometimes inverted, apparently sailing in the clouds near the horizon. Over the Great Lakes, trees, boats,

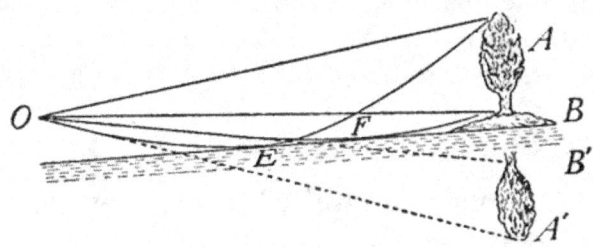

FIG. 375.—Diagram of a mirage.

and towns on the opposite shore, sixty or seventy miles away, can sometimes be plainly seen, apparently but a few miles out. In this case the images are erect, the total reflection being from warm, still layers of air over colder layers near the water.

Important Topics

(A) Refraction: cause, illustration, two principles.
(B) Index of refraction, meaning.
(C) Plates, prisms, lenses, action of each.
(D) Total reflection, uses.

Exercises

1. Compute the speed of light in water, the index of refraction being 1.33.
2. If one wished to shoot a fish under water, should he aim at the apparent location of the fish as viewed from the air? Explain, using a diagram.
3. Define refraction. Mention two illustrations of this action that you have observed out of school.
4. Why does the moon look larger near the horizon?
5. Is your reflection seen in a pool of water upside down? Why?
6. Why does it whiten molasses candy to pull it?
7. When looking at a building through the ordinary glass of a window why do straight lines of the building appear to be so distorted? What makes them appear to move as you move your head slightly?

8. Explain the phenomenon which one observes when looking at an object through the air arising from a hot stove or radiator.

9. Frequently the horizontal diameter of the setting sun appears to be greater than the vertical. Explain.

10. Explain why one observes several images of a luminous body like a lighted candle when the reflected light from a thick glass mirror enters the eye, the angle of reflection being large.

(5) The Formation of Images by Lenses

382. Uses of Lenses in Optical Instruments.—The use of instruments that employ lenses in their operation, such as spectacles, reading and opera glasses, and the camera, microscope, and telescope, is familiar to most students of physics. The part played by the lenses, however, is not generally understood. Consequently the study of the formation of images by lenses is of general interest and importance.

383. Forms of Lenses.—While a lens may be formed from any transparent solid it is commonly made of glass.

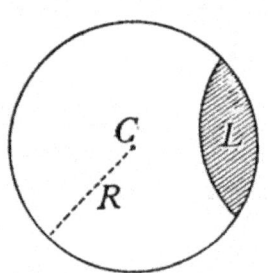

FIG. 376.—Formation of a spherical lens.

It may have two curved surfaces or one curved and one plane surface. Most lenses are *spherical lenses*, since their curved surfaces form a part of the surface of a sphere. Fig. 376 represents a spherical lens with a curved surface coinciding with that of a sphere whose center is at C This center is called the *center of curvature*, while the radius of the sphere R, is the *radius of curvature*.

There are two classes of lenses: those thick in the middle are called *convex*, while those thick at the edges are *concave*. The mode of constructing the six forms of spherical lenses is shown in Fig. 377. These are named as follows: (1) double convex, (2) plano convex, (3) concavo-convex, (4) double concave, (5) plano concave, (6) convexo-concave.

384. Effect of Lenses upon Light.—The most important characteristic of a lens is its effect upon a beam of light. Most persons have seen a "burning glass," a double convex lens, used to bring to a point, or focus, a beam of sunlight. To show the action of a burning glass send a beam of

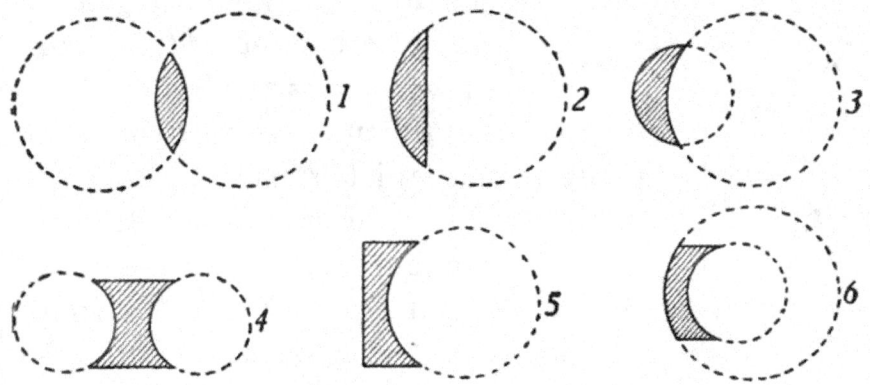

FIG. 377.—Forms of Lenses. 1. double convex; 2. plano convex: 3. concavo convex; 4. double concave; 5. plano concave; 6. convexo concave.

light into a darkened room, and place in its path a double convex lens. (See Fig. 378.) If two blackboard erasers are struck together near the lens, the chalk particles in the path of the light are strongly illuminated, showing that the light after passing through the lens it brought to a focus and that it spreads out beyond this point. This point

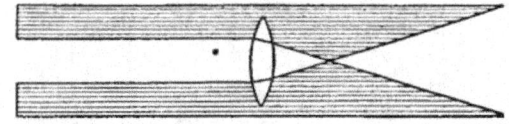

FIG. 378.—The action of a burning glass.

to which the cone of light rays converges after passing through the convex lens is called the *principal* focus of the lens. The distance from the principal focus to the center of the lens is the *focal length* or *principal focal distance* of the lens. *The focal length of double convex*

27

lenses of crown glass is about the same as the radius of curvature of either surface. The action of a convex or converging lens upon light may be better understood by studying Fig. 379 in which light is passing from S to F.

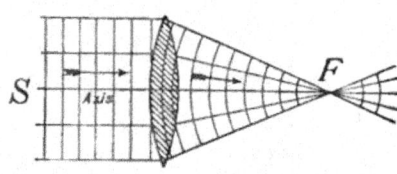

FIG. 379.—Wave diagram of light passing through a convex lens.

The successive positions and shape of the advancing light waves are indicated by lines drawn across the beam. The light being retarded more in the thicker part of the lens, the light waves on leaving the lens have a concave front. Since light waves tend to move at right angles to the front of the wave, the light is brought to a focus. After passing the focus the waves have a convex front, forming a diverging cone.

385. Concave Lenses.—When sunlight passes through a *concave* lens a diverging cone of light is formed. (See Fig. 380.) This is caused by the edges of the wave being retarded more than the center, producing a convex wave front. This diverging cone of light acts as if it had proceeded from a luminous point at F.

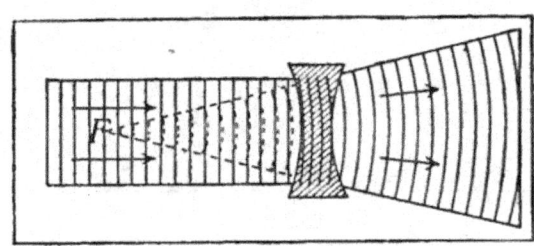

FIG. 380.—Wave diagram of light passing through a concave lens.

This point is called a *virtual* focus and is nearly at the center of the curvature of the nearer surface.

386. The Formation of Images by Lenses.—If a beam composed of *parallel* rays of light, as sunlight, is sent in turn through three convex lenses of the same diameter but of different thickness, it is found that the *thicker the lens the greater is its converging power, or the shorter is its focal* length. (See Fig. 381.) Now if a luminous body, such as a lighted candle, be placed near the convex lens but *beyond its focal length,* the light will be brought to a focus upon the other side of the lens and an image of the candle may be clearly seen upon the screen placed at this point. (See Fig. 382.) *The two points so situated on opposite sides of a lens that an object at one will form an image at the other arc called conjugate foci.*

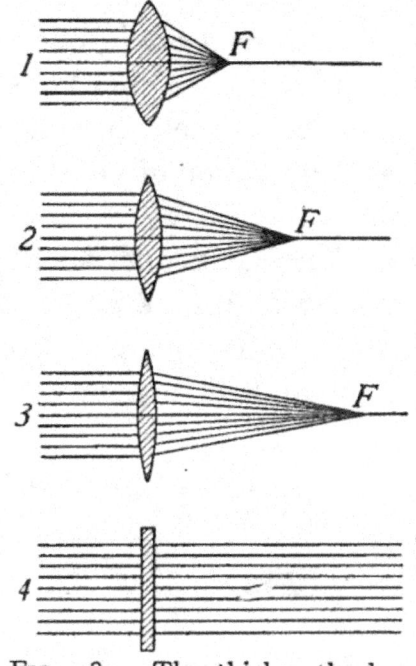

FIG. 381.—The thicker the lens, the shorter is its focal length.

It will be helpful to compare the images formed of a

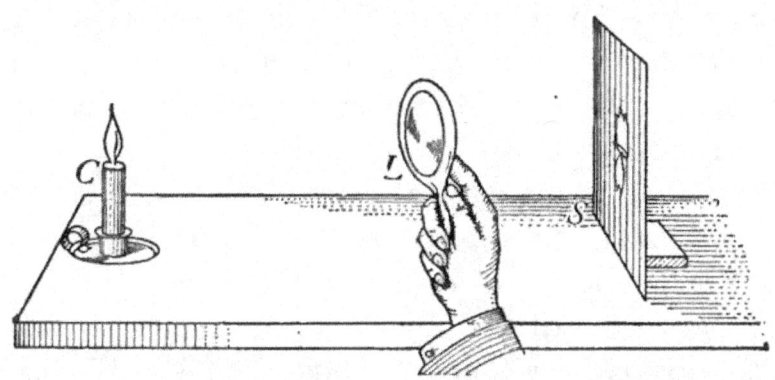

FIG. 382.—*C* and *S* are at conjugate foci.

candle by an *aperture* and by a *convex* lens. Rays of light from each point of the luminous body pass through the aperture in straight lines and produce upon the screen a lighted space of the same shape as the candle. This image is rather *hazy* in outline. Each cone of rays from luminous points of the flame is brought by the lens to a focus on the screen, producing a *sharp image*. It is the converging power of convex lenses that enables them to produce clear images.

387. The Construction of Diagrams to Represent the Formation of Images by Lenses.—Just as the earth has an axis at right angles to its equator to which are referred positions and distances, so a lens has a *principal axis* at

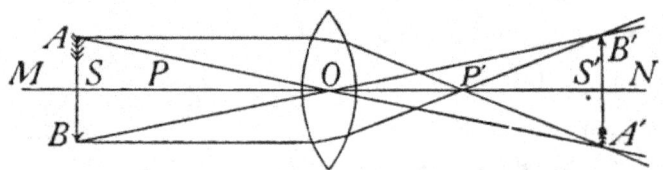

FIG. 383.—Construction of a real image by a convex lens

right angles to its greatest diameter and along this axis are certain definite positions as shown in Fig. 383. Let *MN* be the *principal axis* of a convex lens, *P* and *P'* are *principal foci* on either side of the lens, *S* and *S'* are *secondary foci*. These are at points on the principal axis that are twice as far from *O*, the center of the lens, as are the principal foci. In the formation of images by a convex lens, several distinct cases may be noticed:

(A) If a luminous body is at a *great distance* at the left, its light is brought to a *focus* at *P*, or its *image is formed at P*. (B) As the *object approaches* the lens the *image gradually recedes* until the object and image are at *S* and *S'*, *equally distant from O and of equal size* (as in Fig. 383). The object and image are now said to be at the *secondary foci* of the lens. (C) As the *object moves from S to P* the image recedes, rapidly increasing in size until (D) when the object is at P the rays become parallel and no image is formed. (E) When the object is between P and the lens, the rays *appear to proceed from points back of the object*, thus forming an *erect, larger, virtual image* of the object. (See Fig. 384.) This last arrangement illustrates the *simple microscope*.

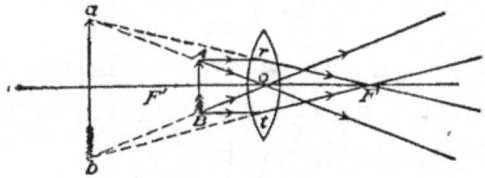

FIG. 384.—Construction of a virtual image by a convex lens.

With a concave lens but one case is possible, that corresponding to the one last mentioned with convex lenses; since the rays from a body are divergent after passing through a concave lens they appear to proceed from points *nearer* the lens than the object and hence a *virtual, erect,*

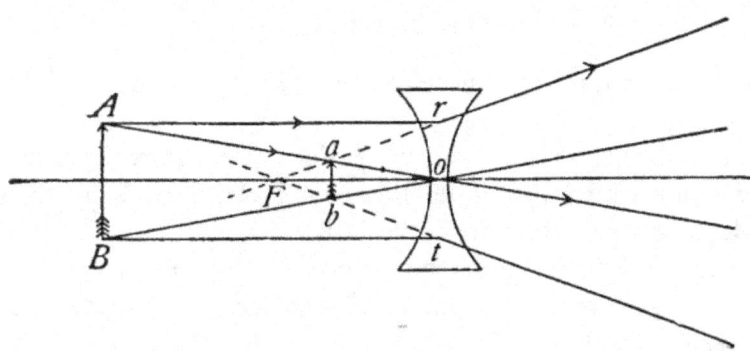

FIG. 385.—Construction of a virtual image by a concave lens.

smaller image of the object is formed. This virtual image may be seen by looking *through* the lens toward the object. (See Fig. 385.)

388. The Lens Equation.—The location of either the object or of the image upon the principal axis of the lens may be calculated if the position of one of these and the focal length are known. This is accomplished by the use of a formula $\frac{1}{F} = \frac{1}{D_0} + \frac{1}{D_1}$ in which F represents the focal length and D_0 and D_1 the distance from the lens of the object and the image respectively. Thus if an object is placed 30 cm. from a lens of 10 cm. focal length, where will the image be formed? Thus: $\frac{1}{10} = \frac{1}{30} + \frac{1}{D}$, and $3D_1 = D_1 + 30$, or $2D_1 = 30$ $D_1 = 15$. This result indicates that a real image will be 15 cm. from the lens. A minus value would indicate a virtual image.

Important Topics

(A) Lenses: convex, concave, six forms, center and radius of curvature.
(B) Principal focus, focal length, virtual focus, conjugate foci.
(C) Principal axis, images formed when object is in various locations.
(D) Computation of location of images.

Exercises

1. Why is an image of a candle formed by an aperture, not sharply defined?
2. When a photographer takes your picture and moves the camera nearer you, must he move the ground glass screen toward the the lens or away from it? Explain.
3. How can you find the principal focal length of a lens?
4. How can you test a spectacle lens to see whether it is convex concave?
5. When will a convex lens produce a virtual image? Have you ever seen one? Where?
6. When a photographer wishes to obtain a full length view of a person, where does he place the camera?
7. The focal length of the lens is 24 cm. How far from the lens must an object be placed in order that a real image may be three times as long as the object?
8. There is a perfect image of an object on the ground glass of a camera. The center of the lens is 20 cm. in front of the image

and the object 75 cm. from the lens. What is the focal length of the lens?

9. An object is 60 cm. from the lens, the image 120 cm. from it. Find the focal length.

10. How can you find experimentally the principal focal length of a lens?

11. A lens is used to project an enlarged image of a candle upon a screen. Which is farther from the lens, the candle or the image? Explain.

(6) Optical Instruments

389. The Eye.—The most common optical instrument is the *eye*. While the structure of the eye it complicated, the principle of it is simple, involving the formation of an image by a double convex lens. (See Fig. 386, in which is shown a front to back, vertical cross-section of the eye.) The eye appears to be made of portions of two spheres, one of which, smaller than the other, is placed in front. This projecting part is transparent, but refracts the light which

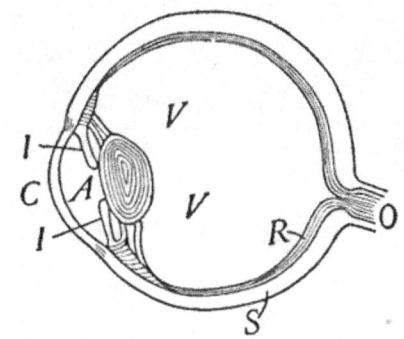

FIG. 386.—Cross-section of the eye.

strikes it obliquely, so as to turn it into the eye. This enables us to see objects at the side when looking straight ahead. Test this by looking directly in front of you and see how far back on each side of the head you can notice a movement of the forefinger of each hand.

390. Action of the Eye in Vision.—When we look at an object, a small, real, inverted image is formed upon the *retina* at the back of the interior of the eye. The retina is an expansion of the optic nerve and covers the inner surface at the back of the eyeball. Seeing is due to the action of light in forming images upon the retina. Our eyes are so constructed that when they are relaxed the lens is adjusted to form clear images of *distant* objects upon the retina. If we look from distant to near objects without changing the shape of the eye lens, a sharp image of the latter cannot be formed and we get a blurred impression. It is difficult, however, to look at objects without automatically adjusting the eye lens so that it will make a sharp image. Test this by looking out of a window at a distant object, then without moving the head or eyes look at the glass of the window; you will notice a slight change of some sort *in* the eye itself as the vision is adjusted. This adjustment is made by muscles that pull or compress the eye lens so as to make it thicker for near objects and thinner for distant ones. The eye ordinarily does not see objects nearer than 10 in. clearly. This means that the greatest possible thickening of lens will not form clear images upon the retina if the object is nearer than 10 in. (25 cm.).

391. The Visual Angle.—To examine objects carefully we usually bring them as close to the eye as possible, for

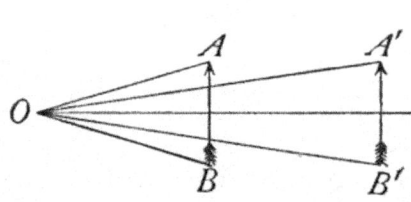

FIG. 387.—The visual angle, *AOB* is greater at *AB* than at *A'B'*.

the nearer to the eye the object is brought, the larger is the visual angle formed by it (see Fig. 387), and the larger is its image upon the retina. *The visual angle of an object is the angle at the eye lens between the rays that have come from the ends of the object.* Consequently the more distant the object, the smaller is its visual angle. Now if we wish to examine small objects with great care, we frequently find that it is necessary to bring them close to the eye so that they have a visual angle of adequate size. If they must be brought closer than 10 in. a double convex lens is placed in front of the eye. This assists the eye lens in converging the light so that a clear image may be formed when the object is close, say an inch or so from the eye. This is the principle of the magnifying glass used by watch-makers and of the *simple microscope*. The action of the latter is illustrated by Fig. 388. The convex lens forms a virtual, enlarged image "*A'–B'''*" of the object "*A–B*" which it observed instead of the object itself.

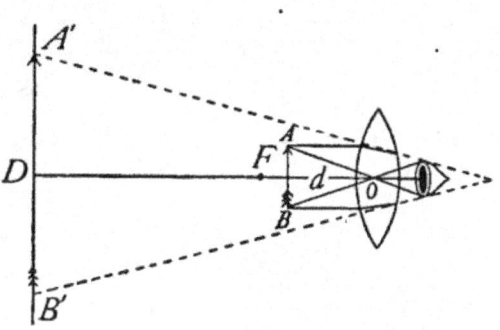

FIG. 388.—Action of the simple microscope.

392. Defects of Vision.—There are several defects of vision that may be corrected by spectacles or eye-glasses. One of these is "near-sightedness." It is due either to an eyeball that is elongated, or to an eye lens that is too convex, or to both conditions. This condition brings light from distant objects to a focus too soon (as shown in Fig. 389). Only light from near objects will focus upon the retina in such cases. With *normal* vision light from *distant* or *near* objects may be focused without unusual

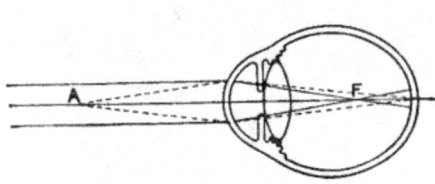

FIG. 389.—"Near sightedness, or myopia. Parallel rays come to a focus at *F*; emerging rays focus at *A*, the far point.

effort upon the retina, see Fig. 390. The remedy for near-sightedness is to use concave lenses which will assist in properly refracting the light so the focus will be formed on the retina (Fig. 391). "Far-sightedness" is the reverse of near-sightedness; the eyeball is either too short, or the lens too flat, or both conditions obtain, so that the light entering the eye is brought to a focus behind the eyeball (Fig. 392). The remedy is convex lenses which will assist in properly converging the light, see Fig. 393. A third defect is called *astigmatism*. This is caused by some irregularity or lack of symmetry in the eye. It is corrected

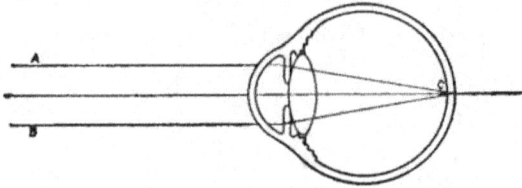

FIG. 390.—The normal eye. The parallel rays *A B* focus without accommodative effort at *C*.

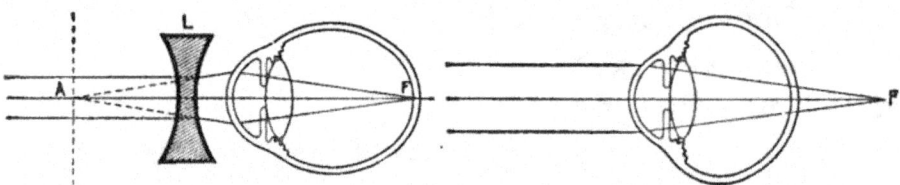

FIG. 391.—Correction of near-sightedness by concave lens.

FIG. 392.—Far-sightedness or hyperopia. Parallel rays focused behind the retina.

by a *cylindrical* lens that compensates for this defect of the eye. A diagram similar to Fig. 394 is used as a test

for astigmatism. If the lines appear with unequal distinctness, some irregularity of refraction (astigmatism) is indicated.

393. The Photographic Cam-

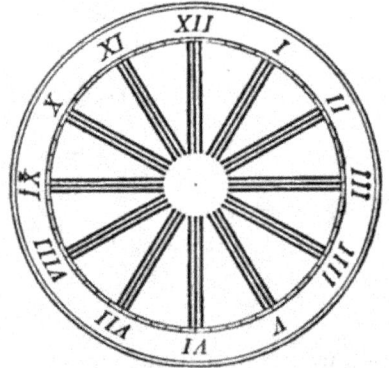

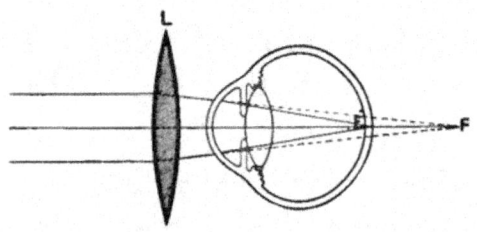

FIG. 393.—Correction of far-sightedness by a convex lens.

FIG. 394.—Test card for astigmatism.

era.—This is a light-tight box, provided with a convex lens in front, covering an aperture and a ground glass screen at the back. The distance between the lens and the screen is adjusted until a sharp image is obtained upon the latter, which is then replaced by a sensitive plate or film. The sensitized surface of the plate or film contains a salt of silver which is changed by the action of light. After the plate has been "exposed" to the action of light, it is "developed" by the use of chemicals producing a *negative* image. From "negative," by the use of sensitized paper, "positive" prints may be secured which resemble the object photographed.

394. The projecting lantern (see Fig. 395) employs a strong source of light, as an electric arc lamp L, to strongly

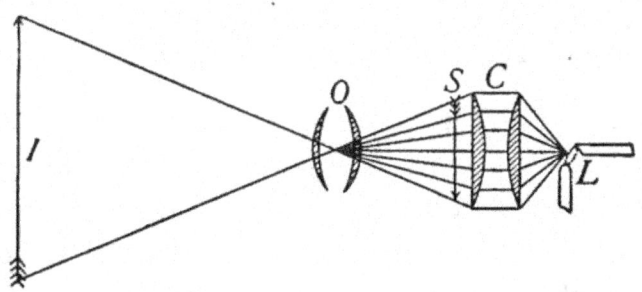

FIG. 395.—Diagram of the projecting lantern.

illuminate a transparent picture, or *lantern slide, S,* a real image (I) of which is formed upon a large screen. Two large plane-convex lenses (C), called condensing lenses, are placed near the lamp to concentrate the light upon the "slide" S. The convex lens forming the image is called the "objective" (O).

395. The compound microscope consists of two lenses. One called the *objective* is placed near the object to be viewed. This lens has a short focal length usually less than a centimeter. It forms a *real image* of the object. $A'-B'$. The other lens, the *eyepiece* forms a virtual image of this real image. $A''-B''$. (See Fig. 396.)

396. The telescope consists of two lenses, the eyepiece and the objective. As in the compound microscope, the objective of the telescope forms a real image of the distant object, the eyepiece forming an enlarged virtual image of the real image. It is the virtual image that is viewed by

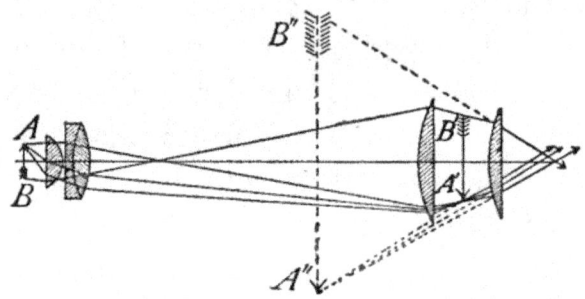

FIG. 396.—Formation of an image by a microscope. *A–B* is the object. *B'–A'* the real image formed by the "objective." *B''–A''* is the virtual image formed by the eyepiece. The eye sees the virtual image.

the observer. (See Fig. 397.) In order to collect sufficient light from distant stars the objective is made large, sometimes 50 in. in diameter.

The length of the telescope tube depends upon the focal length of the objective, since the distance between the two lenses must equal the *sum* of their focal lengths.

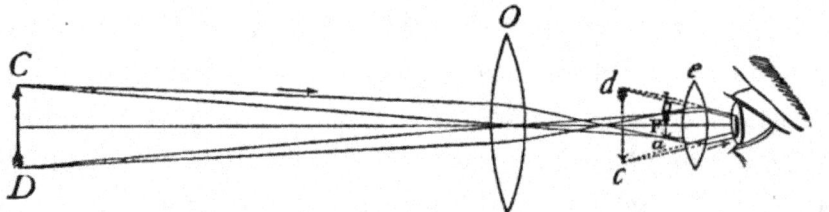

FIG. 397.—Formation of an image by a telescope. *b–a* is the real image; *d–c* is the virtual image seen by the observer.

397. The opera glass consists of a convex lens as objective and a *concave* lens as an eyepiece. The former tends to form a real image but the latter diverges the rays before a real image can be formed, the action of the two lenses producing an enlarged virtual image (as in Fig. 398) which is viewed by the one using the glass. The compact size of the opera glass is due to the fact that the distance between the two lenses is the *difference* of the focal lengths.

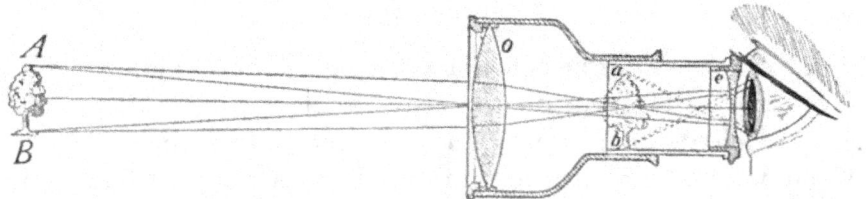

Fig. 398.—Formation of an image by an opera-glass. *a–b* is the virtual image.

398. The Prism Field Glass or Binocular.—This instrument has come into use in recent years. It possesses the wide field of view of the spy glass but is as compact as the opera glass. This compact form is secured by causing the light to pass back and forth between two right-angle

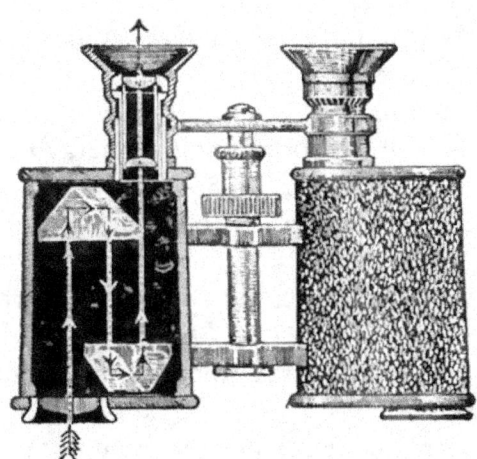

Fig. 399.—Diagram of the Zies binocular or prism field glass.

prisms (as shown in Fig. 399). This device permits the use of an objective lens with a focal length three times that of the tube, securing much greater magnifying power than the short instrument would otherwise possess. A further advantage is secured by the total reflection from the two prisms, one of which is placed so as to reverse the image right for left and the other inverts it, so that when viewed in the eyepiece it is in its proper position.

Important Topics

1. The eye: parts, formation of image, kind, how, where.
2. Eye defects, how remedied.   Visual angle.
3. Simple microscope, camera; images, kind, how formed.
4. Compound microscope, telescope and opera glass; images, action of each lens.

Exercises

1. Name three instruments in which lenses form virtual images and three in which *real* images are formed.
2. In what direction is an oar in water apparently bent?   Explain by a diagram.
3. What optical instruments have you used?   Is the *visible* image formed by each of these *real* or *virtual?*
4. The focal length of a copying camera lens is 14 in.   Where must a drawing be placed so that an image of the same size may be formed upon the ground glass screen?   What must be the distance of the screen from the lens?
5. What are two methods by which you can determine the focal lengths of the lens of a photographic camera?
6. The critical angle for water is 48½ degrees.   Show by a diagram how much of the sky can be seen by a diver who looks upward through the water.
7. How is near-sightedness caused?   How is it corrected?   Illustrate by a diagram.
8. How is the eye accommodated (focused) as an object gradually approaches it?
9. Explain why a simple microscope assists in looking at the parts of a flower or insect.
10. Why do people who have good eyesight when young require glasses as they grow old?

(7) Color and Spectra

399. Color.—Much of the pleasure experienced in gazing at beautiful objects is due to the *color* shown by them.

Guglielmo Marconi (Italy). Inventor of wireless telegraphy.

GUGLIELMO MARCONI
"Copyright by Underwood & Underwood, N. Y."

Alexander Graham Bell, Washington, D. C. Inventor of the telephone.

ALEXANDER GRAHAM BELL
"Copyright by Underwood & Underwood, N. Y."

The blue sky, the green grass, and the varied tints of flowers, and of the rainbow all excite our admiration   The study of color begins naturally with the production of the *spec-*

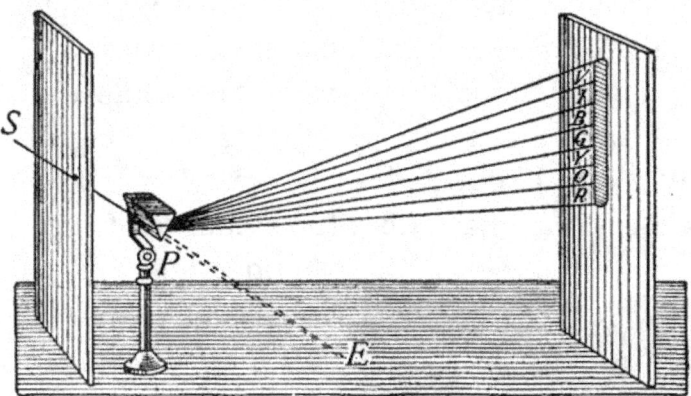

FIG. 400.—Formation of the spectrum by a prism.

trum, the many-colored image upon a screen produced by passing a beam of light through a prism.   The spectrum is best shown when the light enters by a narrow slit (Fig. 400).   The spectrum was first produced by Sir Isaac Newton in 1675 by the means just described.   The names usually given to the more prominent colors of the spectrum are violet, indigo, blue, green, yellow, orange, and red. The initials of these names, combined, spell *vibgyor*, a word without meaning except to assist in remembering the order of the colors in a spec-trum.   If the light that has passed through a prism is sent through a second prism placed in reverse position (see Fig. 401), the light passing through both prisms is found to be white.   This experiment

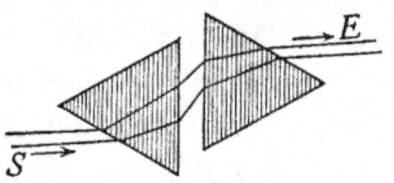

FIG. 401.—The colors of the spectrum recombine to form white light.

indicates that white light is composed of light of all colors.

400. Dispersion.—The separation of the colors by a prism is called dispersion.   In experimenting to find a

28

reason for dispersion, it has been learned that lights of
different colors are of different wave lengths. Color in
light is therefore analogous to pitch in sound. We hear
through many octaves, but we see through about one
octave. That is, the shortest visible waves of violet
light are about 0.000038 cm. in length while the longest
visible red rays are 0.000076 cm., or the longest visible
light waves are about twice the length of the shortest
visible ones. It appears from the evidence of experiments
upon dispersion that *light
waves of different lengths
are refracted differently.*
This causes the images
formed by refraction
through simple glass lenses
to be fringed with color
and to lose some of their
sharpness and definiteness
of outline, since the violet
light is brought to a focus

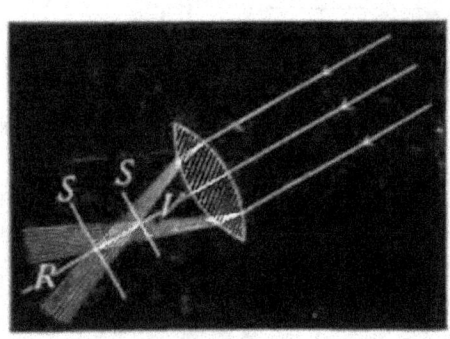

Fig. 402.—Violet light comes to a
focus sooner than red.

sooner than the red. (See Fig. 402.) This seriously affects
the value of such lenses for optical purposes. Fortunately
it is found that *different kinds of glass have a different rate
of dispersion for the same amount of refraction.*

401. The Achromatic Lens.—The existence of these
different kinds of glass makes possible a combination of
lenses in which dispersion is entirely overcome with the
loss of only about one-half of the refraction. Such a combination is shown in Fig. 403. It is called an *achromatic lens*, since images formed by it are not colored
but white (*a* = without, *chroma* = color). *The achromatic
lens consists of a double convex lens of crown glass combined
with a plano-concave lens of flint glass.* Achromatic lenses
are used in all high-grade optical instruments such as

telescopes and microscopes.   The colored images that are sometimes seen in cheap opera glasses show the result of not using achromatic lenses.

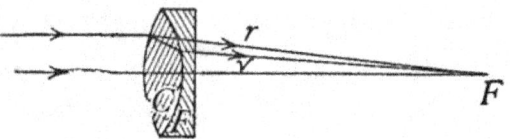

FIG. 403.—An achromatic lens.   C is of crown glass; F, of flint glass.

402. The Color of Bodies.—Project the spectrum of sunlight upon a white surface in a darkened room.

Now place in different parts of the spectrum objects of various colors.   Red objects will show brilliant red when at the red end of the spectrum but look black at the blue end, while blue objects appear blue only at the blue end.

These facts indicate that the color of an object depends upon two things: (a) *the light that falls upon it and* (b) *the light which it sends to the eye.*   A *black* surface absorbs all color while a *white* one reflects all wave lengths to the eye in the same proportion that they come to it.   A white object will appear red in red light, and blue in blue light since it reflects both of these.   A *colored* object reflects light of its own color but absorbs all others.   The color then of a body is due to the light which it does not absorb, but which comes from it to the eye.

403. The color of transparent bodies, such as colored glass, is due to the presence of a *dye* or *pigment* contained in the body.   This pigment absorbs a part of the light, the part transmitted giving the color.   This may be shown by holding a sheet of colored glass in a beam of light either before or after it has passed through a prism.   Some colors, as red, may be found to be nearly *pure*, only the red passing through, while green glass often transmits in addition to the green some yellow and some red light.

404. Complementary Colors.—If two prisms are placed in reversed position near each other (see Fig. 401), a beam of light dispersed by one is recombined into white light by the other. If now a card is held between the two prisms so as to cut off some of the colored light, say the red, the remaining light will be found to form a *greenish blue*. If the card is removed, the light becomes *white* again. That is, red and *peacock blue* light together form white. Any two colors that together form white light are called *complementary*. Other complementary colors are light yellow and blue, green and crimson, orange and greenish blue, violet and greenish yellow. We must not confuse the combining of colors (light) and the combining of *pigments*, the latter consisting of bodies that absorb light. Yellow pigment absorbs all but yellow and some green, while blue pigment absorbs all but blue and some green. Mixing these two pigments causes the absorption of all colors but *green*. Blue and yellow *paint* mixed produce *green*, while blue and yellow *light* give white.

405. The solar spectrum, as the spectrum of sunlight is called, may be observed in the *rainbow*. The latter is produced through the dispersion of light by spherical raindrops. Its formation may be imitated by sending a small circular beam of light through a screen against a round glass flask filled with water. (See Fig. 404.) The light passes through the water and is dispersed when it enters and when it leaves, producing a color upon the screen at $R-V$. The course of the light within the drop is indicated in Fig. 405. The violet ray comes to the eye more nearly horizontal and is therefore below red, as we look at the rainbow.

406. Fraunhofer Lines.—Some of the most important features of the solar spectrum are not seen in the rainbow or in the band of light usually observed upon a screen.

By the use of a narrow slit and a convex lens to carefully focus the slit upon a white screen it is seen that the solar spectrum is crossed by many *dark* lines. These are called

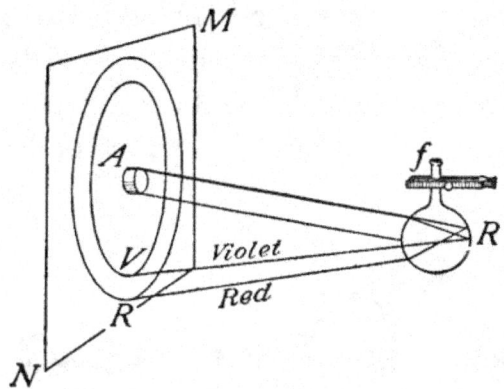

FIG. 404.—A rainbow formed by a beam of light striking a flask of water.

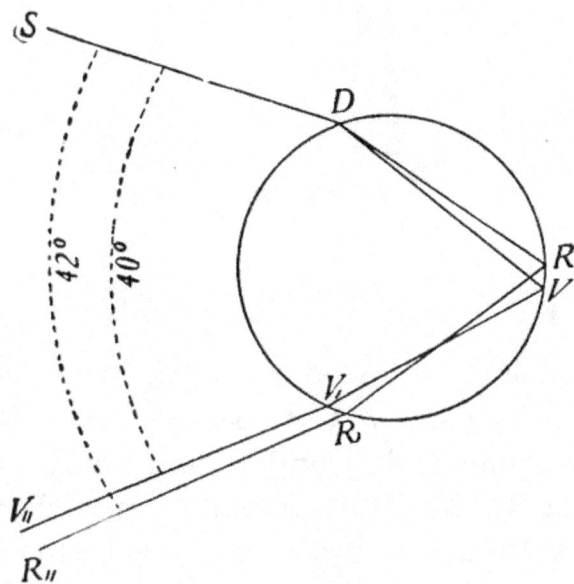

FIG. 405.—The course of a beam of light within a drop of water.

Fraunhofer lines, to honor the German scientist who in 1814 first accurately determined *their* position. Two experiments *with a spectroscope* will help to make clear the meaning of the Fraunhofer lines.

407. The Spectroscope and Its Uses.—The spectroscope (Fig. 406) is an instrument for observing spectra. It consists of a prism, a slit, and a convex lens T for focusing an image of the slit accurately upon a screen (Fig. 407) where the spectrum is observed through the eyepiece E.

(A) A Bunsen flame is placed in front of the slit and a heated platinum wire which has been dipped in common salt or some sodium compound placed in the Bunsen flame;

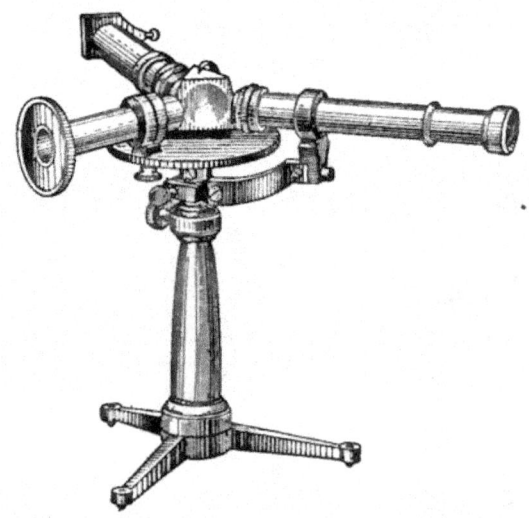

FIG. 406.—The spectroscope.

the latter becomes yellow and a vivid yellow line is observed on the screen in the spectroscope. Other substances, as barium and strontium salts, when heated to incandescence in the Bunsen flame, give characteristic bright lines. In fact each *element* has been found to have its own characteristic set of colored lines. This fact is made use of in *spectrum analysis*, by which the presence of certain elements in a substance can be definitely proved upon the appearance of its particular lines in the spectrum.

(B) If light from, for example, an arc light is sent over a gas flame containing *sodium* vapor, a *dark line* appears in the spectrum—in the exact position in which the yellow sodium line appeared. It seems that the sodium vapor removes from white light the same wave lengths that it itself produces. This absorption is supposed to be due to sympathetic vibration; just as a tuning fork it set in vibra-

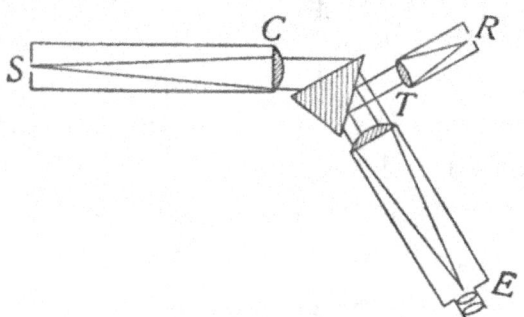

FIG. 407.—Diagram of a spectroscope.

tion by the waves of another fork in unison with it, at the same time absorbing the wave energy, so in the gas flame the sodium particles absorb the wave motion of the · same vibration rate as that emitted by them. The fact

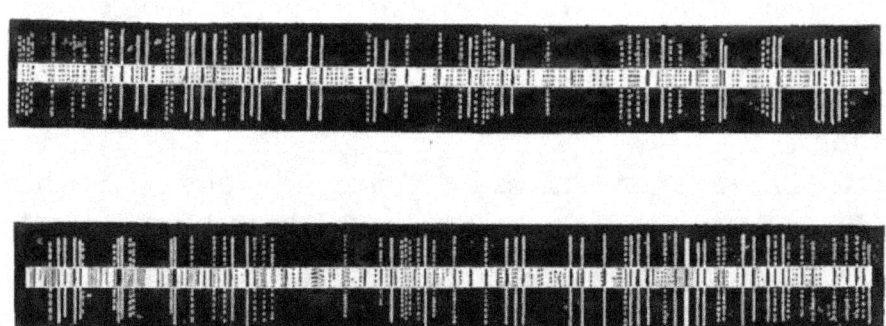

FIG. 408.—The bright line spectrum of iron and its coincidences with some of the dark lines of the solar spectrum.

that the spectrum of sunlight contains a great many dark lines is believed to indicate that the sun is surrounded by clouds formed by the vaporization of the various substances in the sun itself. By comparing the dark lines of the solar spectrum with the *bright-line spectra* of various substances found in the earth, such an exact correspondence of the lines is found that the presence of the vapor of these substances about the sun is considered proved. (See Fig. 408 which shows the exact correspondence between the bright-line spectrum of iron vapor and the dark lines appearing in a portion of the sun's spectrum.) The spectra of the stars also contain certain dark lines. Thus the presence of the corresponding substances in distant stars is considered as determined.

408. Theory of Color Vision.—By combining light of the *three colors red, green* and *blue-violet* in proper proportions, it has been found possible to produce any color effect, even white. This leads to the conclusion that in the retina of the eye are three different kinds or sets of sensitive nerve endings, sensitive respectively to red, to green, and to blue light. This idea is given corroboration by some facts of color blindness. Thus some persons have no sensation of *red*, this color not being distinguished from green. Others are color blind to green or blue. It is supposed that in color blind persons one of the sets of nerve endings sensitive to one of these three colors is lacking.

409. Three-color Printing.—Since all colors may be produced by mixing the three colors, light red, green, and blue-violet, these are called *the three primary colors*. The so-called primary pigments or paints are simply the complements of the three primary colors. They are, in order, peacock blue, crimson, and light yellow. The three pigments when mixed yield black, since combined they absorb all kinds of visible light. The process of three-color printing, now so generally employed in printing colored pictures for books, calendars, etc., consists in combining upon white paper three colored impressions, using successively the three primary pigments (yellow, crimson and blue) from plates prepared as follows:

Three photographs of a given colored object are taken, each through a different sheet of gelatine called a filter, stained the color of one of the primary colors. From these photographs half-tone blocks are made in the usual way. The colored picture is made by carefully superposing impressions from these blocks, using in each case an ink whose color is the complement of the "filter" through which the original picture was taken. An illustration of the process is given upon the plate in the frontispiece of this book.

Important Topics

1. Color, due to wave length; dispersion by prism, sphere in rainbow, complementary colors, color of opaque and transparent bodies.
2. Spectra, solar; formation of rainbow; bright-line spectra, how formed, how used; dark-line, how formed, used.
3. Theory of color vision. Three color printing.

Exercises

1. How does a white flower look when viewed through a blue glass? Through a red glass? Through a red and blue glass at the same time?
2. Why does a red ribbon appear black when seen by blue light and red when seen by red light?
3. In what part of the sky must you look to see a rainbow in the morning? In the afternoon? Explain.
4. How would you arrange two similar prisms so as to produce double the deviation produced by one?
5. The color of an object depends upon what two things?
6. What kind of a spectrum should moonlight give? Why?
7. A mixture of green and red lights gives a sensation of yellow. Can you suggest why a mixture of blue and yellow lights gives the sensation of white?

(8) Nature of Light, Interference, Polarization

410. The Corpuscular Theory.—The theory of the nature of light that was most generally accepted until about the year 1800, held that light consists of streams of minute particles, called corpuscles, moving at enormous velocities. This *corpuscular theory* was in accord with the facts of reflection and the *rectilinear* motion of light, but was abandoned after the discovery of the *interference of light*, as it could not account for the latter phenomenon.

411. The Wave Theory of Light.—The theory that *light is* a *form of wave motion* was first advanced by Huygens, a Dutch physicist, in the seventeenth century. This theory was opposed at the start since (A) *no medium* was known to exist which would convey wave motion through space, as from the sun to the earth, and (B) the *rectilinear motion* of light was *unlike* that of any *other* form of known wave motions, such as that of water or of sound waves which are able to bend around corners. In answer to the first objection, Huygens assumed the presence of a medium which he named *ether*, while the second objection has been completely overcome during the past century by the discovery that *light may deviate from* a *straight line*. It is now known that the *excessive shortness* of light waves is the reason for its straight-line motion. Further, long ether waves, as those of wireless telegraphy, are found to bend around obstacles in a manner similar to those of water or sound.

412. The interference of light is one of the phenomena for which the wave theory offers the only satisfactory explanation. Interference of light may be shown by taking two pieces of plate glass and forcibly pressing them together by a screw clamp, as shown in Fig. 409. After a certain pressure has been reached, colored rings will appear about the compressed spot when viewed by light *reflected* from the upper surface of the glass. If light of one color, such as that transmitted by red glass, falls upon the apparatus, the rings are seen to be alternately red and dark bands. The explanation of this phenomenon according to the wave theory is as follows: The two sheets of glass, although tightly pressed together, are separated in most places by a thin wedge of air (see Fig. 410), which represents in an exaggerated form the bending of the plates

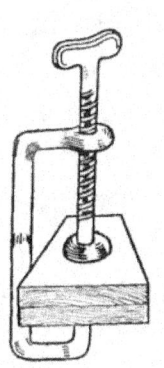

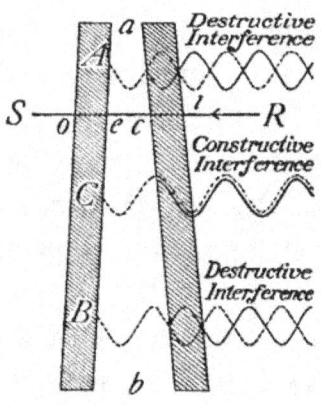

FIG. 409.—Two plates pressed together by a screw clamp.

FIG. 410.—Illustrating the interference of light by a thin film of air.

when pressed by the clamp. Several waves are represented as coming from the right and entering the glass. Now the wave moving from R to the plates has some of its light reflected from each glass surface. Consider the two portions of the wave reflected at each of the surfaces between the plates, *i.e.*, from the two surfaces of the wedge of air. If the portion of the wave reflected from the second surface of the air wedge combines with that reflected from the first surface, in the *same phase* as at C, the two reflected waves strengthen each other. While if the two reflected portions of the wave meet in opposite phases as at A and B, a decrease or a complete extinction of the light results.

This is called *interference*. If light of one wave length is used, as red light, the regions of reinforcement and interference are shown by red and dark rings, while if white light is used, the ring where red light interferes, yields its complementary color, greenish blue. Where interference of greenish blue occurs, red is found, etc. Many phenomena are due to interference, such as (A) the color of thin films of oil on water, where the portions of light reflected from the two surfaces of the oil film interfere resulting in the production of color; (B) the color of soap bubbles. When first formed, soap-bubble films are not thin enough to show interference well, but as the bubbles increase in size or become thinner on standing, the conditions for interference are reached and, as the film becomes thinner, a regular succession of colors is noticed.

413. Differences Between Light and Sound.—Among the important differences between light and sound that have been considered are the following: the former are (a) *waves* in the ether, (b) *of very short wave length*, and (c) their *motion is in straight lines*. Another difference (d) is in *the mode of vibration*.

Sound waves are *longitudinal, while light waves are transverse*. Light waves consist of vibrations of the ether at right angles to the line of motion. To illustrate the reasoning that has led to this conclusion, suppose a rope to be passed through two vertical gratings. (See Fig. 411, 1.) If the rope be set in *transverse* vibration by a hand, the waves produced will readily pass through to the gratings P and Q and continue in the part extending beyond Q. If, however, Q is at right angles to P, no motion will be found beyond Q. Now if a stretched coiled spring with longitudinal vibrations should take the place of the rope, it is evident that the crossed position of the two gratings would offer no obstacles to the movement of the vibration.

In other words, crossed gratings offer no obstruction to longitudinal vibrations, while they may completely stop transverse vibrations.

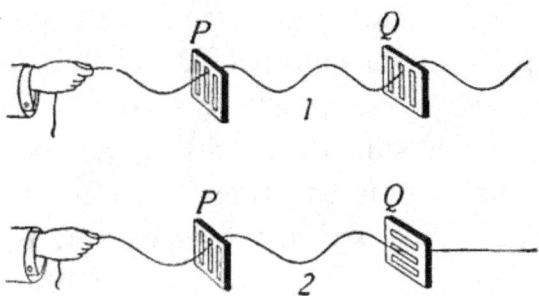

FIG. 411.—Transverse waves will pass through both gratings in (1) where the openings in the two gratings are at right angles. The waves passing P are stopped by Q (2).

414. Polarization of Light.—It is found that two crystals of tourmaline behave toward light just as the two gratings behave with respect to the transverse waves of the rope. Thus, if a small opening in a screen is covered with a *tourmaline* crystal, light comes through but slightly diminished in intensity. If a second crystal is placed over the first one so that the two axes are in the same direction as in Fig. 412P, light is as freely transmitted through the second crystal as through the first, but if the crystals are crossed (Fig. 412S) no light passes the second crystal.

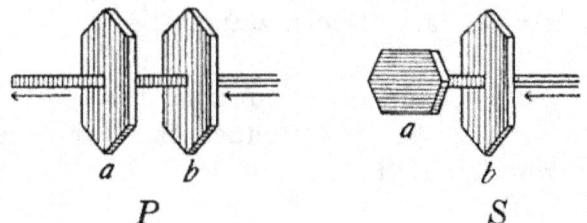

FIG. 412.—Effect of tourmaline crystals on light.

This experiment shows that the light which has passed through one tourmaline crystal will pass through another only when the latter is held in a certain position, hence it is believed that a tourmaline crystal is capable of transmitting light that is vibrating in one particular plane. The direct conclusion from this is that *light waves* are *transverse rather than longitudinal.* The exper ment just described illustrates what is called *polarization of light.* The beam that after passing through *a* (Fig. 412) is unable to pass through *b,* if the two axes are crossed, is called a *polarized beam.* The conclusion that light waves are transverse is therefore based upon the phenomenon of the po'arization of light. This was first discovered by Huygens in 1690.

Important Topics

1. Interference of light: evidence, reasoning involved, illustration.
2. Polarization of light: evidence, reasoning involved.
3. Nature of light, differences between sound and light.

Exercises

1. Make a list of the differences between sound and light and state briefly the evidence upon which the knowledge of these differences is based.
2. Why will a thickness of film that will produce interference of red light be different from that producing interference for green or blue?
3. Using the formula $n = \dfrac{v}{l}$ compute the vibration rate for violet light if its wave length is considered as 0.00004 cm.
4. Explain how the fact of polarization affects the wave theory of light.
5. Show how it is possible by comparing the spectrum of the sun with that of a star to tell whether the star is approaching or receding from the earth.

Review Outline: Light

Light; speed, source, medium.
Straight Line Motion; shadow, umbra, penumbra, eclipse, image.
Photometry; Law of intensity, candle power, foot-candle.
Mirrors; Law of reflection; image—real, virtual; plane, curved, parabolic, mirrors.

Refraction; cause and effects; plate, prism, lens; total reflection.

Lenses; six forms, principal focus, center, lens equation, $1/F = 1/D_o + 1/D_i$.

Optical instruments; eye, defects and correction, camera, microscope, etc.

Spectra; 3 kinds, dispersion, production of color effects, spectroscope, uses.

Nature of Light; wave theory, interference, polarization, significance.

CHAPTER XVII

INVISIBLE RADIATIONS

(1) ELECTRIC WAVES AND RADIO-ACTIVITY

415. Oscillatory Nature of the Spark from a Leyden Jar.—In studying sound (Art. 339), the sympathetic vibration of two tuning forks having the same rate of vibration was given as an illustration of resonance. The conditions for obtaining *electrical resonance* by the use of two Leyden jars are given in the following experiment.

Join the two coats of a Leyden jar (Fig. 413) to a loop of wire L,

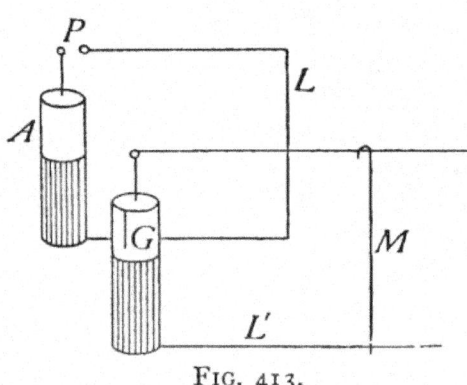

FIG. 413.

the sliding crosspiece M being arranged so that the length of the loop may be changed as desired. Also place a strip of tinfoil in contact with the inner coating and bring it over to within about a millimeter of the outer coating as indicated at G. Now join the outer coating of another exactly similar jar A to a wire loop of fixed length, the end of the loop being separated from the knob connected to the inner coating, a short distance at P. Place the jars near each other with the wire loops parallel and connect coatings of A to the terminals of a static machine or an induction coil. At each discharge between the knobs at P, a spark will appear in the other jar at G, if the crosspiece M is so adjusted that the areas of the two loops are exactly equal. When the wire M is moved so as to make the areas of the two loops quite unequal, the spark at G disappears.

The experiment just described shows that two electrical circuits can be *tuned* by adjusting their lengths, just as two tuning forks may be made sympathetic by adjusting their lengths. This fact indicates that the discharge of the Leyden jar is *oscillatory*, since resonance can plainly not be secured except between bodies having natural periods of vibration. This same fact is also shown by examining the discharge of a Leyden jar as it appears when viewed in a rapidly revolving mirror. (See Fig. 414.)

The appearance in the mirror shows that the discharge is made up of a number of sparks, often a dozen or more, vibrating back and forth until they finally come to rest. The

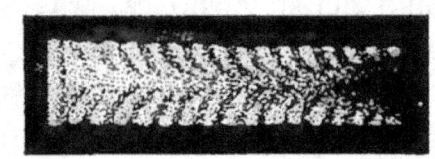

FIG. 414.—Photograph of the oscillatory discharge of a Leyden jar.

time of one vibration varies from one millionth to one-hundred millionth of a second, depending on the space between the discharging balls and the size of the jars.

The discharge of a Leyden jar or of another condenser sets up ether waves that have the speed of light. Heinrich Hertz in Germany first proved this in 1888. These waves are now known as Hertzian waves. The length of these varies from 3 cm. to several miles, depending upon the size and conditions of the discharging circuit.

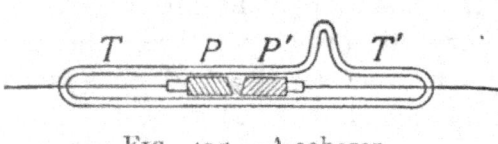

FIG. 415.—A coherer.

416. The Coherer.— The coherer is a device for detecting electric waves. It consists of a glass tube with metal filings loosely packed between two metal plugs that fit the tube closely. (See Fig. 415.) These filings offer a *high* resistance to the passage of an electric current, but when electric waves pass through the filings these *cohere* and allow a weak current to pass through. This current may be strong enough to operate a relay connected with a sounder or bell that gives audible signals. If the tube be tapped the filings will be disturbed and the resistance again made so high that no current can pass through.

417. Wireless Telegraphy.—In 1894 Marconi, then a young man of twenty, while making some experiments with electrical discharges discovered that the coherer would detect electrical waves at a considerable distance from their source and that by the use of a telegraph key the "dots and dashes" of the telegraph code could be reproduced by a sounder attached to a relay. At present the coherer is used principally in laboratory apparatus, as much more sensitive detectors are now available for commercial work. The essential parts of a modern wireless telegraph apparatus as used in many commercial stations are shown in Fig. 416.

Alternating current at 110 volts is sent into the primary, P, of a transformer, the secondary, S, of which produces a potential of 5000 to 20,000 volts. The secondary charges a condenser until its potential becomes high enough to produce a discharge across a spark gap, SG. This discharge is oscillatory, the frequency being at the rate of about one million a second, depending upon the capacity of the condenser and the induction of the circuit.

These oscillations pass through the primary of the oscillation transformer, inducing in the secondary, electric oscillations which surge back and forth through the antennæ, or aerial wires, A. These oscillations set up the "wireless waves." The production of these waves is explained as follows: An electric current in a wire sets up a magnetic field spreading out about the conductor; when the current stops the field returns to the conductor and disappears. The oscillations in the antennæ, however, have such a high frequency, of the order of a million a second, that when one surge of electricity sets up a magnetic field, the reverse surge immediately following sets up an opposite magnetic field before the first field can return to the wire. Under these conditions a succession of oppositely directed magnetic fields are produced which move out from the antennæ with the speed of light and induce electric oscillations in any conductors cut by them.

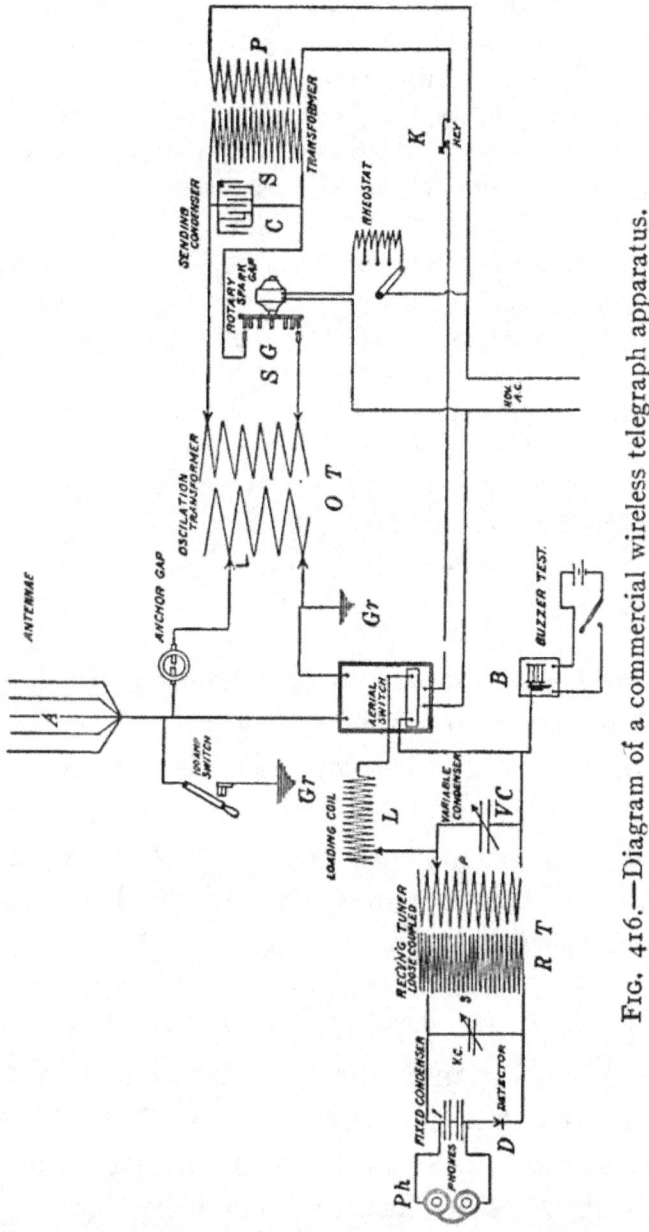

FIG. 416.—Diagram of a commercial wireless telegraph apparatus.

While the electric waves are radiated in all directions from the aerial, the *length* of the waves set up is approximately four times the combined length of the aerial wires and the "lead in" connection to the oscillation transformer.

The electric waves induce effective electrical oscillations in the aerial of the receiving station, even at distances of hundreds of miles, provided the receiving transformer, RT, is "tuned" in resonance with the transmitting apparatus by adjustments of the variable condenser, VC, and the loading coil, L. The *detector* of these oscillations in the receiving transformer is simply a crystal of silicon or carborundum, D, in series with two telephone receivers, Ph. The crystal detector permits the electric oscillations to pass through it in one direction only. If the crystal did not possess this property, the telephone could not be used as a receiver as it cannot respond to high frequency oscillations. While one spark passes at SG, an intermittent current passes through the receiver in one direction. Since some 300 to 1200 sparks pass each second at SG while the key, K, is closed, the operator at Ph hears a musical note of this frequency as long as K is depressed. Short and long tones then correspond to the dots and dashes of ordinary telegraphy. In order to maintain a *uniform tone* a *rotary spark gap*, as shown, is often used. This insures a tone of fixed pitch by making uniform the rate of producing sparks.

The *Continental* instead of the *Morse* code of signals is generally employed in wireless telegraphy, since the former employs only *dots* and *dashes*. The latter code employs, in addition to dots and dashes, *spaces* which have sometimes caused confusion in receiving wireless messages. The United States government has adopted the regulations of the *International Radio Congress* which directs that commercial companies shall use wave lengths between 300 and 600 or above 1600 meters. Amateurs may use wave lengths less than 200 meters and no others, while the government reserves the right to wave lengths of 600 to 1600 meters. See p. 459 for Continental telegraph code.

418. Discharges in Rarefied Air.—Fig. 417 represents a glass tube 60 or more centimeters long, attached to an air pump. Connect the ends of the tube to the terminals of a static machine or of an induction coil, a–b. At first no sparks will pass between a and f, because of the high resistance of the air in the tube. Upon exhausting the air in the tube, however, the discharge begins to pass through it instead of between *a* and *b*. This shows that an electrical discharge will pass more readily through a partial vacuum than through air at ordinary pressure. As the air becomes more and more exhausted, the character of the discharge changes. At first it is a faint spark, gradually changing until it becomes a glow extending

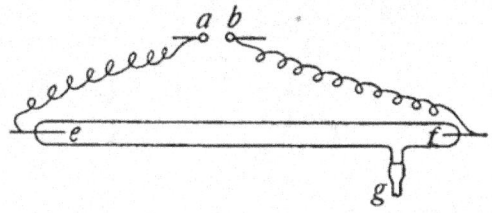

FIG. 417.—An Aurora tube.

from one terminal to the other and nearly filling the tube.

Geissler tubes are tubes like the above. They are usually made of different kinds of glass twisted into various shapes to produce beautiful color effects. The *aurora borealis* or northern light is supposed to be electric dis-

FIG. 418.—Aurora Borealis.

charges through rarefied air at the height of from 60 to 100 miles above the earth's magnetic poles. (See Fig. 418.)

419. Cathode Rays.—When the tube in Art. 420 is exhausted to a pressure of 0.001 mm., or a little less than one millionth of an atmosphere, the character of the discharge is entirely changed. The tube becomes filled with a yellowish green phosphorescent light. This is produced by what are called cathode rays striking the glass walls of

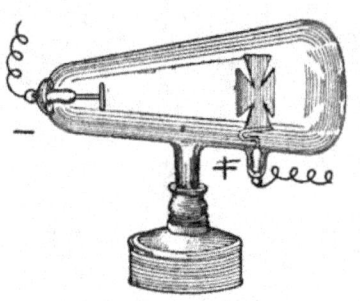

the tube. These rays are called cathode rays because they come from the cathode of the tube. They are invisible and that they travel in straight lines is shown by the shadow obtained by using a tube with a screen (Fig. 419).

FIG. 419.—A cathode ray tube.

420. "X" Rays.—In 1895, Professor Röntgen of Wurtzburg, Germany, discovered that when the cathode rays strike the walls of the tube or any solid within it they excite a form of invisible radiation. This radiation is called Röntgen rays, or more commonly, "X" rays. Careful experiments show that they travel in straight lines, and that they can not be reflected or refracted as light waves are. They pass through glass and opaque objects such as flesh, cardboard, cloth, leather, etc., but not through metallic substances. The tube in Fig. 420 has a screen covered with crystals which become luminous when struck by the cathode rays. On bringing a magnet near the tube the luminous line is raised or lowered showing that the magnetic field affects the stream of cathode rays, attracting it when in one position but repelling it when in the reverse direction. The cathode rays which cause the bright line possess a negative charge of electricity. They are now believed to be electrons shot off from the surface of the cathode with speeds that may reach 100,000 miles a second. "X" rays possess no electrical charge whatever and cannot be deflected by a magnet. They produce the same effect on a photograph plate as light does, only more slowly. Hence, they can be used in taking "X" ray photographs. Certain crystals, like barium platinum cyanide, fluoresce when struck by the "X" rays. The

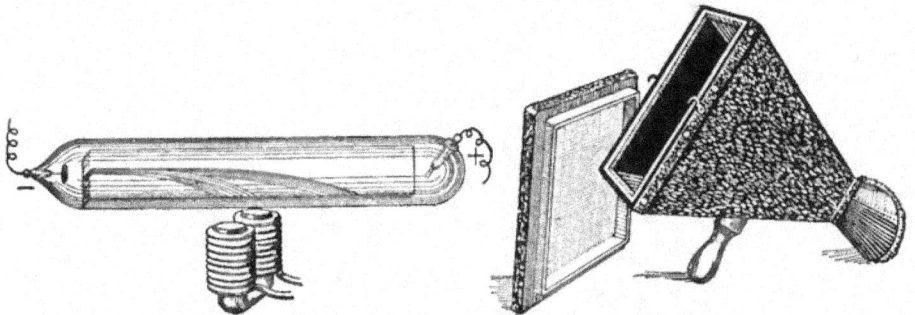

FIG. 420.—The stream of cathode rays is deflected by a magnet.

FIG. 421.—A fluoroscope.

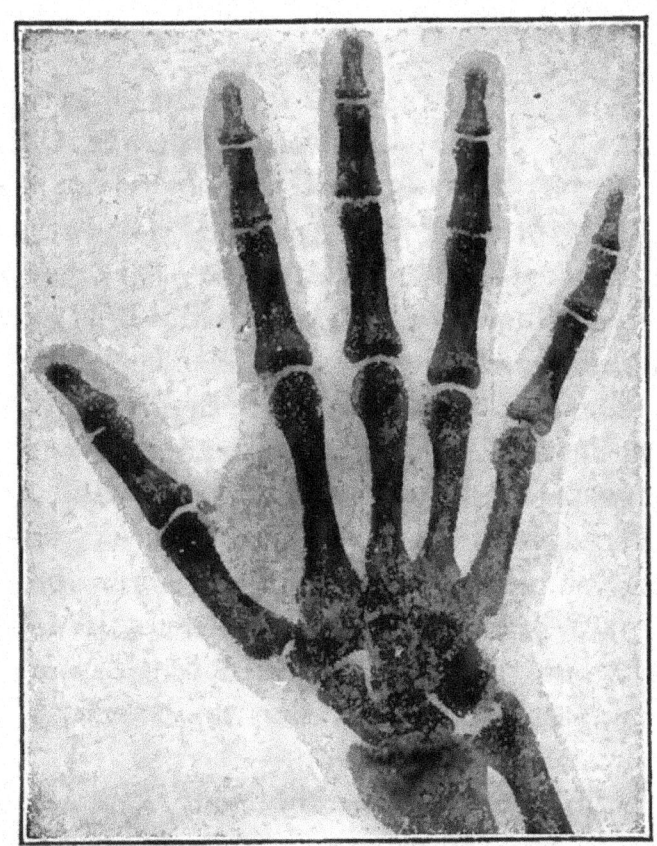

FIG. 422.—A view of the "shadow" of a hand as seen in a fluoroscope.

fluoroscope is the name given to a light-tight box closed at one end by a cardboard covered with these crystals (Fig.

421). On looking into the fluoroscope with an opaque object such as the hand placed between the screen and the "X" ray tube, a shadow of the bones of the hand can be seen upon the screen of the fluoroscope. (See Fig. 422.)

A special form of the tube is used. (See Fig. 423.) In this tube a platinum disc is placed at the focus of the concave cathode. This concentrates the "X" rays in one direction. It is now generally believed that "X" rays are waves in the ether set up by the sudden stoppage of the cathode rays at the platinum anode.

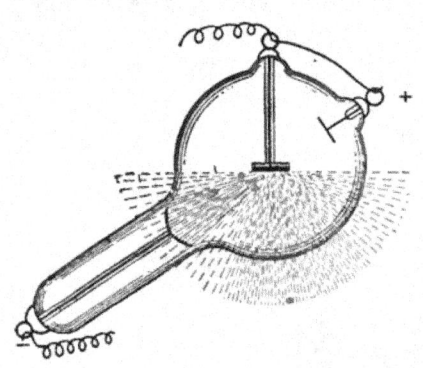

FIG. 423.—An "X" ray tube.

421. The Electromagnetic Theory of Light.—The study of electric waves has shown that they are similar to light waves in many respects: (a) they have the same velocity; (b) they can be reflected and refracted. The main difference is in their length, light waves being very much shorter. In 1864 James Clark Maxwell, an English physicist, proposed the theory that ether waves could be produced by electrical means and that light waves are electromagnetic. In 1888 Hertz proved by his experiments that ether waves having the same velocity as light could be produced in this way. It is now the general belief that light waves are ether waves produced by the vibrations of the electrons within the atoms and that they consist of electromagnetic waves in the ether.

422. Radio-activity.—In 1896 Henri Becquerel of Paris discovered that uranium and its compounds emit a form of radiation that produces an effect upon a photographic plate that is similar to that resulting from the action of "X" rays. These rays are often called *Becquerel* rays in honor of their discoverer.   The property of emitting such rays is called **radio-activity,** and the substances producing them are called **radio-active.**

In 1898, Professor and Mme. Curie after an investigation of all the elements found that *thorium*, one of the chief constituents of incandescent gas mantles, together with its compounds, was also radio-active.   This may be shown by the following experiment:

Place a flattened gas mantle upon a photographic plate and leave in a light tight-box for several days.   Upon developing the plate in the usual way a distinct image of the mantle will be found upon the plate.

423.   Radium.—Mme. Curie discovered also that pitchblende possessed much greater radio-active power than either thorium or uranium.   After prolonged chemical experiments she obtained from several tons of the ore a few milligrams of a substance more than a million times as active as thorium or uranium.   She called this new substance *radium*.   Radium is continually being decomposed, this decomposition being accompanied by the production of a great deal of heat.   It has been calculated that it will take about 300 years for a particle of radium to be entirely decomposed and separated into other substances.   It is also believed that radium itself is the product of the decomposition of uranium, atomic weight 238, and that the final product of successive decompositions may be some inert metal, like lead, atomic weight 207.

The radiation given off by radio-active substances consists of three kinds: (A) Positively charged particles of helium called *alpha* rays: (B) negatively charged particles called *beta* rays: (C) *gamma* rays.

The alpha rays have little penetrating power, a sheet

of paper or a sheet of aluminum 0.05 mm. stopping them. Upon losing their charges they become atoms of helium. Their velocity is about $\frac{1}{10}$ of that of light or 18,000 miles a second. The *spinthariscope* is a little instrument devised by Sir Williams Crookes in 1903 to show direct evidence that particles are continually being shot off from radium. In this instrument (Fig. 424), a speck of radium R is

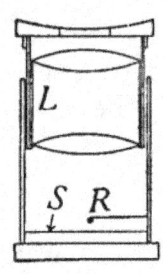

placed on the under side of a wire placed a few millimeters above a screen S covered with crystals of zinc sulphide. Looking in the dark at this screen through the lens L, a continuous succession of sparks is seen like a swarm of fireflies on a warm summer night. Each flash is due to an alpha particle

FIG. 424.—A spinthariscope.

striking the screen. The beta rays are supposed to be cathode rays or electrons with velocities of from 40,000 to 170,000 miles a second. The gamma rays are supposed to be "X" rays produced by the beta rays striking solid objects.

424. The discovery of radio-activity has revolutionized the ideas of the constitution of matter. Further, the results of experiments upon radio-active materials reveals the presence of immense quantities of sub-atomic energy. If man ever discovers a means of utilizing this, he will enter a storehouse of energy of far greater extent and value than any of which he has as yet made use. A consideration of this unexplored region gives zest to the work of those who day by day are striving to understand and control forces of nature.

Important Topics

1. Oscillatory nature of discharge of Leyden jar. Proofs.
2. Wireless telegraphy and telephony.
3. Electrical discharges in rarefied gases.

4. Cathode and "X" rays.
5. Electromagnetic theory of light.
6. Radio activity and radium.

CONTINENTAL TELEGRAPH CODE

A .—	J .———	S ...
B —...	K —.—	T —
C —.—.	L .—..	U ..—
D —..	M ——	V ...—
E .	N —.	W .——
F ..—.	O ———	X —..—
G ——.	P .——.	Y —.——
H	Q ——.—	Z ——..
I ..	R .—.	

PERIOD	INTERROGATION	EXCLAMATION
......	..——..	——..——
1 .————	2 ..————	3 ...——
4 ....—	5	6 —.... 7 ——...
8 ———..	9 .	0 —————

CHAPTER XVIII

WIRELESS TELEPHONY AND ALTERNATING CURRENTS

The developments in Wireless communication have been so rapid during recent years that a more extended account, than that given in Art. 417 of the apparatus and methods used at the present time, seems desirable. The study of Alternating Currents is also included with the idea that it will make the text more complete and of wider usefulness.

WIRELESS TELEPHONY

425. The Wireless Telephone.—One of the most important developments in wireless communication in recent years has been in wireless telephony. We realize its possibilities, when we hear of the achievements of talking across an ocean or between airplanes and the ground.

The wireless telephone can be best understood by comparing it with the common telephone. When the latter is in use, a direct current flows continually through the instrument. (See Arts. 312–316.) When a person speaks into the transmitter, the sound waves of the voice cause the diaphragm to vibrate, this action causes rapid changes in the *resistance* of the transmitter, which in turn causes the direct current to fluctuate just in step with the pulses of the voice waves. This fluctuating direct current passes through the primary of an induction coil, producing in the secondary an intensified alternating current. This passes over the line wires to the receiver where it produces variations in the magnetic field affecting the receiver diaphragm, causing the latter to reproduce the voice of the person speaking in the transmitter. Now to make the comparison clear, two facts must be noted with regard to the wire telephone: first, there must be an action in the transmitter which causes variations in a current through the instrument; second, this fluctuating current produces a more intense alternating current which flows over the line and affects the receiver diaphragm, producing there sound vibrations of greater intensity than those used at the transmitter. This added energy comes from the current flowing through the transmitter. The case is analogous to that of an electric bell. The armature of the bell vibrates with greater energy than is required to push the button, the extra energy being derived from the battery.

426. The Action of the Wireless Telephone.—In the wireless telephone we have a continuous stream of electric waves of high frequency. (See Fig. 425A.) This stream of electric waves corresponds to the current that flows through the transmitter in the wire telephone. These waves are of such high frequency that even though we had a receiver diaphragm vibrating in step with the waves, we could not hear the sound because the human ear cannot hear a sound which consists of more than about 40,000 vibrations per second. The sound waves act upon this stream of waves very much, as in the wire telephone, the transmitter acts to modify the line current. The impulses caused by the voice are much slower than the electric waves first mentioned and these slower impulses are reproduced in the receiver. Not only are these slower impulses reproduced but they are *amplified*, that is, produced with greater energy than the impulses impressed on the stream of waves. Fig.

425*A* represents as nearly as is possible in a diagram the continuous stream of electric waves. Fig. 426 *B*, repre-

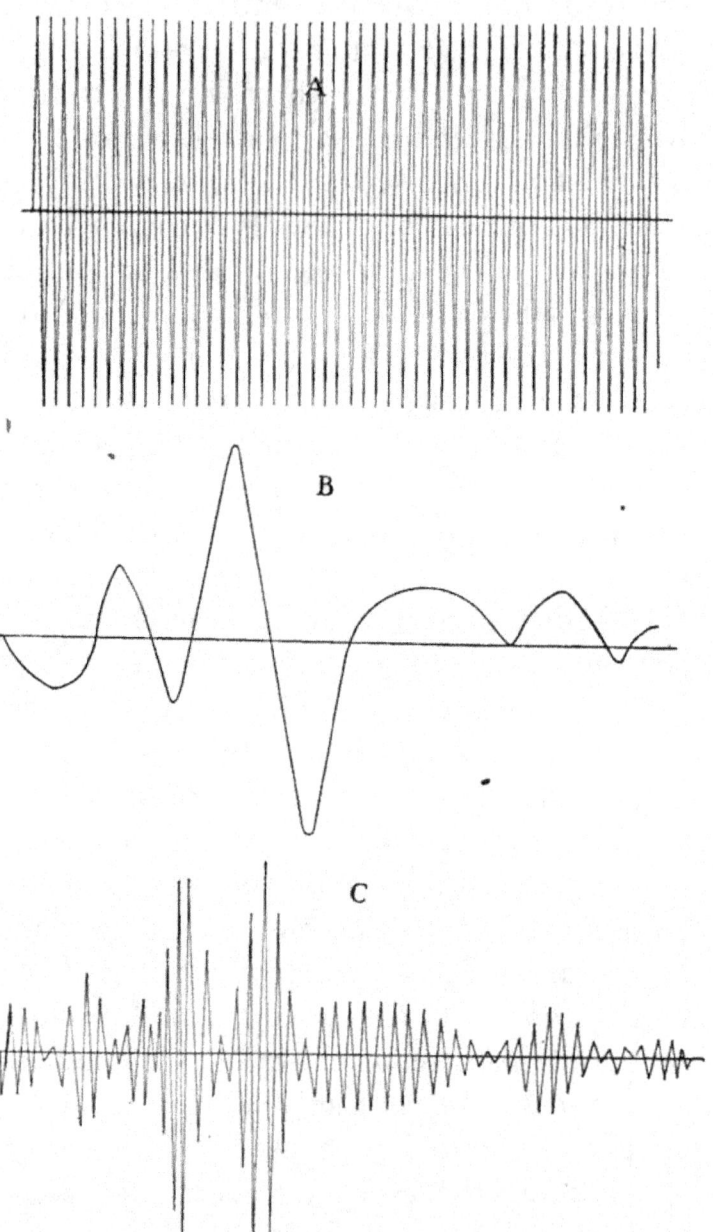

FIG. 425.—*A*, unmodified high frequency waves; *B*, waves of voice frequency; C, high frequency waves modified by waves of voice frequency.

sents the impulses produced by the sound alone, and Fig.

426 C, shows how these voice impulses are impressed on the stream of waves.

427. The Vacuum Tube or Audion.—The device by which all of this is accomplished is the *vacuum tube*. (See Fig. 426.) This tube contains three electrodes. *First*, a *filament* (*F*, in Fig. 428) which is heated by a

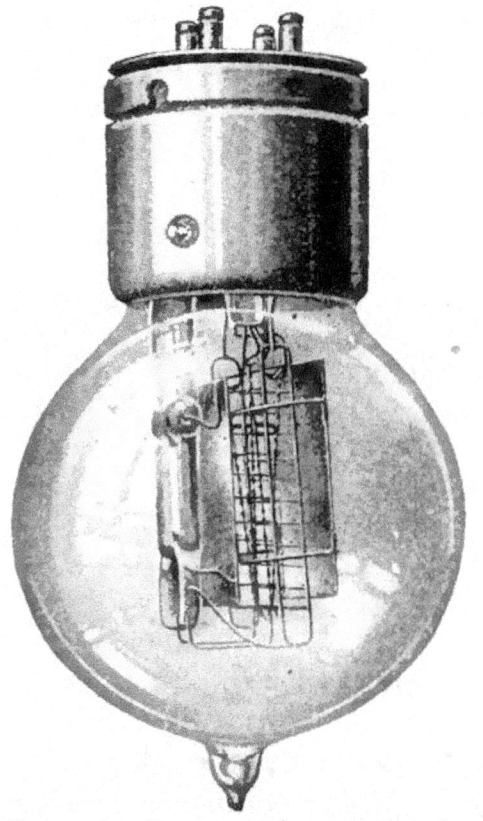

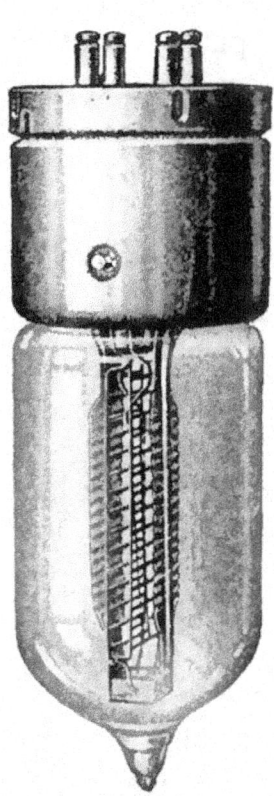

FIG. 426.—Vacuum tube, transmitting type. (*Western Electric Co.*)

FIG. 427.—Vacuum tube, receiving type. (*Western Electric Co.*)

current from a battery (B_1, Fig. 428) and because it is heated, sends out a stream of electrons. *Second*, the *plate* which forms the anode of the circuit from battery, B_2. This plate receives the electrons which are thrown off by the heated filament, hence a current flows through the circuit of B_2; the discharge through the tube depending on the e.m.f. between the filament and the

plate. *Third*, a *grid* is placed between the filament and
the plate and is connected to the *secondary* of the induction
coil, the primary of which is connected to the transmitter.
When the transmitter diaphragm is vibrating, the e.m.f.
induced in the secondary of the induction coil causes a
variation in the potential of the grid. This means a
variation in the electric field between the filament and
the plate. (See Fig. 428.) The changing electric field
causes a variation in the discharge of electrons through

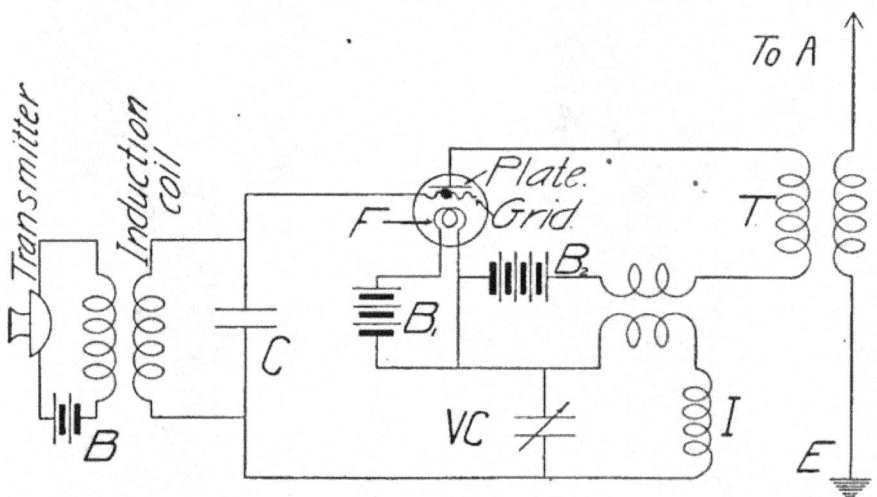

FIG. 428.—Diagram of wireless telephone transmitting set.

the tube; the variation corresponds to the vibrations
of the transmitter diaphragm. This produces a surging
current of the frequency of the sound waves in the primary
of the transformer (*T*, Fig. 428). The secondary of this
transformer is connected to the antennæ (*A*) and the earth
(*E*). By means of the transformer, rapid surgings are set
up in the antennæ and these surgings produce a continuous
stream of electromagnetic waves which goes out in space.
(Like Fig. 426*C*.) These electromagnetic waves produce oscillations in the antennæ of a receiving station. The antennæ transmit the impulses to a *tube* (Fig. 427) which acts

as a *detector*, and makes possible the reproduction of the sound by an ordinary telephone receiver.

The *vacuum tube* in the transmitting circuit also *amplifies* the impulses, that is, the energy of the waves given out is greater than that of the impulses which produce them, the additional energy being derived from the battery

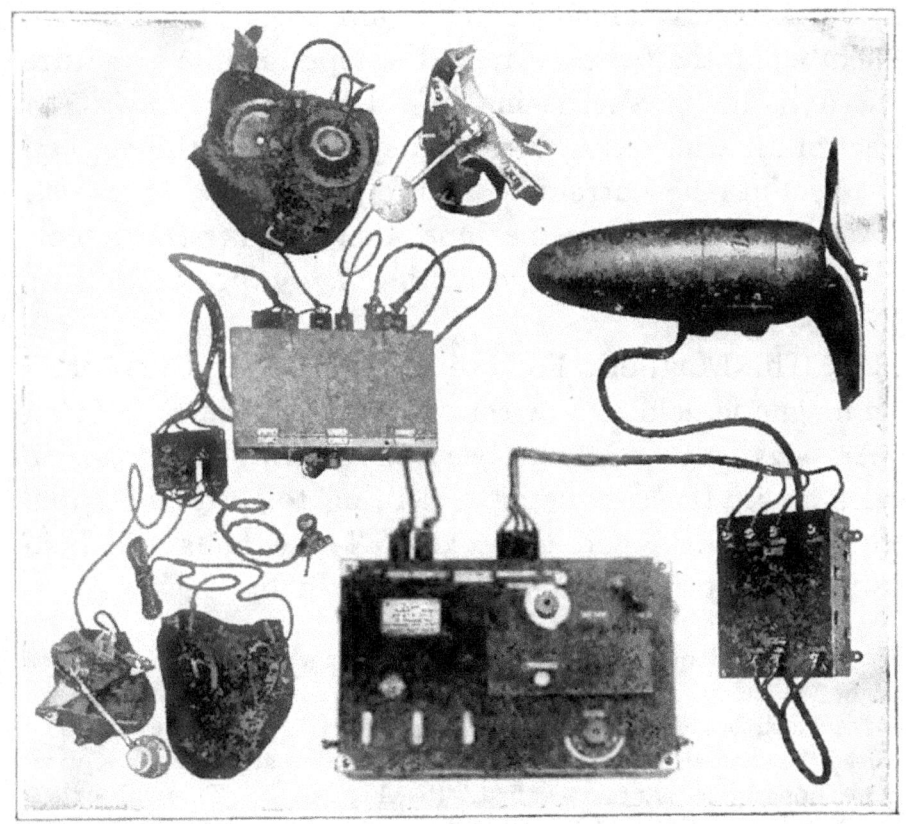

FIG. 429.—View of wireless telephone set.

sending current through the plate and filament. In operation, the filament and the plate are connected to a battery with a *condenser* (*VC*) and an *inductance coil* (*I*) in the circuit, as shown in Fig. 428. Photograph of a complete modern wireless telephone set is shown in Fig. 429.

30

ALTERNATING CURRENTS

428. Alternating currents are of interest to us because of their general commercial uše. To understand the reason for the extensive application of alternating currents it is necessary to learn the fundamental principles which pertain to them. The production of such currents has already been explained in Arts. 300–304. It should be remembered that the current developed in the armature of a dynamo is alternating. A dynamo may *deliver* a direct or an alternating current, depending on the method of collecting the current from the armature. If a *commutator* is used, the machine delivers *direct* current, if *slip rings* are employed, an *alternating* current is delivered.

429. The Magnetic Field of an Alternating Current.— The magnetic field of a direct current has been considered in Arts. 255–256. It has been shown to be arranged in circles about the conductor, according to the *Right Hand Rule*. (See Figs. 229 and 230.) These facts will help one to understand the following experiment:

If a number of magnetic compasses be arranged in a circle about a straight vertical wire carrying a direct current, the compass needles will point out a circle about the wire. (See Fig. 430, *A*.) If now the current be reversed the compass needles will reverse themselves and point in a direction just opposite to that taken at first. (See Fig. 430, *B*) This will be clear if you imagine yourself walking around the wire in the direction the compass needles pointed at first, and then walking around the wire in the reverse direction. This illustrates what happens in the field of an alternating current. The field reverses each time the current reverses.

The magnetic field of an alternating current not only rapidly reverses itself, but also continually *changes in intensity*. At the instant when the current reverses, the force of the magnetic field is zero since the current at that instant is zero. As the current begins flowing and increases to its maximum intensity, the magnetic field appears and increases in intensity; and as the current decreases to zero, the magnetic field changes in a similar manner. The field as it grows in strength extends farther and farther from the wire, as it decreases in strength it contracts or draws closer to the wire. Thus the magnetic field may be said to expand and contract. We may picture the lines of force as continually moving. In a typical a.-c. circuit, the complete series of changes takes place in a

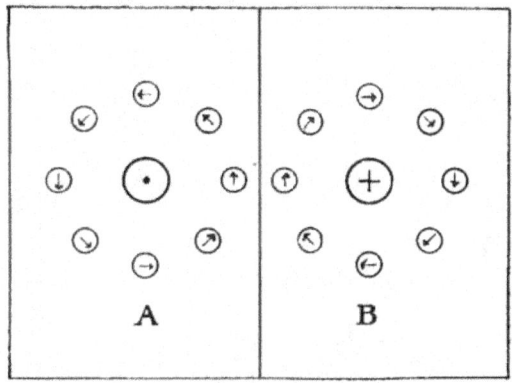

FIG. 430.—Arrangement of compasses about a wire carrying an alternating current.

small fraction of a second, and is repeated many times over in a second. Contrast this with the magnetic field of a constant direct current. Here the magnetic field has the same direction as long as the current flows and does not change in strength. This comparison is important because most of the differences between direct and alternating currents depend on differences in the action of their magnetic fields.

430. Transformers.—The transformer has been described in Arts. 309–310. The principle of the transformer may be illustrated by the following experiment:

A coil having several hundred turns of No. 18 d.c.c. copper wire is placed over one arm of a "U" shaped iron core (see Fig. 431) and then connected to a 110 volt a.-c. lighting circuit. Another coil (S) having about 50 turns of No. 22 d.c.c. copper wire is connected to an electric bell or buzzer, or a low voltage electric light bulb. When the small coil is held over the other arm of the "U" shaped iron core, the bell rings or the bulb glows. It is evident that the electromotive force developed in the small coil (S) is due to the alternating magnetic field surging back and forth through the iron core. In Fig. 431 the core is "open" since the magnetic field must pass through the air from one end of the core to the other. A typical transformer has a *closed core* to provide a *closed magnetic circuit*. To secure this, take a suitable bar of iron and lay across the end of the "U" shaped core, and notice any change in the induced current produced in the small coil, due to increased movement of magnetism through the closed iron core.

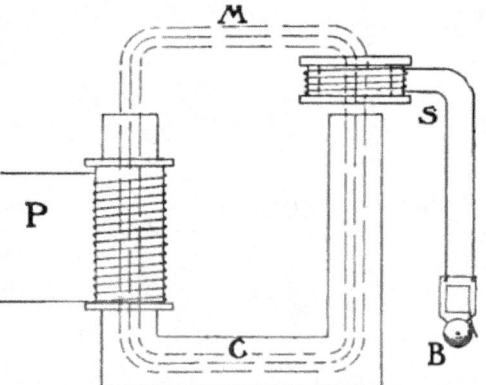

Fig. 431.—Diagram of a transformer.

This experiment illustrates the construction and action of a transformer. In a commercial transformer, the two windings are on a closed magnetic circuit. (See Figs. 304 and 305, p. 346.) To keep the coils insulated, the transformer is placed in an iron "housing" and covered with oil. These "housings," or transformer cases are generally attached to poles near buildings in which alternating current is used.

431. Voltage Relation in a Transformer.—In the experiment described above, a bell was rung by an induced current produced in the secondary coil. The induced e.m.f. was less than the voltage of the primary coil partly because there was some magnetic leakage, but mainly because there were fewer turns of wire on the secondary. In a commercial transformer the magnetic leakage is practically zero. In such a case, the ratio of the number of turns on the primary coil to the number on the secondary equals the ratio of the e.m.f. induced in the primary to the e.m.f. induced in the secondary. Suppose, for example, we wish to make a bell ringing transformer to use on a 110 volt lighting circuit, 10 volts being required for the bell; the secondary will then need one-eleventh of the number of turns of the primary. So that if 550 turns are on the primary, then 50 turns will be needed for the secondary. This will be a "step-down" transformer. On the other hand, suppose we wish to "step-up" the voltage as is done in a certain power station where the voltage of the generators is 6000 volts, the voltage being stepped up to 44,000 by means of large transformers. This means that the secondary coils have approximately $7\frac{1}{3}$ times as many turns as the primary.

432. Power Loss in a Transformer.—When the voltage is "stepped up" in a transformer, do we gain power? To answer this question we must remember that electric power does not depend on voltage alone but on the *product* of e.m.f. and current intensity. (See Art. 291.) By tests with a.-c. voltmeters and ammeters, we find that when the secondary e.m.f. is *greater* than the primary e.m.f., the secondary current intensity is *less* than that in the primary. It is also found that the *power* developed is less than the power received by the transformer, *i.e.*, the "output" is less than the "input" as we would expect from the law of machines. The power loss is mainly due to the work required to reverse the magnetism, that is, to continually reverse the position of the iron molecules. (See Art. 205.) The energy lost in this manner is known as "core loss" since it occurs in the

iron core. The lost energy appears as heat. So much
heat is developed in large transformers that special means
of cooling are provided. In order to make the heat
developed as small as possible, the cores are "laminated"
(see Fig. 305, p. 346), that is, built up of thin sheets of
iron, because if the iron cores were solid, the changing
magnetic fields would induce electric currents in the
iron cores, which would produce an excessive amount of
heat with a correspondingly large power loss.

433. Choke Coils and Inductance.—If we refer to Fig.
432 we see that the primary winding of the bell ringing

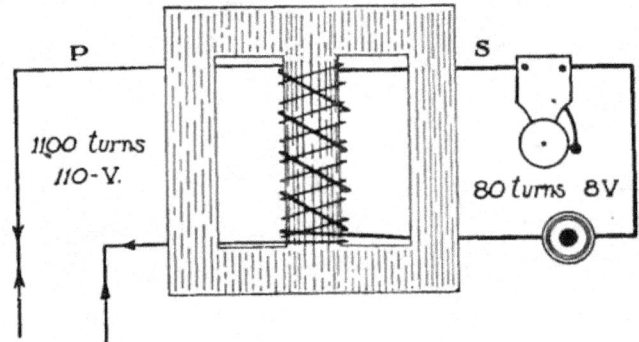

FIG. 432.—Diagram of "bell-ringing" transformer.

transformer is connected across the line. This winding
forms a closed circuit whether the bell is ringing or not.
The resistance of this winding is small. Let us assume
it to be one ohm. With a one ohm resistance connected
across a 110 volt line we might expect a current of 110
amperes. This is certainly what we should get if we were
to connect a one ohm resistance across a line having 110
volts direct. The primary would form a short circuit
if the current were direct. But the fact is that practically
no current flows through the primary winding when the
bell is not ringing. Herein lies one of the important differences between alternating and direct currents. With
an alternating current the primary winding of our trans-

former acts as a *choke coil* and "chokes" down the current almost to zero. Let us see how this is done.

Let Fig. 433 represent a choke coil. Since alternating current is used, the magnetic field is continually changing. Each turn of wire has its own magnetic field. The lines of force of turn number 1 expand and contract and as they do so they move across turns 2, 3 and so on. In like manner the lines of force from each turn of wire move across the other turns. In other words the coil is cutting its own lines of force. Now whenever an electric conductor cuts magnetic lines of force an electromotive

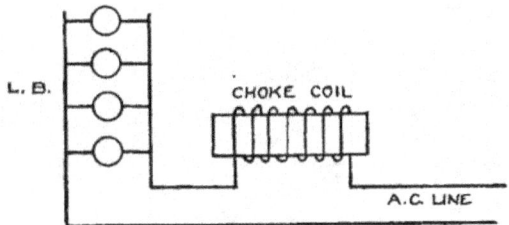

FIG. 433.—A circuit containing a choke coil.

force is induced in the conductor. There is then an e.m.f. induced in the coil by its own magnetic field. This induced e.m.f. on the whole opposes the applied e.m.f.; in the primary of our bell ringing transformer the induced e.m.f. opposes the e.m.f. of the line to such an extent as to reduce the current almost to zero. *Inductance* is the action of an alternating current in inducing an opposing e.m.f. in the coil in which the current is flowing. Since this opposing e.m.f. is induced in the coil by its own magnetic field this action is also called *self-induction.* In a transformer the action of the field of the primary upon the secondary is *mutual induction*; while the action of the field of the primary in choking the current in the primary itself is self-induction or inductance. A coil having a single winding and used to introduce inductance in a circuit is called a *choke coil*. A choke coil inserted in a lamp circuit in series with the lamps dims the lamps because it reduces the intensity of the current.

Self-induction causes the current to *lag*, that is, the current does not quite reach its maximum at the instant the voltage reaches its maximum. Fig. 434 shows graphically an e.m.f. and a lagging current. In this figure the maximum current is shown following the maximum voltage at an interval of 30 degrees. In other words the armature in

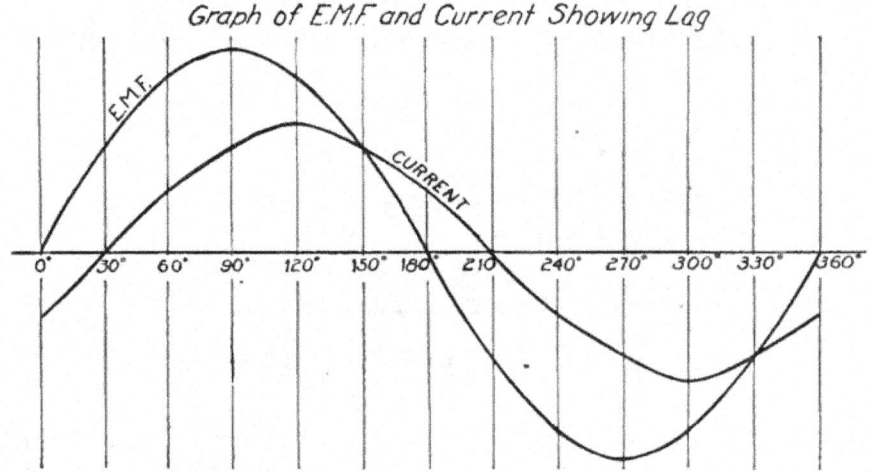

FIG. 434.—Diagram showing graphically an alternating current with a "lag" of 30° behind its electromotive force.

a two-pole field must turn 30 degrees from the position of maximum voltage before the current in the coil, where the self-induction occurs, reaches its maximum.

434. Reactance and Impedance.—A choke coil has resistance as well as inductance. Its resistance can be found by the voltmeter-ammeter method, using a direct current. (See Art. 278.) Let us take for example the primary winding of a bell ringing transformer. Using a direct current and testing the coil with a voltmeter and ammeter we find its resistance to be, let us say, one ohm. If we connect the same coil across a 110 volt a.-c. line we find the current to be very small, say 0.05 ampere. The coil now has resistance and *reactance*. Reactance is the effect of self-induction in hindering the flow of current. It is measured in ohms. The combined effect of resistance and reactance is called *impedance*. In the example above, the coil has $\dfrac{110 \text{ (volts)}}{0.05 \text{ (ampere)}} = 2200$ ohms of impedance. In applying Ohm's law to an alternating current circuit, impedance must be substituted for resistance. Ohm's law as applied to an a-c. circuit should be stated:

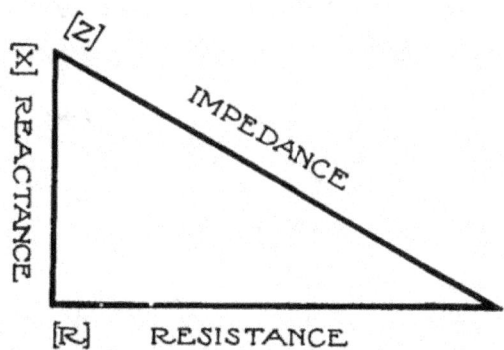

FIG. 435.—The relation between resistance, reactance and impedance.

"Current intensity equals e.m.f. divided by impedance", or $I = E/Z$. (Z = impedance.)

Impedance, however, does not equal the *sum* of resistance and reactance. The relation between these three quantities is similar to that between the three sides of a right triangle, in which the impedance represents the hypotenuse, and the resistance and reactance the other two sides. See Fig. 435 which indicates that Resistance^2 + Reactance^2 = Impedance^2, or ($R^2 + X^2 = Z^2$). (X = reactance.) To illustrate this relation; suppose the primary of a transformer has 10 ohms impedance and 8 ohms resistance, then the reactance equals $10^2 - 8^2 = 6^2$, or the reactance is 6 ohms.

Exercises

1. Find the reactance of a choke coil having a resistance of 10 ohms, when its impedance is 50 ohms. How great a current flows through this coil if the terminal voltage is 110 volts?

2. When the bell is ringing, the primary of a bell ringing transformer has an appreciable current. Suppose this current is 0.2 ampere. What is the impedance if the voltage of the line is 115 volts? What is the reactance if the resistance is 1 ohm?

3. The primary of a large transformer has a terminal voltage of 6000 volts and a current of 600 amperes. What is the impedance? If the resistance is 6 ohms, what is the reactance?

435.—The electric condenser (see Art. 231) is a very useful device in a.-c. circuits; *e.g.*, in telephone sets

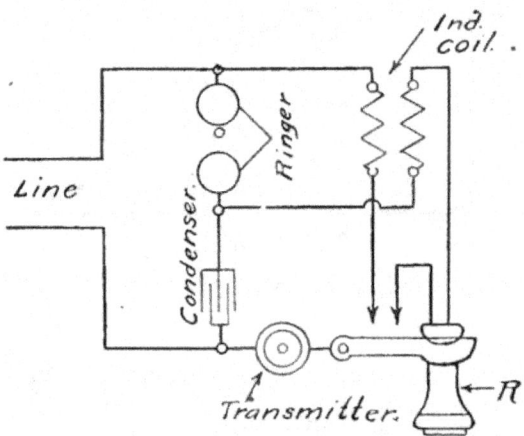

Fig. 436.—A telephone set showing a condenser used in the circuit of the "ringer."

used in cities, a condenser is used in the ringing circuit, as shown in Fig. 436. Alternating current is required to ring such a bell and a condenser permits an a.-c. current to act through it, although it entirely prevents the flow of a direct current. This peculiar action will now be explained.

436. The action of a condenser in an alternating current circuit may be illustrated by the following experiment. Connect twelve, 1 m.f. (microfarad) condensers, in parallel, and then attach them to a 110 volt a.-c. line so that an incandescent lamp is in circuit as shown in Fig. 437. The lamp will be found to glow brightly, although there is no electrical connection between the two sets of condenser plates. If the same arrangement is connected to a 110 volt direct current circuit, the lamp does not glow because it is really an open circuit. The lamp glows on an a.-c. circuit because, although no electricity flows *through* the condenser, it does flow *into and out of* the condenser, surging back and forth through the lamp with sufficient intensity to cause it to glow brightly. When the a.-c. current moves one way in the circuit, one set

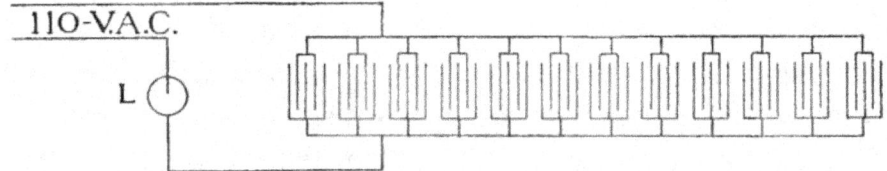

FIG. 437.—Twelve condensers in circuit with an incandescent lamp.

of plates of the condensers becomes charged positively, the other, negatively. When the a.-c. current reverses, the charges on the condenser plates reverse. In the ordinary lighting circuit 120 reversals take place each second, so that electricity rapidly flows into and out of the condensers. On removing one condenser after another from the circuit, the lamp is found to glow less and less, till when but one condenser is left, no glowing is observed, since one small condenser does not have sufficient *capacity*.

The unit of capacity is the *Farad*. Capacity is defined as the quantity of electricity per second that flows into a condenser when the voltage at the terminals changes at the rate of one volt per second. If a change of one volt per second causes one coulomb to flow per second, that is, a current of one ampere, the capacity is one *farad*. The condensers used in the above experiment have a capacity of one microfarad, or one millionth of a farad.

A condenser, on account of its capacity, causes an a.-c. current to *lead* the voltage, that is the current reaches its maximum value before the voltage does. In this respect a condenser has an effect opposite to that of the self-induction of a choke coil (the latter causing the current to "lag"). (See Fig. 435.)

437. Transmission of Electric Power.—A field of peculiar usefulness for a.-c. currents is in the economical transmission of electric power. This fact is due to the following reasons: (*a*) The loss of electrical power in a transmission line is due to the production of heat; the heat produced being proportional to I^2R, or to the *square* of the *current intensity*. Any lessening of the current flow required to transmit a given power will therefore increase the efficiency of transmission. (*b*) In order to employ a small current in transmitting a large amount of power, we must use a very high e.m.f. Such high electromotive forces, say from 60,000 to 100,000 volts, can be obtained only by the use of a.-c. transformers, since it is not practicable to build a direct current generator capable of producing 60,000 volts. In large power transmission systems, a.-c. generators are used to produce powerful alternating currents. The e.m.f. is then stepped up to a suitable voltage (2300–100,000) by transformers and sent over transmission lines to the various places where the power is to be used; at these places suitable transformers "step-down" the e.m.f. to a convenient or safe voltage for use. (See Fig. 442 of a transmission line and Fig. 438 of a large power transmission system, and Fig. 439 of an a.-c. generator and power plant.)

438. Power Factor.—The *power factor* is a matter of interest and importance in the use of a.-c. machines. It meaning and use may be learned from the following explanation: In a direct current circuit, watts equals

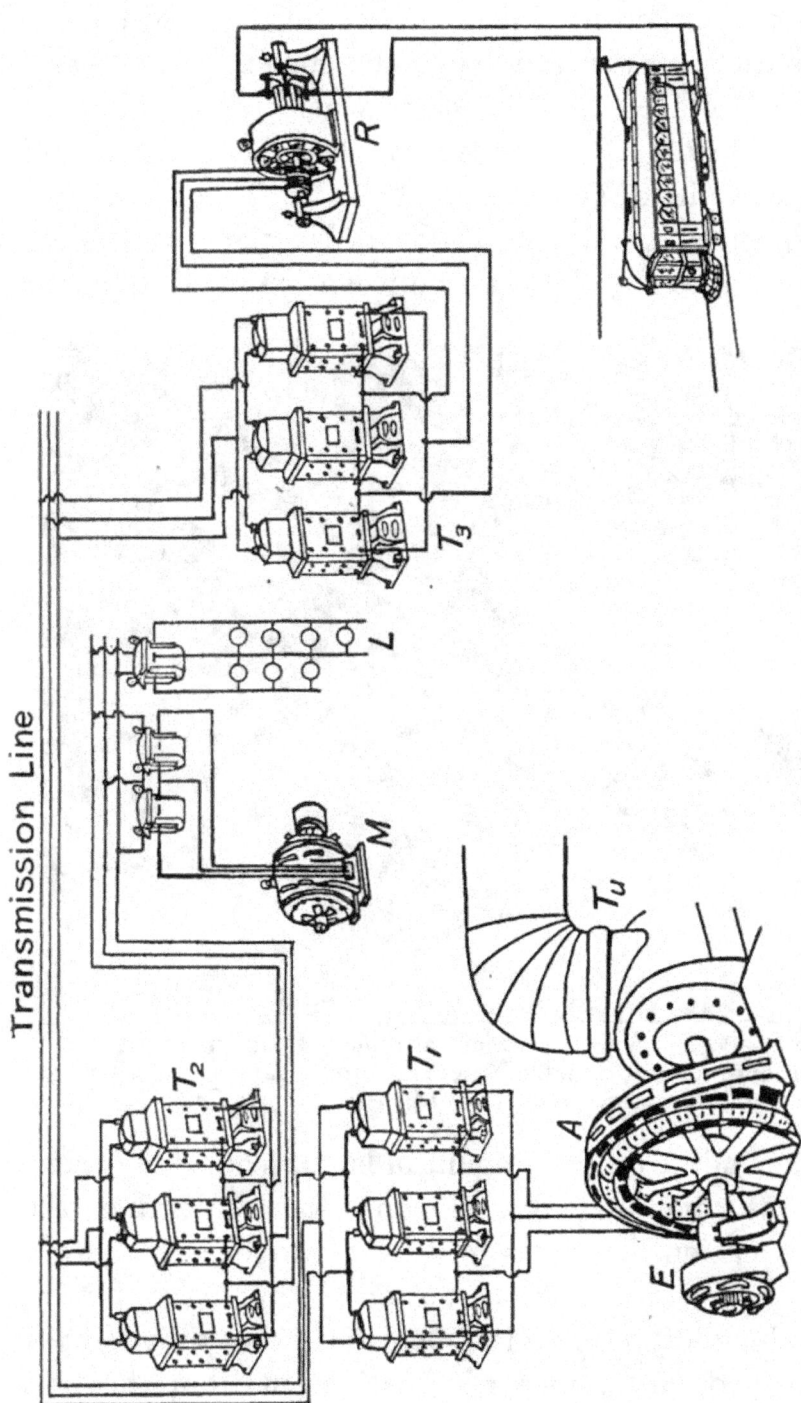

FIG. 438.—Diagram of an alternating current high tension power system. (A) Alternator, (T_4) water turbine, direct connected to alternator, (E) exciter, (T_1) step-up transformers in power station, (T_2) step-down transformers in substation, (M) motor, (L) lamps, single-phase, three-wire system, (T_3) step-down transformers delivering three-phase current to rotary converter (R) which delivers direct current to the trolley line.

Transmission Line

volts times amperes. In an alternating current circuit, this equation is true only when the current is "in step" with the voltage, that is, only when there is no *inductance* or *capacity* in the circuit. If current and voltage are out of step, *i.e.*, if there is *lag* or *lead* (see Fig. 434), the product of volts and amperes gives only the *apparent power*, the ratio between true and apparent

FIG. 439.—Power house showing alternators, direct connected to horizontal hydraulic turbines. Note the direct current "exciter" on end of shaft of alternator. (*Courtesy of General Electric Co.*)

power depending on the amount of lag or lead. This ratio is called the power factor. In an a.-c. circuit, then, the power equation is: watts = volts × amperes × power factor, or power factor = true power/apparent power. The products of volts and amperes is the *apparent power* and is called volt-amperes in distinction from the true power or watts. Therefore the following is true: power factor = true watts/volt-amperes.

439. Single-phase Currents.—There are several kinds of a.-c. currents. One of the most common is the *single-phase*. It is simply the common a.-c. current used for light and power in the average home, and uses a two-wire circuit around which the current is rapidly alternating. Fig. 440 illustrates the changes of e.m.f. in an a.-c. single-phase current. It may be produced by a single coil rotating in a magnetic field. The curve of Fig. 440 represents one *cycle*, that is, one complete series of changes in

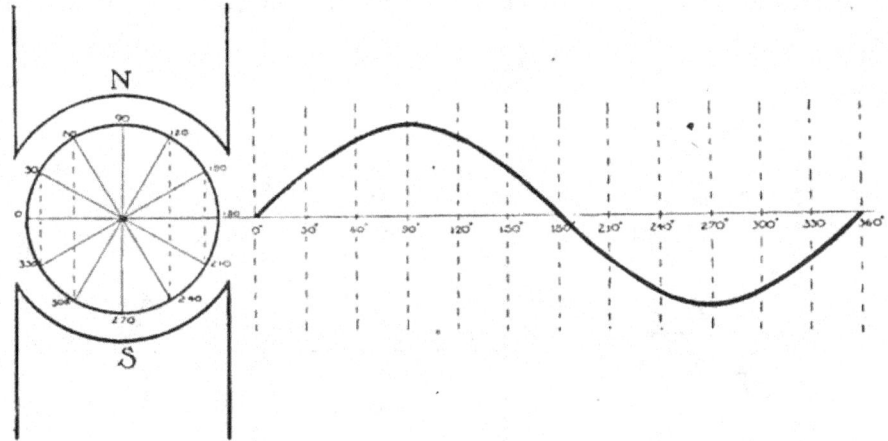

FIG. 440.—Graph showing the e.m.f. changes of a single-phase current for one "cycle."

the electromotive forces. At the end of the cycle the armature is in the same condition as at the beginning so far as the magnetic field is concerned. It then begins a new cycle. The ordinary commercial alternating current has a frequency of 60, that is 60 cycles per second. One rotation produces as many cycles as there are pairs of poles. For example, if there are 48 poles in the generator field, one rotation produces 24 cycles.

440. Three-phase Currents.—Now suppose we have three coils as in Fig. 441, the coils being evenly spaced, or 120 degrees apart, at *A*, *B*, and *C*. If the coils are rotated in a magnetic field, each will produce an electro-

motive force. The result produced by three such coils
is called a *three-phase* current. Ordinarily six wires, or
three circuits, would be required to carry the current produced by three separate coils; for when coil "*C*" is in the
90 degree position, where its e.m.f. is a maximum, coil "*B*"
is 120 degrees past its maximum, and coil "*A*" is 240 degrees past its maximum. The graph (Fig. 441) shows the
maximum points of the three e.m.f.'s. separated by intervals of 120 degrees. In practice, however, it is found
possible to use *three wires* instead of six, as explained in
Art. 441.

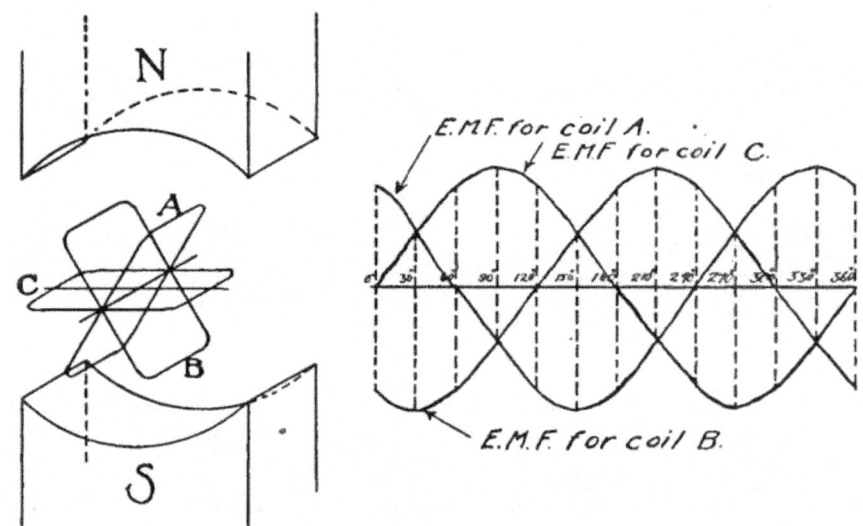

FIG. 441.—Graph showing the e.m.f. changes of a three-phase current
for one "cycle."

441. Three-wire Transmission.—The currents produced
in the three coils just described undergo precisely the
same changes as those represented in the *graph* (Fig. 441)
for the three electromotive forces. Careful examination
of the graph will show that at any point the sum of the
plus e.m.f.'s. equals the sum of the *minus* e.m.f.'s. In
other words the algebraic sum of the three e.m.f.'s. is
zero. Therefore if we properly connect a transmission
line of three wires to the generator, the sum of the currents

leaving the generator will equal the sum of the currents returning to it. Since the algebraic sum of the currents produced by the three coil combination described in Art. 440 is always zero, it is possible to use three wires on three-phase transmission lines. Fig. 442 shows a "tower" carrying three, three-wire transmission lines. Long

FIG. 442.—A "tower" supporting three, three-phase circuits of a high tension transmission line.

distance, high tension transmission lines are generally three-wire lines carrying three-phase a.-c. currents.

442. Alternators.—A dynamo which delivers alternating current is known as an *alternator*. Commercial alternators have many pairs of poles in the field and as a rule the field rotates while the armature is stationary. The field must be supplied with *direct* current for the polarity of each coil in the field must remain unchanged. Usually

31

a separate "exciter" is used, which is a small direct current generator. The current from this exciter is fed into the rotating field by means of slip rings. Fig. 439 shows a d.-c. (direct current) exciter on the end of the armature shaft of the large alternator.

443. The A.-C. Series Motors.—The only type of motor that will run on either alternating or direct current is the *series motor*. The "universal" motor used in

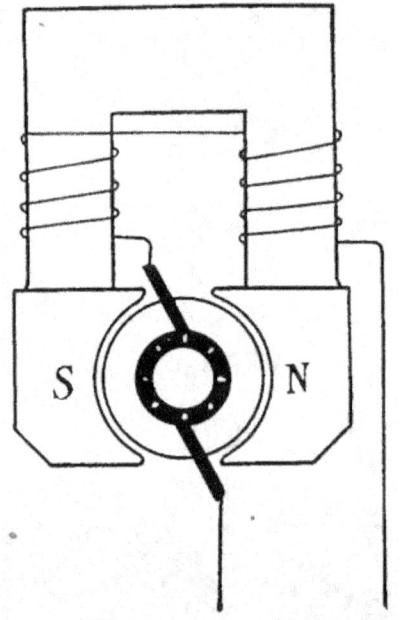

Fig. 443.—Diagram of a "Series Motor."

household appliances such as electric fans, vacuum cleaners, etc., is a series motor. The reason a series motor will run on either direct or alternating current is because the direction of rotation of the armature of a motor depends on (*a*) the direction of the current in the armature, and (*b*) the polarity of the field. Reversing either of these alone, reverses the direction of rotation of the armature, while reversing both at the same instant leaves the direction of rotation unchanged. Fig. 443 is a diagram of a series motor since the field coils and armature are connected in series. On an a.-c. line, both field and armature current must therefore reverse at the same instant. In a shunt motor (similar to Fig. 286) we have a divided circuit, and the greater self-induction of the field coils causes an a.-c. current through these coils to lag behind that flowing in the armature so that the two currents do not reverse at the same instant.

444. The Induction Motor.—Another common type of a.-c. motor is the *induction motor*. Its advantage lies in its simplicity. It has neither commutator nor brushes,

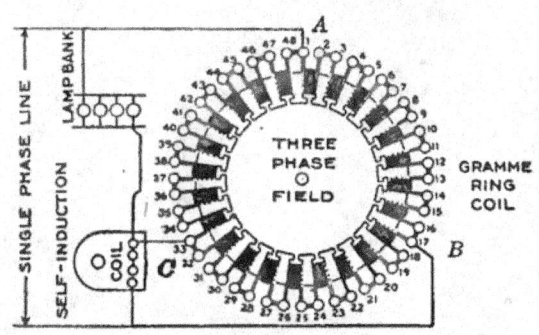

THREE PHASE FROM SINGLE PHASE

FIG. 444.—Diagram of a gramme ring. It is shown connected to a single-phase current so as to produce a rotating magnetic field, similar to that obtained with a three-phase current. (*Ahrens, Harley and Burns.*)

the armature having no connection with an external circuit. If the wires of a three-phase line be connected to a coil wound in the form of a *gramme ring*, the connections being 120 degrees apart as in Fig. 444, the magnetic field within this coil will change in the same manner as if a magnet were spinning upon a pivot at the center of the coil. Suppose the *N* pole at one instant is at *A*, in one-third of a cycle it moves to *B*, in another third to *C*, and in one cycle it makes a complete revolution. Thus we have a *rotating magnetic field*. If a cup of some non-magnetic metal such as aluminium or copper be placed on a pivot in the center of this coil, the cup is cut by the moving lines of force and currents are induced in it. Because of these currents, the cup has a magnetic field of its own, and the action of the two magnetic fields is such as to pull the cup around and cause it to rotate in the same direction as that in which the field of the coil rotates. The coil represents the stationary part, the *stator* (Fig. 445) and the cup the rotating part, the *rotor*, of an induction

FIG. 445.—The "stator" of an induction motor.

motor.   While the cup rotates in the same direction, it does not rotate so rapidly as the magnetic field.   If it should it is plain that it would not cut the lines of force.   The difference between the rate of rotation of the rotor and that of the magnetic field is called the "slip."   The rotating part in small induction motors is frequently made in a single casting.   In large motors, it is built up of heavy copper bars.   Thus, from its appearance the common form of rotor is known as the "squirrel cage" rotor. (See Fig. 446.)

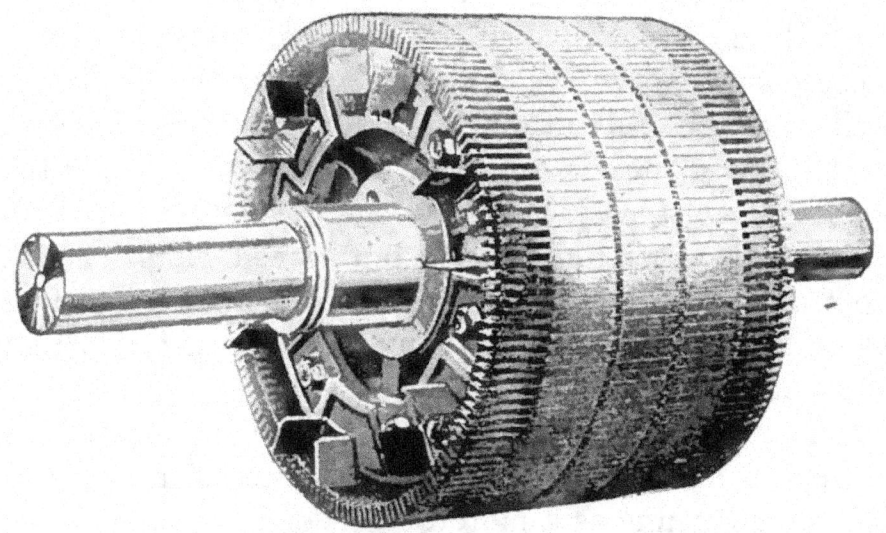

FIG. 446.—The "rotor" of an induction motor.

445. A synchronous motor is one that keeps step with the alterations of an alternating current. The line current is fed into the armature by means of two slip rings and

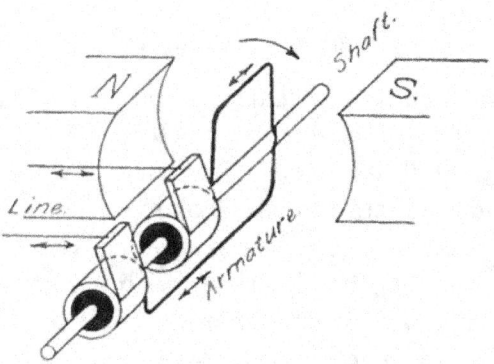

FIG. 447.—Diagram illustrating the principle of the synchronous motor. The armature coil passes the position shown in the figure at the instant the current in the line reverses. Thus the armature keeps with the line current, making one revolution with each "cycle."

brushes. The principle of the synchronous motor is illustrated in Fig. 447. This shows a motor having a two-pole field. The armature current must be reversed twice in each revolution. The reversal must take place when the armature winding is perpendicular to the lines of force of the field. In a direct current motor this reversal is brought about by the commutator. In a synchronous motor the armature reaches the 90 degree position at the exact instant at which the current reverses in the line. Thus in the case of a two-pole motor the armature must make exactly one revolution for each cycle; it is, therefore, a constant speed motor. Such motors are frequently employed in converter stations where alternating current is converted into direct current by what are called *rotary converters*.

In practice the synchronous motor has a number of pairs of field poles. It is essentially an alternating current generator running as a motor. One of the principal uses of the synchronous motor is that of a converter, receiving alternating current and delivering direct current. Synchronous motors are also used in transmission lines to aid in maintaining constant voltage.

Important Topics

The wireless telephone, essential parts, action, arrangement.
Alternating currents, alternating fields.
Transformers, voltage relation of coils, power and core losses.
Self-induction, inductance, and coke coils, uses, applications.
Impedance, reactance, and resistance; relation and effects.
Condensers, uses and applications with a-c. circuits.
Alternating current power transmission; uses, advantages.
Power factor, lag, lead, volt-amperes, true watts.
Single- and three-phase currents; uses and nature of each.
Three-wire transmission systems, alternators, construction, and action.
A-c. motors, series, induction, synchronous.

INDEX

www.ingramcontent.com/pod-product-compliance
Lightning Source LLC
Chambersburg PA
CBHW081254170526
45165CB00011B/3306